Ion Beam Analysis

Ion Beam Analysis

Fundamentals *and* Applications

Michael Nastasi
James W. Mayer
Yongqiang Wang

CRC Press
Taylor & Francis Group
Boca Raton   London   New York

CRC Press is an imprint of the
Taylor & Francis Group, an **informa** business

CRC Press
Taylor & Francis Group
6000 Broken Sound Parkway NW, Suite 300
Boca Raton, FL 33487-2742

© 2014 by Taylor & Francis Group, LLC
CRC Press is an imprint of Taylor & Francis Group, an Informa business

First issued in paperback 2019

No claim to original U.S. Government works

ISBN-13: 978-0-367-44584-3 (pbk)
ISBN-13: 978-1-4398-4638-4 (hbk)

Library of Congress Cataloging-in-Publication Data

Nastasi, Michael Anthony, 1950-
    Ion beam analysis : fundamentals and applications / Michael Nastasi, James W. Mayer, and Yongqiang Wang.
        pages cm
    "A CRC title."
    Includes bibliographical references and index.
    ISBN 978-1-4398-4638-4 (alk. paper)
    1. Ion bombardment. I. Mayer, James W., 1930- II. Wang, Yongqiang (Nuclear physicist) III. Title.

QC702.7.B65N35 2014
539.7'3--dc23   2014007466

Visit the Taylor & Francis Web site at
http://www.taylorandfrancis.com

and the CRC Press Web site at
http://www.crcpress.com

Dedicated to the memory of Jim Mayer—
innovator, educator, colleague, and friend

Brief Contents

Contents

SECTION I — Fundamentals

SECTION II — Applications

Preface

Materials are the cornerstone of many modern technological innovations today. Materials' properties and performance are strongly coupled with their composition and structure. This is especially true of surface layers and thin films. These materials may be made from a variety of techniques, such as physical and chemical vapor deposition techniques, molecular beam epitaxial growth, ion implantation and laser surface modification, and sol-gel coatings. Materials systems may include metals, semiconductors, and insulators, both in thin film and bulk form. A rapid, nondestructive, and quantitative way of analyzing these materials is with ion beam analysis.

In this book we focus on the fundamentals and applications of ion beam methods of materials characterization. These techniques all rely on the use of energetic ions to interrogate the material. In most instances the ion interaction with the material is Coulumbic; at low enough incident energies, the interaction of charged particles with an atomic nucleus strictly follows classical laws that depend only on the respective charges. The repulsive Coulomb forces effectively shield the incident ion from short-range nuclear forces. As the incident projectile energy increases, the projectile can penetrate to within the range of the attractive nuclear force, which modifies the simple Rutherford scattering results, and nuclear reactions occur.

This book is divided into two sections: fundamentals and applications. The fundamentals section includes Chapters 1 through 9, focusing on fundamentals of ion–solid interactions (e.g., derivations of basic parameters used in ion beam analysis). Chapter 1 provides an overview of the topics. Chapter 2 focuses on classical collision theory or kinematics, the energy transfer from the incident particle to target atoms, used to identify target elements. Chapter 3 focuses on cross sections, the probability of collision events taking place, used to quantify the concentration of the target elements. Chapter 4 deals with ion stopping and energy loss, which provide depth information in ion beam analysis.

Chapters 5 through 9 deal with the fundamentals of five specific ion beam analysis techniques. Chapter 5 focuses on backscattering spectrometry, both Rutherford and non-Rutherford applications. Chapter 6 presents a discussion on elastic recoil detection. Nuclear reaction analysis is discussed in Chapter 7. Particle-induced x-ray emission is discussed in Chapter 8. Finally, ion channeling is presented in Chapter 9.

The applications section includes Chapters 10 to 15, presenting specific applications and examples of ion beam analysis in traditional and emerging research fields. Chapter 10 discusses applications in thin film analysis including simple elemental diffusion, reaction kinetics, intermixing of multilayers, corrosion, etc. Chapter 11 details specific examples of how ion channeling is used to measure defects information in single crystalline solids. Chapter 12 gives application examples in nuclear energy research including unique advantages for light elements analysis, such as He and C, etc. Chapter 13 summarizes applications in art and archaeology and describes the importance of external ion beam analysis (nonvacuum) in cultural heritage studies. Chapter 14 provides an overview of applications in biomedicine and emphasizes key contributions of microbeam analysis to the life science in general. Finally, Chapter 15 compares the most frequently used ion beam analysis computer codes and their algorithms, as well as providing real examples of how these codes are used to convert acquired energy spectra into elemental depth profiles.

This book is written for researchers interested in the use of ion beam analysis techniques for materials characterization. It is primarily aimed at college seniors, graduate students, and researchers who have had previous training in classical and modern physics. At the end of each chapter, there are a set of problems and references and/or suggested readings.

We are indebted to the following authors who have written the applications chapters: Chris Jeynes and Richard Thompson (UK), Lin Shao and Amit Misra (US), Thomas Calligaro and Jean-Claude Dran (France), Harry Whitlow (Switzerland), Min-Qin Ren (Singapore), Nuno Barradas and M. A. Reis (Portugal), M. Mayer (Germany), and F. Schiettekatte (Canada). We also want to thank our spouses, Holly, Betty, and Minhua, for their constant and enthusiastic support during the development of this textbook. We are grateful to CRC Press for undertaking the formidable task of publishing this book. Senior editor Luna Han and her staff offered valuable assistance in guiding the book to completion. Initial support for the writing of this book was provided by the Materials Sciences and Engineering Division of the US Department of Energy, Office of Basic Energy Sciences. Partial support was also provided later by the Center for Integrated Nanotechnologies (CINT), a DOE nanoscience user facility jointly operated by Los Alamos and Sandia National Laboratories.

About the Authors

Michael Nastasi is director of the Nebraska Center for Energy Sciences Research (NCESR) and Elmer Koch Professor in the Department of Mechanical and Materials Engineering at the University of Nebraska-Lincoln. Prior to this appointment he was director of the Department of Energy (DOE) Energy Frontier Research Center on Materials at Irradiation and Mechanical Extremes and nanoelectronics and mechanics thrust leader at the Center for Integrated Nanotechnologies (CINT). He served as team leader for the Nanoscience and Ion–Solid Interaction Team and as Fellow at Los Alamos National Laboratory. He earned his PhD in materials science and engineering at Cornell University. Dr. Nastasi is an elected fellow of the American Physical Society (APS) and the Materials Research Society (MRS).

James W. Mayer (1930–2013) was a pioneer in the application of ion beam techniques for materials analysis. He received his PhD from Purdue University, followed by appointments at California Institute of Technology (1967–1980) and Cornell University (1980-1992), as Francis Norwood Bard Professor of Materials Science and Engineering and director of the Microscience and Technology Program. He was appointed to the faculty at Arizona State University (ASU) in 1992, where he became Regents' Professor and P.V. Galvin Professor of Science and Engineering, as well as director of the Center for Solid State Science until his retirement. His research contributions were in many areas of solid-state engineering, especially ion implantation and Rutherford backscattering spectrometry. Among his many accolades, Dr. Mayer was recipient of the Materials Research Society's Von Hippel Award, a Fellow of the American Physical Society and Institute of Electrical and Electronic Engineers, and an elected member of the National Academy of Engineering.

Yongqiang Wang is the team leader of the Ion Beam Materials Laboratory (IBML) in Los Alamos National Laboratory. Dr. Wang received his PhD in nuclear physics and technology at Lanzhou University in China and has worked in the field of ion beam applications in materials research for more than 25 years. Over the years, he has used and maintained a variety of electrostatic accelerators and implanters for research and education. He is a coauthor of nearly two hundred peer-reviewed publications, three invited book chapters, two US patents, and the *Handbook on Modern Ion Beam Materials Analysis* (2nd edition). He is a member of the International Conference Committee for Ion Beam Analysis. He was a cochair of the 21st International Conference on Ion Beam Analysis in Seattle, Washington in June 2013. He currently serves as a cochair of the Biennial International Conference on Application of Accelerators in Research and Industry (CAARI).

Contributors

Nuno P. Barradas
Instituto Superior Técnico/ITN
Universidade Técnica de Lisboa
Sacavém, Portugal

Thomas Calligaro
Centre de Recherche et de
    Restauration des Musées de
    France
Palais du Louvre
Paris, France

Jean-Claude Dran
Centre de Recherche et de
    Restauration des Musées de
    France
Palais du Louvre
Paris, France

Chris Jeynes
University of Surrey Ion Beam
    Centre
Guildford, England

Matej Mayer
Max Planck Institute for Plasma
    Physics
EURATOM Association
Garching, Germany

Amit Misra
Department of Materials Science
    and Engineering
University of Michigan
Ann Arbor, Michigan

Miguel A. Reis
Instituto Superior Técnico/ITN
Universidade Técnica de Lisboa
Sacavém, Portugal

Min-Qin Ren
Centre for Ion Beam Applications
Department of Physics
National University of Singapore
Singapore

François Schiettekatte
Département de Physique
Université de Montréal
Montréal, Québec, Canada

Lin Shao
Department of Nuclear
    Engineering
Texas A&M University
College Station, Texas

Richard L. Thompson
Department of Chemistry
Durham University
Durham, England

Harry J. Whitlow
Haute Ecole Arc Ingénierie
University of Applied Sciences
    (HES-SO)
La Chaux-de-Fonds, Canton of
    Neuchâte
Switzerland

Fundamentals

Overview

1.1 INTRODUCTION

The fundamental physics of the interactions between incident particles and target atoms provides the underlying science for ion beam analysis. These inter-particle interactions depend on many parameters, including ion velocity and energy, ion and atom size, atomic number and mass, and the distance of closest approach between ion and atoms in the solid. Interactions include scattering, inner-shell ionization, and nuclear reactions. Emission products from these interactions are used to derive information on material composition and structure and provide the basis for *elemental* ion beam analysis. Other interactions control the slowing down of the ion (energy loss) and perturb its trajectory (multiple scattering) and, more importantly, bring to the methods of ion beam analysis a unique capability for deriving depth information without physically removing/sputtering the surface target material and provide the basis for ion beam *depth profiling*. The fundamental details about the interactions between incident ions and target atoms are presented in Chapters 2–9, while the applications of ion beam analysis will be covered in Chapters 10–15. Some common parameters and data tables are included in Appendices at the end of the book.

1.2 ATOMIC AND PLANAR DENSITIES

The natural unit of composition in ion beam analysis is areal density. To have a better sense of this it helps to know the atomic density, interplanar distance between planes, and the number of atoms per square centimeter on a given plane of a material. In cubic systems with an atomic density of N atoms per cubic centimeter, the crystal lattice parameter, a_c, is given by

$$(1.1) \qquad a_c = \left(\frac{\text{atoms/unit cell}}{N} \right)^{1/3}$$

where, for systems with one atom per lattice point, there are four atoms per unit cell for a face-centered cubic lattice (Al, Cu, Ag, Au, Pd, Pt), two for a body-centered cubic lattice (V, Fe, W), and eight for the common semiconductors like germanium (Ge) and silicon (Si), which have the diamond cubic structure. Aluminum has an atomic density of 6.02×10^{22} atoms/cm³, so that the lattice parameter is

$$(1.2) \qquad a_c = \left(\frac{4}{6.02 \times 10^{22}} \right)^{1/3} = 4.05 \times 10^{-8} \, \text{cm}$$

The atomic volume can be calculated without the use of crystallography. The atomic density N of atoms per cubic centimeter is given by

$$(1.3) \qquad N = \frac{N_A}{A} \rho$$

where N_A is Avogadro's number, ρ is the mass density in grams per cubic centimeter, and A is the atomic mass number. Taking Al as an example, where ρ is 2.7 g/cm³ and A is 27, the atomic density is $N = (6.02 \times 10^{23} \times 2.70)/27 = 6.02 \times 10^{22}$ atoms/cm³. The semiconductors Ge and Si have atomic densities of about 4.4×10^{22} and 5.0×10^{22} atoms/cm³, respectively. Metals such as Co, Ni, and Cu have densities of about 9×10^{22} atoms/cm³. The volume Ω_v occupied by an atom is given by

$$(1.4) \qquad \Omega_v = \frac{1}{N}$$

with a typical value of 20×10^{-24} cm³.

The average areal density of a monolayer, N_s atoms/cm², also can be estimated without the use of crystallography by taking the atomic density N to the 2/3 power:

$$(1.5) \qquad N_S \cong N^{2/3}$$

Equation (1.5) gives the average areal density of one monolayer for a material with an atomic density N. Based on this equation, the average areal density N_s for Al is approximately 1.54×10^{15} atoms/cm².

1.3 ENERGY AND PARTICLES

In the SI (or MKS) system of units, the joule (J) is a unit of energy, but the electron volt (eV) is the traditional unit used in ion-solid interactions. We can define 1 eV as the kinetic energy gained by an electron accelerated through a potential difference of 1 V. The charge on the electron is 1.602×10^{-19} C, and a joule is a coulomb-volt, so that the relationship between these units is given by

(1.6)
$$1 \text{ eV} = 1.602 \times 10^{-19} \text{ J}$$

Commonly used multiples of the electron volt are the kiloelectron volt (keV or 10^3 eV) and megaelectron volt (MeV or 10^6 eV).

In ion–solid interactions it is convenient to use centimeter–gram–second (cgs) units rather than SI units in relations involving the charge on the electron. The usefulness of cgs units is clear when considering the Coulomb force between two charged particles with Z_1 and Z_2 units of electronic charge separated by a distance r,

(1.7)
$$F = k_c \frac{Z_1 Z_2 e^2}{r^2}$$

where the Coulomb law constant $k_c = \frac{1}{4\pi\varepsilon_0} = 8.988 \times 10^9$ m/F in the SI system (where 1 farad $\equiv$ 1 ampere sec./volt) and is equal to unity in the cgs system.

The conversion factor follows from

$$e^2 k_c = (1.6 \times 10^{-19} \text{ C})^2 \times 8.988 \times 10^9 \text{ m/F} = 2.3 \times 10^{-28} \text{ C}^2 \text{ m/F}$$

The conversions 1 coulomb $\equiv$ 1 ampère sec. and 1 joule $\equiv$ 1 coulomb-volt lead to the units of the farad:

$$1 \text{ farad} \equiv 1 \text{ ampère sec. per volt}$$

so that

$$\frac{1 \text{ C}^2 \text{m}}{F} \equiv \frac{1 \text{ C}^2 \text{ Vm}}{As} \equiv 1 \text{ Jm} \equiv 10^9 \text{ Jnm} \equiv \frac{10^9 \text{ Jnm}}{\left(1.6 \times 10^{-19} \text{ J/eV}\right)} = \frac{10^{28}}{1.6} \text{ eV nm}$$

and

$$e^2 k_c = 2.31 \times 10^{-28} \frac{\text{C}^2 \text{m}}{F} = \frac{2.31}{1.6} \text{ eV nm} = 1.44 \text{ eV nm}$$

In this book we will follow the cgs units for e^2 with $k_c = 1$, so that

(1.8)
$$e^2 = 1.44 \text{ eV nm}$$

Each nucleus is characterized by a definite atomic number, Z, and mass number, A; for clarity, we use the symbol M to denote the atomic mass in kinematic equations. The atomic number Z is the number of protons, and hence the number of electrons, in the neutral atom; it reflects the atomic properties of the atom. The mass number A gives the number of nucleons: protons and neutrons; isotopes are nuclei (often called nuclides) with the same Z and different A. The current practice is to represent each nucleus by the chemical name with the mass number as a superscript (e.g., ^{12}C). For greater clarity, Z is sometimes written as a subscript, as in $^1_1H, ^2_1H, ^4_2He$, and so on.

The chemical atomic weight (or atomic mass) of elements as listed in the periodic table gives the average mass (i.e., the average of the stable isotopes weighted by their abundance). Carbon, for example, has an atomic weight of 12.011, which reflects the 1.1% abundance of ^{13}C.

The masses of particles may be expressed as given in Table 1.1 in terms of energy through the Einstein relation

(1.9)
$$E = Mc^2$$

which associates 1 J of energy with $1/c^2$ of mass, where c is the velocity of light, $c = 2.998 \times 10^8$ m/s. The mass of an electron, m_e, is 9.11×10^{-31} kg, which is equivalent to an energy of

(1.10) $E = (9.11 \times 10^{-31} \text{kg})(2.998 \times 10^8 \text{m/s})^2 = 8.188 \times 10^{-14} \text{J} = 0.511 \text{ MeV}$

The Einstein relation also is useful when calculating the velocity, v, of an ion of mass M and energy E for a nonrelativistic energy domain where ion beam analysis energies are

(1.11)
$$v = \left(\frac{2E}{M}\right)^{1/2} = c\left(\frac{2E}{Mc^2}\right)^{1/2}$$

TABLE 1.1   Mass Energies of Particles and Light Nuclei

Particle	Symbol	Atomic Mass [u]	Mass [10^{-27} kg]	Mass Energy [MeV]
Electron	e or e$^-$	0.000549	9.1095×10^{-4}	0.511
Proton	p or ^{1}H$^+$	1.007276	1.6726	938.3
Atomic mass unit (amu)	u	1.00000	1.6606	931.7
Neutron	n	1.008665	1.6747	939.6
Deuteron	d or ^{2}H$^+$	2.01410	3.3429	1875.6
Alpha	α or ^{4}He^{2+}	4.00260	6.6435	3727.4

For example, the velocity of a 2 MeV ^{4}He ion is

$$v = 3 \times 10^8 \text{ m/s} \left(\frac{2 \times 2 \times 10^6 \text{ (eV)}}{3727 \times 10^6 \text{ (eV)}} \right)^{1/2} = 9.8 \times 10^6 \text{ m/s}$$

1.4   THE BOHR VELOCITY AND RADIUS

The Bohr atom provides useful relations for simple estimates of atomic parameters. The Bohr radius of the hydrogen atom is given by

(1.12) $$a_0 = \frac{\hbar^2}{m_e e^2} = 0.5292 \times 10^{-8} \text{ cm} = 0.05292 \text{ nm}$$

and the Bohr velocity of the electron in this orbit is

(1.13) $$v_0 = \frac{\hbar}{m_e a_0} = \frac{e^2}{\hbar} = 2.188 \times 10^8 \text{ cm/s}$$

where $\hbar = h/2\pi$ with Planck's constant $h = 4.136 \times 10^{-15}$ eV s. For comparison with the Bohr radius, the radius of a nucleus is given by the empirical formula

(1.14) $$R = R_0 A^{1/3}$$

where A is the mass number and R_0 is a constant equal to 1.4×10^{-13} cm. The nuclear radius is about four orders of magnitude smaller than the Bohr radius.

PROBLEMS

1.1.   Aluminum is a face-centered-cubic metal with lattice parameter a = 4.05 × 10⁻⁸ cm. Calculate the density of atoms/cm² on a {200} planar face using
a. The planar spacing, Equation (1.5), and $d_{200} = 0.203$ nm
b. The relation

$$N_s = \frac{\text{surface atoms/unit cell}}{\text{surface cell area}}$$

1.2.   For the canonical value of 10¹⁵ atoms/cm² in a monolayer on a cubic crystal, estimate
a. The bulk density
b. The volume Ω_V occupied by an atom

1.3. Nickel is a face-centered-cubic metal with an atomic weight of 58.7 and mass density of 8.91 g/cm³.
  a. What are the atomic density of nickel and the atomic volume, Ω_V?
  b. What is the lattice parameter, a_c?

1.4. Silicon has a diamond cubic lattice structure with an atomic density of 5×10^{22} atoms/cm³ and an atomic weight of 28.09.
  a. What are silicon's mass density and average areal density?
  b. What are silicon's lattice parameter and atomic volume, Ω_V?

1.5. Calculate the energy in electron volts of a proton moving at the Bohr velocity. What is the velocity of a 35 keV As ion?

1.6. What is the value of the K-shell radius for silicon ($Z = 14$), germanium ($Z = 32$), and gold ($Z = 79$)?

1.7. Estimate the nucleus size of ^{1}H, ^{4}He, ^{12}C, and ^{79}Au.

SUGGESTED READING

Cullity, B. D. 1978. *Elements of x-ray diffraction.* Boston, MA: Addison–Wesley Publishing Company, Inc.

Feldman, L. C., and Mayer, J. W. 1986. *Fundamentals of surface and thin film analysis.* New York: North–Holland.

Krane, K. S. 1988. *Introductory nuclear physics.* New York: John Wiley & Sons.

Omar, M. A. 1975. *Elementary solid state physics: Principles and applications.* Boston, MA: Addison–Wesley Publishing Company.

Weidner, R. T., and Sells, R. L. 1980. *Elementary modern physics,* 3rd ed. Boston, MA: Allyn & Bacon.

Kinematics

2.1   KINEMATICS OF ELASTIC COLLISIONS

The energy transfers and kinematics in elastic collisions between two isolated particles can be solved fully by applying the principles of conservation of energy and momentum. We consider those collisions in which the kinetic energy is conserved to be *elastic*. An *inelastic* collision does not conserve kinetic energy; an example is the promotion of electrons to higher energy states in collisions where substantial K-shell overlap occurs. The energy lost in promoting the electrons is not available in the particle–atom kinematics after the collision. In this chapter we consider only elastic processes.

For an incident energetic particle of mass M_1, the values of the velocity and energy are v_0 and E_0 ($E_0 = 1/2\, M_1 v_0^2$), while the target atoms of mass M_2 are at rest. After the collision, the values of the velocities v_1 and v_2 and energies E_1 and E_2 of the projectile and target atoms, respectively, are determined by the scattering angle θ and recoil angle ϕ. The notation and geometry for the laboratory system of coordinates are given in Figure 2.1. Table 2.1 is a list of symbols used in kinematic expressions.

Conservation of kinetic energy and conservation of momentum parallel and perpendicular to the direction of incidence are expressed by the equations

(2.1) $$E_0 = \frac{1}{2}M_1 v_0^2 = \frac{1}{2}M_1 v_1^2 + \frac{1}{2}M_2 v_2^2$$

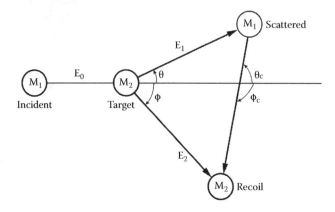

FIGURE 2.1   Elastic collision diagram between two unequal masses as seen in the laboratory reference frame with reference to CM reference frame scattering angles.

TABLE 2.1   Definitions and Symbols Used in Collision Kinematics

E_0	Energy of the incident projectile
E_c	Total kinetic energy in the center-of-mass system
E_1	Laboratory energy of the scattered projectile
E_2	Laboratory energy of the recoiling target
T	Energy E_2 transferred to the target atom
θ_c	Center-of-mass variable scattering angle defined in Figure 2.1
K	Backscattering kinematic factor E_1/E_0
M_1	Mass of the incident projectile
M_2	Mass of the target particle
M_c	Reduced mass in center-of-mass system
X	Mass ratio M_1/M_2
V_0	Velocity of the incident projectile in laboratory coordinates
V_1	Velocity of the scattered projectile in laboratory coordinates
V_2	Velocity of the recoiling atom in laboratory coordinates
V_c	Velocity of the reduced mass in center-of-mass coordinates
V_{ion}	Velocity of the incident projectile (ion) in center-of-mass coordinates
V_{atom}	Velocity of the target atom in center-of-mass coordinates
θ	Laboratory angle of the scattered projectile
θ_c	Center-of-mass angle of the scattered projectile
θ_m	Maximum laboratory angle for M_1 scattering ($M_1 > M_2$)
ϕ	Laboratory angle of the recoiling target atom
ϕ_c	Center-of-mass angle of the recoiling target atom
π	$\pi = 180° = \theta_c + \phi_c$

(2.2)
$$M_1 v_0 = M_1 v_1 \cos\theta + M_2 v_2 \cos\phi$$

(2.3)
$$0 = M_1 v_1 \sin\theta - M_2 v_2 \sin\phi$$

Equations (2.1)–(2.3) can be solved in various forms; for example, transposing the first term on the right to the left side in Equations (2.2) and (2.3), squaring and adding, will eliminate ϕ, giving

(2.4)
$$\left(M_2 v_2\right)^2 = \left(M_1 v_0\right)^2 + \left(M_1 v_1\right)^2 - 2M_1^2 v_0 v_1 \cos\theta$$

Substituting Equation (2.4) into Equation (2.1) to eliminate v_2, one finds the ratio of the particle's velocity after and before the collision:

(2.5)
$$\frac{v_1}{v_0} = \frac{M_1}{M_1 + M_2} \cos\theta \pm \left[\left(\frac{M_1}{M_1 + M_2} \right)^2 \cos^2\theta + \frac{M_2 - M_1}{M_1 + M_2} \right]^{1/2}$$

Equation (2.5) can be used with Equation (2.4) to determine v_2 and E_2, and it can be used with Equation (2.2) to find the angle of recoil, ϕ, of the scattered target atom.

If $M_1 > M_2$, the quantity under the radical in Equation (2.5) will be zero for $\theta = \theta_m$, where θ_m is found from

(2.6)
$$\cos^2\theta_m = 1 - \frac{M_2^2}{M_1^2}, \qquad 0 \le \theta_m \le \frac{\pi}{2}$$

For $\theta > \theta_m$ (and θ), v_1/v_0 is either imaginary or negative, neither of which is physical, so θ_m represents the maximum angle through which M_1 can be scattered.

For the condition $M_1 < M_2$, all values of θ from 0 to π are possible, and a positive value for v_1/v_0 results if the plus sign in Equation (2.5) is chosen. Choice of the minus sign in this equation leads to negative values of v_1/v_0, which is physically unrealistic. The ratio of the projectile energies for $M_1 \le M_2$, where the plus sign holds, is

(2.7)
$$\frac{E_1}{E_0} = \left[\frac{\left(M_2^2 - M_1^2 \sin^2\theta\right)^{1/2} + M_1 \cos\theta}{M_2 + M_1} \right]^2$$

This allows us to define the *kinematic factor* K,

(2.8)
$$K \equiv E_1/E_0$$

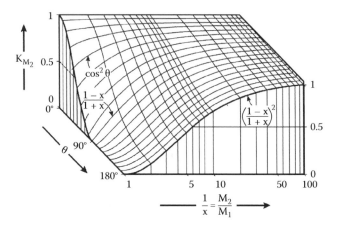

FIGURE 2.2   The kinematic factor K plotted as a function of scattering angle θ and mass ration $x^{-1} = M_2/M_1$. (Adapted from Chu, W.-K., Mayer, J. W., and Nicolet, M -A. 1978. *Backscattering spectrometry*, NY: Academic Press.)

Equation (2.7) can be rewritten as

$$(2.9) \qquad K_{M_2} = \left[\frac{[1-(M_1/M_2)^2 \sin^2 \theta]^{1/2} + M_1/M_2 \cos \theta}{1 + (M_1/M_2)} \right]^2$$

where the subscript has been added to K to indicate the target mass for which the factor applies. Equation (2.9) shows that the kinematic factor depends only on the ratio of the projectile to the target masses and the scattering angle θ.

Figure 2.2 plots Equation (2.9) with the mass ratio $M_1/M_2 = x$. The plot of K versus $M_2/M_1 = x^{-1}$ and θ shows that for any combinations of projectile and target mass, K always has its lowest values at 180°. In addition, the value of K at θ = 180° is the square of its value at θ = 90°. Also, for the condition of $M_1 = M_2$ (i.e., x = 1), the value of K is zero for θ > 90°. The implication is that a projectile colliding with a target equal to its own weight cannot be scattered backward but only forward. This is also true for the condition $M_1 > M_2$.

The kinematic factor, as expressed by Equations (2.7)–(2.9), is the term in the applications of backscattering spectrometry for sensing the mass of target atoms. In backscattering spectrometry experiments the energy E_0 and mass M_1 of the projectile are known and the energy of the backscattered projectile E_1 is measured at angle θ. The only unknown is the mass of the target atom M_2 that prompted the backscattering event.

When a target contains two different atoms with mass difference ΔM_2 it is important that small differences in masses produce as large a change in ΔE_1 as possible if these masses are to be observed separately. Figure 2.2 shows that for fixed M_1 a change in ΔM_2 produces the largest change in K for θ = 180° for all but the smallest values of M_2.

In quantitative terms, for the condition that $\theta \sim 180°$, ΔM_2 and ΔE_1 are related to each other by (Chu, Mayer, and Nicolet 1978)

$$(2.10) \qquad \Delta E_1 = E_0 \frac{1-x}{(1-x)^3} [4(1-x\delta^2) - \delta^2(1-x^2)]x \frac{\Delta M_2}{M_2}$$

where $\delta = \pi - \theta$. For $M_2 \gg M_1$, this reduces to

$$(2.11) \qquad \Delta E_1 = E_0 [4(1-x\delta^2)(M_1/M_2^2)\Delta M_2$$

Both Equations (2.10) and (2.11) show that ΔE_1 increases with E_0 and ΔM_2. This is demonstrated in Figure 2.3, where the backscattered ($\theta \sim 180$) energies for an He projectile (E_1) on targets of C, O, Si, Ti, and Bi for E_0 equal to 2 and 5 MeV are plotted separately. The normalized heights of these elements are indicative of their scattering strength or—more precisely, scattering cross section—that is discussed in Chapter 3.

Alternately, increases in ΔE_1 can also be accomplished by increasing M_1 (using a heavier incident ion beam such as Li). Keep in mind that M_2 values less than M_1 can produce forward scattering but will not produce any backscattering.

Additional equations related to Figure 2.1 are listed in Table 2.2.

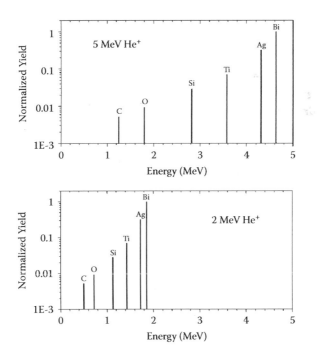

FIGURE 2.3   Energies for a He projectile (E_1) on targets of C, O, Si, Ti, and Bi for E_0 equal to 2 and 5 MeV in backscattering geometry ($\theta \sim 180$).

TABLE 2.2   Relationships between Energy and Scattering Angles

Center-of-mass energy	$E_c = \dfrac{M_2 E_0}{M_1 + M_2} = \dfrac{E_0}{1+x}; \quad x = \dfrac{M_1}{M_2}$
Laboratory energy of the scattered projectile for $M_1 \leq M_2$	$K = \dfrac{E_1}{E_0} = \dfrac{\left[x \cdot \cos\theta + \left(1 - x^2 \sin^2\theta\right)^{1/2} \right]^2}{(1+x)^2}$ When $M_1 = M_2, \theta \leq \dfrac{\pi}{2}$
Laboratory energies of the scattered projectile for $M_1 > M_2$	$\dfrac{E_1}{E_0} = \dfrac{\left[x \cdot \cos\theta \pm (1 - x^2 \sin^2\theta)^{1/2} \right]^2}{(1+x)^2}$ $\theta \leq \sin^{-1}(1/x)$
Laboratory energy of the recoil nucleus	$\dfrac{E_2}{E_0} = 1 - \dfrac{E_1}{E_0} = \dfrac{4 M_1 M_2}{\left(M_1 + M_2\right)^2} \cos^2\varphi =$ $\dfrac{4x}{(1+x)^2} \cos^2\varphi = \dfrac{4x}{(1+x)^2} \sin^2\left(\dfrac{\theta_c}{2}\right)$ where $\phi \leq \dfrac{\pi}{2}$
Laboratory angle of the recoil nucleus	$\phi = \dfrac{\pi - \theta_c}{2} = \dfrac{\phi_c}{2} \qquad \sin\varphi = \left(\dfrac{M_1 E_1}{M_2 E_2}\right)^{1/2} \sin\theta$
Laboratory angle of the scattered projectile	$\tan\theta = \dfrac{M_2 \sin\theta_c}{M_1 + M_2 \cos\theta_c}$
Center-of-mass angle of the scattered projectile	$\theta_c = \pi - 2\phi = \pi - \phi_c$ When $M_1 \leq M_2 \Rightarrow x \leq 1$, θ_c is defined for all $\theta \leq \pi$ and $\theta_c = \theta + \sin^{-1}(x \sin(\theta))$. When $M_1 > M_2 \Rightarrow x > 1$, θ_c is double valued and the laboratory scattering angle is limited to the range $\theta \leq \sin^{-1}(1/x)$. In this case: $\theta_c = \theta + \sin^{-1}(x \sin\theta)$ or $\theta_c = \pi + \theta - \sin^{-1}(x \sin\theta)$

Source: Weller, R. 2009. Appendix 4 in *Handbook of Modern Ion Beam Materials Analysis*, 2nd ed., ed. Y. Wang and M. Nastasi, Warrendale, PA: MRS Publisher.

2.2   CLASSICAL TWO-PARTICLE SCATTERING

The collision and scattering problem defined by Figure 2.1 will now be restated in terms of center-of-mass (CM) coordinates. The motivation for this transformation will be obvious when we discuss scattering in a central force field later in this chapter. Through the use of CM coordinates it will be shown that no matter how complex the force is between the two particles, so long as it acts only along the line joining them (no transverse forces), the relative motion of the two particles can be reduced to that of a single particle moving in an interatomic potential centered at the origin of the center-of-mass coordinates. By introducing the CM system, the mutual interaction of the two colliding particles can be described by a force field, V(r), which depends only on the absolute value of the interatomic separation, r. The motion of both particles is given by one equation of motion. This equation has r as the independent variable and describes a particle moving in the central force field V(r).

The CM coordinates for a two-particle system are defined in a zero-momentum reference frame. In the frame, the total force on two particles that interact only with each other is zero. We can define the total force of two interacting particles as

(2.12) $$\mathbf{F_T} = \mathbf{F_1} + \mathbf{F_2} = \frac{d\mathbf{p_T}}{dt}$$

where $\mathbf{F_T}$ = total force, $\mathbf{F_1}$ and $\mathbf{F_2}$ are the individual forces on particles 1 and 2, respectively, and $\mathbf{p_T}$ is the total linear momentum of the two-particle system. For $\mathbf{F_T} = 0$, $d\mathbf{p_T} = 0$, indicating that the total momentum is unchanged or conserved during the interaction process.

One of the consequences associated with observing elastic collisions in the CM coordinates is that the individual particles' kinetic energies are unchanged by the collision process. Thus, the CM velocities of the two colliding particles are the same before and after the collision process. In addition, the CM scattering angle of particle 1 will equal the scattering angle of particle 2. Finally, all scattering angles in the CM system are allowed, unlike the scattering angles in the laboratory reference frame, where the allowed scattering angles depend on the ratio $x = M_1/M_2$.

The collision processes described in Figure 2.1 is represented in Figure 2.4. For CM coordinates, Figure 2.4(b), we define the system velocity, v_c, such that in this coordinate system there is no net momentum change, so

(2.13) $$M_1 v_0 = (M_1 + M_2) v_c$$

We also define in CM coordinates a reduced mass, M_c, given by the relation

(2.14) $$\frac{1}{M_c} = \frac{1}{M_1} + \frac{1}{M_2}$$

or

$$(2.15) \qquad M_c = \frac{M_1 M_2}{M_1 + M_2}$$

From Equations (2.13) and (2.15) we can represent the CM velocity in terms of reduced mass as

$$(2.16) \qquad v_c = v_0 \frac{M_c}{M_2}$$

From the velocity vector diagram in Figure 2.4 and Equation (2.16), the ion and target atom velocities in CM coordinates are

$$(2.17) \qquad v_{ion} = v_0 - v_c = v_0 \frac{M_c}{M_1}$$

$$(2.18) \qquad v_{atom} = v_c = v_0 \frac{M_c}{M_2}$$

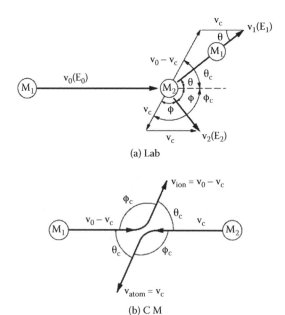

(a) Lab

(b) C M

FIGURE 2.4   Schematic representation of classical two-particle scattering in (a) laboratory reference frame and (b) CM reference frame.

Equation (2.18) shows that the target atom, which has zero velocity before the collision in the laboratory reference frame, has the system velocity v_c before and after the collision in the CM reference frame.

Equations (2.17) and (2.18) show the advantage of the CM reference frame. The system velocity, v_c, and the atom and ion velocities, v_{atom} and v_{ion}, remain constant and are independent of the final scattering angle between the two particles (Figure 2.4b). Thus, regardless of whether the collision is elastic or inelastic, the total momentum is unchanged in a collision. In addition, from Equations (2.17) and (2.18), we see that the ratio of the ion to atom velocities is inversely proportional to the ratio of their masses:

$$(2.19) \qquad \frac{v_{ion}}{v_{atom}} = \frac{v_0 - v_c}{v_c} = \frac{M_2}{M_1} = \frac{1}{x}$$

Another advantage to the CM reference frame is that the CM total energy, E_c, is equal to the CM initial kinetic energy:

$$(2.20a) \qquad E_c = \frac{1}{2} M_c v_0^2$$

$$(2.20b) \qquad E_c = \frac{1}{2} \frac{M_1 M_2}{M_1 + M_2} v_0^2 = \frac{M_2}{M_1 + M_2} E_0$$

where

$$E_0 = \frac{1}{2} M_1 v_0^2.$$

The conversion of scattering angles from the laboratory system to the CM system is determined from the scattering diagrams given in Figures 2.1 and 2.4. Examining the target atom (M_2) trajectory portion of Figure 2.4(a), we see that the final target velocity in the laboratory, v_2, is related to the CM atom velocity, $v_{atom} = v_c$, by the difference vector, v_c. Since the triangle formed by these velocity vectors is isosceles, we have

$$(2.21) \qquad \phi_c = 2\phi$$

From the CM diagram, Figure 2.4(b), we have $\theta_c + \phi_c = \pi$, which allows us to rewrite Equation (2.21) in the form

$$(2.22) \qquad \phi = \frac{\pi - \theta_c}{2}$$

which relates the target atom scattering angle in the laboratory to the CM ion scattering angle.

Another important relationship is the energy transferred to the target atom as a function of the target atom scattering angle θ_c or θ. Again, from the velocity vector diagram in Figure 2.4(a), and the law of cosines, we have

$$(2.23) \qquad v_2^2 = v_c^2 + \left[v_c^2 - 2v_c^2 \cos(\pi - \phi_c) \right]$$

Using Equations (2.21) and (2.22) to recast ϕ_c in terms of θ_c, we obtain

$$(2.24) \qquad v_2^2 = 2v_c^2 (1 - \cos \theta_c)$$

which relates the target atom recoil velocity in the laboratory to the CM velocity and the CM ion scattering angle. Equation (2.24) can be simplified by using Equations (2.18) and (2.21) to obtain

$$(2.25) \qquad v_2 = 2v_0 \frac{M_c}{M_2} \cos \phi$$

which gives the laboratory recoil velocity, v_2, as a function of the initial ion velocity, v_0, and the laboratory recoil angle. This equation can now be used to obtain the energy transferred to the target atom by the incident ion through the kinetic energy velocity relationship,

$$(2.26) \qquad E_2 = \frac{1}{2} M_2 v_2^2$$

In many books, the energy *transferred* to the target atom, E_2, is referred to as T. Substituting Equation (2.25) into Equation (2.26) gives

$$(2.27) \qquad T \equiv E_2 = \frac{M_2}{2} \left(\frac{v_0 M_c \cos \phi}{M_2} \right)^2$$

The transferred energy, T, can be related to the ion scattering angle, θ_c, by Equation (2.22) to yield

$$(2.28) \qquad T = \frac{2}{M_2} \left(v_0 M_c \sin \frac{\theta_c}{2} \right)^2 = \frac{4E_c M_c}{M_1} \sin^2 \frac{\theta_c}{2}$$

From the description of reduced mass, Equation (2.15), we rewrite Equation (2.28) to obtain

$$(2.29) \qquad T = E_0 \frac{4 M_1 M_2}{(M_1 + M_2)^2} \sin^2 \frac{\theta_c}{2} = T_M \sin^2 \frac{\theta_c}{2}$$

where T_m is the maximum energy transferable in a head-on collision when $\theta_c = 0$.

As an example, to determine the energy transferred in a binary collision where a 1 MeV boron ($M_1 = 10$) ion incident on Si ($M_2 = 28$) is scattered through a laboratory angle $\theta = 45°$, one first determines the corresponding CM angle θ_c from the expression given in Table 2.1, $\theta_c = \theta + \sin^{-1}(x \sin\theta)$, which gives $\theta_c = 60°$. Next we calculate the ratio T_M/E from Equation (2.29), which gives $T_M = 0.78E_0$. Finally, for $E_0 = 1$ MeV, $T = 0.195$ MeV.

Additional relationships between the CM and the laboratory reference frames are summarized in Table 2.2.

2.3   THE CLASSICAL SCATTERING INTEGRAL

In this section we will derive an expression for the CM scattering angle θ_c. We will see that θ_c will depend on the interatomic potential V(r), the ion energy E, and the impact parameter, b.

We begin by defining the scattering trajectory of a moving particle in a central force field, assuming that the force between the two particles acts only along the line joining them and that there are no transverse forces. The use of CM coordinates then reduces any two-body problem to a one-body problem—namely, the interaction of a particle with mass M_c and velocity v_c with a static potential field, V(r), centered at the origin of the CM coordinates. This simplification occurs because in the CM system the *total linear momentum of the particles is always zero*, the paths of the two particles are symmetric (as shown in Figure 2.4), and the evaluation of the path of one particle (scattering angle) directly gives the path of the other particle. The conversion from CM scattering angles to laboratory angles is then achieved with the equations summarized in Table 2.2.

In Figure 2.5 we represent the scattering process between an atom moving with initial velocity v_0 and energy E_0 and a stationary target atom with the details of the scattering trajectories displayed for both the laboratory and the CM reference frames. The distance b in the figure is the *impact parameter* and defines the perpendicular distance between the initial position of the target atom and the incident trajectory of the ion. This parameter will be shown to be an important quantity in the scattering process and will define the hardness of the collision. The dashed lines in Figure 2.5 represent the asymptotes of the ion and target atom trajectories. The parameter r_{min} is the distance of closest approach during the scattering event.

Since we are dealing with two particles only and no transverse forces, the problem is two dimensional in the plane defined by the initial velocity vector for the ion and the initial position of the target atom. Since we are dealing with conservative central forces defined by an ion–atom interaction potential V(r), conservation of energy in the center-of-mass system will be

$$(2.30) \qquad E_c = \frac{1}{2}M_c\left(\dot{r}^2 + r^2\dot{\Theta}_c^2\right) + V(r)$$

where the first term is the system kinetic energy. The variable r is defined in Figure 2.5 as

$$(2.31) \qquad r = r_1 + r_2$$

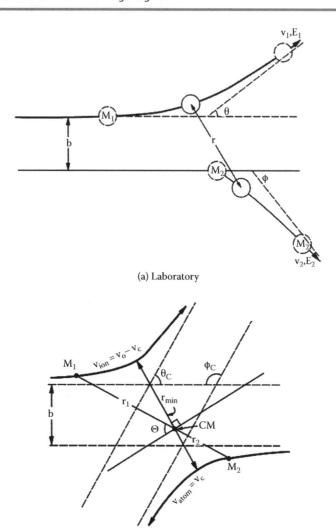

(a) Laboratory

(b) Center-of-Mass

FIGURE 2.5   The collision trajectories at an impact parameter, b, for an elastic collision between two unequal masses, as seen in the (a) laboratory reference frame and (b) CM reference frame.

with the CM distances r_1 and r_2 defined by

(2.32a)
$$r_1 = \frac{M_2}{M_1 + M_2} \, r$$

(2.32b)
$$r_2 = \frac{M_1}{M_1 + M_2} \, r$$

The variable r is the CM separation distance between M_1 and M_2, and r_1 and r_2 represent the distance from the center of mass to the ion (M_1) and the target atom (M_2), respectively. The value $\dot{\Theta}_c$ is the time rate of change in the scattering angle, $d\Theta_c/dt$, and Θ_c is defined as the angle between line $r_1 + r_2$ and the line perpendicular to r_{min} and is different from the CM scattering angle θ_c. The energy E_c is the CM defined in Equation (2.20), and M_c is the CM defined in Equation (2.15).

In addition to the conservation of energy, we have the law of conservation of angular momentum that, during the scattering process in the CM system, is given by

$$(2.33) \qquad\qquad \ell = M_c\, r^2\, \dot{\Theta}_c$$

where ℓ is the constant angular momentum. For large values of r, the angular momentum is simply related to the impact parameter and has the magnitude $M_c\, v_0\, b$. Since angular momentum is conserved, we have

$$(2.34) \qquad\qquad \ell = M_c\, r^2\, \dot{\Theta}_c = M_c\, v_0\, b$$

from which we obtain

$$(2.35) \qquad\qquad \dot{\Theta}_c = \frac{v_0 b}{r^2}$$

With Equations (2.20a), (2.30), and (2.35), we can now solve for Θ_c as a function of any central force potential, $V(r)$. From these equations we obtain the radial equation of motion,

$$(2.36) \qquad\qquad \dot{r} = v_0 \left(1 - \frac{V(r)}{E_c} - \left(\frac{b}{r}\right)^2\right)^{1/2}$$

Using Equations (2.35) and (2.36), and noting that $\dot{r} = dr/dt$ and that $\dot{\Theta}_c = d\Theta_c/dt$, we obtain

$$(2.37) \qquad \frac{d\Theta_c}{dr} = \frac{d\Theta_c}{dt}\frac{dt}{dr} = \frac{b}{r^2\left[1 - \dfrac{V(r)}{E_c} - \left(\dfrac{b}{r}\right)^2\right]^{1/2}}.$$

The CM scattering angle θ_c is found by integrating Θ_c on the left-hand side of Equation (2.37) over the first half of the orbit, from $\theta_c/2$ to $\pi/2$, which corresponds to the integration limits on the right-hand side of r_{min} to infinity:

$$\int_{\theta_c/2}^{\pi/2} d\Theta_c = \int_{r_{min}}^{\infty} \frac{b \, dr}{r^2 \left[1 - \dfrac{V(r)}{E_c} - \left(\dfrac{b}{r}\right)^2\right]^{1/2}} \tag{2.38}$$

$$\frac{1}{2}(\pi - \theta_c) = \int_{r_{min}}^{\infty} \frac{b \, dr}{r^2 \left[1 - \dfrac{V(r)}{E_c} - \left(\dfrac{b}{r}\right)^2\right]^{1/2}} \tag{2.39}$$

This reduces to

$$\theta_c = \pi - 2b \int_{r_{min}}^{\infty} \frac{dr}{r^2 \left[1 - \dfrac{V(r)}{E_c} - \left(\dfrac{b}{r}\right)^2\right]^{1/2}} \tag{2.40}$$

This final equation is called the *classical scattering integral* and gives the angular trajectory information for two-body central-force scattering. Equation (2.40) allows us to evaluate the scattering angle θ_c in terms of energy, E_c; the interatomic potential, $V(r)$; and the impact parameter, b. The scattering angle of an ion with energy E, moving in a force field defined by $V(r)$, will vary with the impact parameter b. The significance of this will become clear when we discuss the differential scattering cross section in Chapter 3, where Equation (2.40) is solved explicitly for a Coulomb potential (Rutherford scattering). In general, potentials are more complex and numerical solutions are required. Transformations from the CM angle θ_c to the laboratory angles θ and ϕ can be made with the help of Table 2.2.

PROBLEMS

2.1.   a. What is the velocity for a 1 MeV He ion and for a 1 MeV proton?
        b. What is the velocity of a 1 MeV He ion after a 180° backscattering collision with a silicon atom?

2.2.   What is the laboratory energy for 1 MeV He ions scattered from carbon at a scattering angle of 150°? How about if the He ions are replaced with the same energy of protons or lithium ions?

2.3.   For a detector that can resolve a 30 keV energy difference, what change in mass can be detected by 2 MeV He ion scattering at 180°? Can you resolve two isotopes of Li and Cu with this system?

2.4. Suppose 1 MeV He ions interact with silicon atoms in an experiment:
   a. What is the reduced mass in the collision?
   b. Determine the final silicon atom velocity if the collision takes place at 90° (right angle).
   c. Find the center of mass angle for the silicon atom after a 90° (right angle) collision.

2.5. What is the maximum energy transferred in a head-on collision of a 1 MeV He ion with silicon? What are the corresponding energy and scattering angle of He ions after such a head-on collision?

2.6. Derive the expression for the laboratory energy of the recoil nucleus as written in Table 2.2.

2.7. Write a simple expression for E_1/E_0 and E_2/E_0 in backscattering (180°) and right angle scattering (90°) for $M_1 = M_2$, $M_1 > M_2$, and $M_1 < M_2$. What are the allowed solutions?

2.8. What is the maximum energy transferred to electrons, silicon atoms, and copper atoms by incident 1 MeV electrons, silicon ions, and copper ions?

2.9. In the laboratory system, we have He ions at 1 MeV scattered from silicon atoms at $\theta = 10°$:
   a. In the laboratory system what are v_1, v_2, ϕ, and E_2?
   b. In the center-of-mass system, what are v_{ion}, θ_c, and ϕ_c?

2.10. Solve the scattering integral, Equation (2.40) for the unscreened Coulomb potential $V(r) = Z_1 Z_2 e^2/4$.

REFERENCES

Weller, R. 2009. Appendix 4. In *Handbook of modern ion beam materials analysis*, 2nd ed., eds. Y. Wang and M. Nastasi. Warrendale, PA: MRS Publisher.

SUGGESTED READING

Chu, W.-K, Mayer, J. W., and Nicolet, M.-A. 1978. *Backscattering spectrometry*, chap. 2. New York: Academic Press.
Feldman, L. C., and Mayer, J. W. 1986. *Fundamentals of surface and thin film analyses*. New York: North–Holland Science Publishing.
French, A. P. 1971. *Newtonian mechanics*. New York: W. W. Norton and Co.
Goldstein, H. 1959. *Classical mechanics*. Reading, MA: Addison–Wesley Publishing Co.
Johnson, R. E. 1982. *Introduction to atomic and molecular collisions*. New York: Plenum Press.
Symon, K. R. 1953. *Mechanics*. Reading, MA: Addison–Wesley Publishing Co.

Torrens, I. M. 1972. *Interatomic potentials.* New York: Academic Press.

Weller, R. 1995. Appendix 4. In *Handbook of modern ion-beam materials analysis*, ed. J. R. Tesmer and M. Nastasi. Pittsburg: Materials Research Society.

Ziegler, J. F., Biersack, J. P., and Littmark, U. 1985. *The stopping and range of ions in solids.* New York: Pergamon Press Inc.

Cross Section

3

3.1 INTRODUCTION

In ion beam analysis experiments, many ions interact with many target nuclei. Due to the large number of interactions, the questions of how much energy will be transferred in a collision or what the scattering angle will be must be answered. The differential cross section is the fundamental parameter that we will develop. It gives a measure of either the probability of transferring energy T in the range between T and $T + dT$ to a target atom or of the probability of scattering a projectile into some angle between θ_c and $\theta_c + d\theta_c$. We will discuss the scattering cross section in this chapter. Chapter 4 will deal with the stopping cross section that is related to the probability of energy transfer. The differential cross section has units of area, typically centimeters squared. The differential cross section integrated over all angles is the total cross section, often referred to simply as the cross section.

3.2 ANGULAR DIFFERENTIAL SCATTERING CROSS SECTION

It is customary to describe the number of particles scattered through different angles θ_c in terms of a quantity called the angular differential scattering cross section. Imagine the experiment depicted in Figure 3.1, where a beam of ions is incident on a thin foil and is scattered into a detector of area Δa at a polar angle between θ_c and $\theta_c + d\theta_c$. Each of the ions in the incident beam has a different impact parameter b (as described in Chapter 2) and will be scattered through a different angle. We

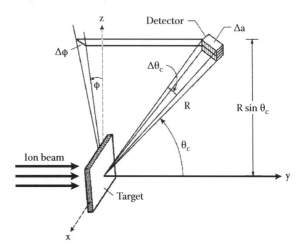

FIGURE 3.1 Experiment for measuring angular differential cross section. The detector area is $\Delta a = (R\Delta\theta_c)(R\sin\theta_c\Delta\varphi)$. By moving the detector to all angular positions for a fixed R, all the scattered particles can be counted, and the detector will have covered an area $4\pi R^2$, or a total solid angle of 4π.

define the differential dn_θ as the number of ions scattered into the detector of area Δa, between angles θ_c and $\theta_c + d\theta_c$, per unit time. We also define I_0 to be the flux of incident particles equal to the number of ions incident on the sample per unit time, per unit area (i.e., ions per second per centimeter squared). The solid angle of the detector, $\Delta\Omega$, is related to the detector area, Δa, and its distance away from the sample, R, and is given by

$$(3.1) \qquad \Delta\Omega = \frac{\Delta a}{R^2} = \frac{(R\Delta\theta_c)(R\sin\theta_c\Delta\varphi)}{R^2} = \Delta\theta_c\Delta\varphi \; \sin\theta_c$$

We now define $d\sigma(\theta_c)$, the differential scattering cross section, to be given by

$$(3.2) \qquad \frac{d\sigma(\theta_c)}{d\Omega} \equiv \frac{1}{I_0}\frac{dn_{\theta_c}}{d\Omega}$$

where, for $\Delta a \to 0$, we have $\Delta\Omega \to d\Omega$. The term $d\sigma(\theta_c)/d\Omega$ is the differential scattering cross section per unit solid angle, and $dn_{\theta_c}/d\Omega$ is the number of particles scattered into the angular regime between θ_c and $\theta_c + d\theta_c$ per unit solid angle, per unit time. Since the solid angle Ω units (steradian) are dimensionless, the differential scattering cross section has units of area.

The cross section is simply the effective target area presented by each scattering center (target nucleus) to the incident beam. At a more microscopic level, the scattering cross section can be shown to be dependent on b, the impact

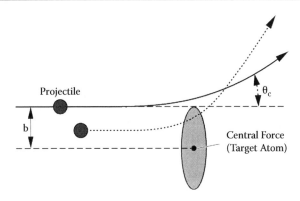

FIGURE 3.2   Scattering of a particle that approaches a nucleus with an impact parameter b. The total cross section is $\sigma = \pi b^2$.

parameter. In Figure 3.2 we present the collision process in which the incident particle is scattered by a target nucleus through an angle θ_c. The projectile moves in a nearly straight line until it gets fairly close to the target nucleus, at which point it is deflected through an angle θ_c. After being deflected, the trajectory of the particle is again nearly a straight line. If there had been no interaction force between the projectile and the target nucleus, the projectile would have maintained a straight trajectory and passed the target nucleus at a distance b.

Examining Figure 3.2, we see that all incident particles with impact parameter b are headed in a direction to strike the rim of the circle drawn around the target nucleus and will be deflected by an angle θ_c. The area of this circle is πb^2, and any particle with a trajectory that strikes anywhere within this area will be deflected by an angle greater than θ_c. The target area defined by the impact parameter is called the total cross section $\sigma(\theta_c)$:

$$(3.3) \qquad\qquad \sigma(\theta_c) = \pi b^2$$

For projectiles moving with small values of b, the cross section defined by Equation (3.3) will be small, but, due to the interaction forces, the scattering angle will be large. Thus, b is proportional to $\sigma(\theta_c)$, while I_0 and $\sigma(\theta_c)$ are inversely related to θ_c. From this discussion we see that $b = b(\theta_c)$.

In addition to the total cross section, there is the differential cross section, $d\sigma(\theta_c)$, and its relationship to b. As shown in Figure 3.3, particles incident with impact parameters between b and b + db will be scattered through angles between θ_c and $\theta_c + d\theta_c$. The differential cross section for this process is found by taking the differential of Equation (3.3) with respect to the impact parameter:

$$(3.4) \qquad\qquad d\sigma(\theta_c) = d(\pi b^2) = 2\pi b \, db$$

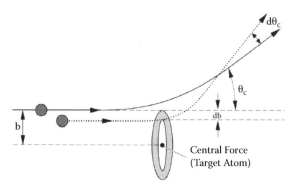

FIGURE 3.3   Nuclear target area for the differential cross section $d\sigma = 2\pi b db$.

From the description given in Equation (3.4) and the schematic presented in Figure 3.3, the differential cross section of each target nucleus is presented as a ring of radius b, a circumference $2\pi b$, and width db. Any incident particles with an impact parameter within db will be scattered into angles between θ_c and $\theta_c + d\theta_c$.

From the examples presented in Figures 3.2 and 3.3, we see that there is a unique connection between the value of I_0 and the scattering angle θ_c. To find the dependence of $d\sigma(\theta_c)$ on the scattering angle, we rewrite Equation (3.4) in the form

$$(3.5) \qquad d\sigma(\theta_c) = 2\pi \, b(\theta_c) \left| \frac{db(\theta_c)}{d\theta_c} \right| d\theta_c$$

We use the absolute value of $db(\theta_c)/d\theta_c$ to maintain $d\sigma(\theta_c)$ as a positive value; θ_c increases as b decreases, indicating that $db(\theta_c)/d\theta_c$ is negative.

To determine an expression for the differential scattering cross section per unit solid angle (Equation 3.1), we note that scattering experiments are performed by observing the number of incident particles that are scattered into a solid angle located at θ_c. Measurements give information in units of the number of scattering particles per element of solid angle. A schematic of this process is presented in Figure 3.4. The annular region represents the solid angle $d\Omega$ subtended between the scattering angles θ_c and $\theta_c + d\theta_c$. The entire area of the sphere of radius R is $4\pi R^2$ and the total solid angle of the sphere is 4π. The shaded area is a ring of radius equal to $R \sin \theta_c$, circumference equal to $2\pi R \sin \theta_c$ and width of $R d\theta_c$. The area of the shaded region is therefore $(2\pi)(R \sin \theta_c)(R \, d\theta_c) = 2\pi R^2 \sin \theta_c \, d\theta_c$. By definition of solid angle, area/R^2, we obtain

$$(3.6) \qquad d\Omega = 2\pi \sin \theta_c d\theta_c$$

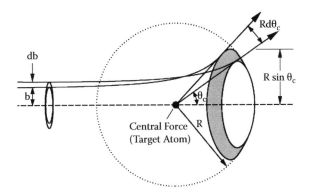

FIGURE 3.4 The solid angle $d\Omega$ subtended at the scattering angle θ_c by the incremental angle $d\theta_c$. By definition, $d\Omega/4\pi$ is the shaded area divided by the entire area of spherical surface; the shaded area is equal to $2\pi(R\sin\theta_c)(Rd\theta_c)$. Then, $d\Omega/4\pi = 2\pi R^2\sin\theta_c d\theta_c/4\pi^2$; therefore, $d\Omega = 2\pi \sin\theta_c d\theta_c$.

The result is equivalent to Equation (3.1), where $\Delta\phi$ has been integrated over 2π. The differential scattering cross section for scattering into a solid angle (Equation 3.1) is obtained by combining Equations (3.5) and (3.6) to produce

(3.7)
$$\frac{d\sigma(\theta_c)}{d\Omega} = \frac{b}{\sin\theta_c}\left|\frac{db}{d\theta_c}\right|$$

Equations (3.5) and (3.7) give the differential scattering cross section in the center of mass. The equivalent expressions in the laboratory reference frame can be obtained for the scattered projectile and scattered target nucleus by using the angular relationships presented in Table 2.2 in Chapter 2.

Integration of Equation (3.7) provides a relationship between the differential scattering cross section and the impact parameter:

$$\int_0^b b(\theta_c)\,db = \int_{\theta_c}^{\pi} \frac{d\sigma(\theta_c)}{d\Omega} \sin\theta_c\,d\theta_c$$

This results in the expression

(3.8)
$$b^2 = 2\int_{\theta_c}^{\pi} \frac{d\sigma(\theta_c)}{d\Omega} \sin\theta_c\,d\theta_c$$

where the dependence of scattering angles on the impact parameter has been omitted for brevity. By linking Equation (3.8) with the expression for θ_c (Equation 2.40 in Chapter 2), an effective means of passing between $V(r)$ and $d\sigma(\theta_c)$ can be established.

3.3   RUTHERFORD DIFFERENTIAL CROSS SECTION

As an example of the use of the angular differential cross section, we consider the condition where the interaction between colliding particles is purely Coulombic, as is the case for Rutherford scattering; for this situation the projectile and target nucleus are treated as pure nuclei, with the projectile described by mass and atomic number M_1 and Z_1 and the target nucleus described by mass and atomic number M_2 and Z_2. The interatomic potential for Coulomb interaction in the centimeter–gram–second (cgs) unit is given by

(3.9)
$$V(r) = \frac{Z_1 Z_2 e^2}{r}$$

where r is the distance of separation between the two nuclei. To put Equation (3.9) into the same form as Equation (2.40), we make the following substitutions:

(3.10a)
$$u \equiv \frac{1}{r}$$

and

(3.10b)
$$\alpha = Z_1 Z_2 e^2$$

leading to

(3.11)
$$V(u) = \alpha\, u$$

With the interatomic potential written in this way, the angular scattering integral (Equation 2.40) becomes

(3.12)
$$\theta_C = \pi - 2 \int_0^{1/r_{min}} \frac{du}{\left[\frac{1}{b^2} - \frac{\alpha u}{E_c b^2} - u^2 \right]^{1/2}}$$

Equation (3.12) can be integrated exactly by noting the following integral solution:

$$\int \frac{dx}{(a + cx + dx^2)^{1/2}} = \frac{-1}{(-d)^{1/2}} \sin^{-1}\left(\frac{c + 2dx}{(q)^{1/2}} \right)$$

where $q = c^2 = 4ad$. For Equation (3.12) these variables are equal to

$$a = \frac{1}{b^2}; \quad c = \frac{-\alpha}{E_c\, b^2}; \quad d = -1$$

$$q = \frac{4}{b^2}\left(1 + \frac{\alpha^2}{4E_c^2\, b^2}\right)$$

and

$$c + 2dx = -\left(2u + \frac{\alpha}{E_c\, b^2}\right)$$

Carrying out these substitutions, the solution to Equation (3.12) is now given by

$$(3.13) \qquad \theta_c = \pi - 2\left[\sin^{-1}\left(\frac{-\left(bu + \frac{\alpha}{2E_c b}\right)}{\left(1 + \frac{\alpha^2}{4E_c^2 b^2}\right)^{1/2}}\right)\right]_0^{1/r_{min}}$$

To complete the integration, a value for r_{min} must first be obtained. It can be shown from the law of conservation of angular momentum that the distance of closest approach, r_{min}, is related to the ion energy, E_c, and the form of the inter-atomic potential, $V(r_{min})$, through the following expression (Nastasi, Mayer, and Hirvonen 1996)

$$0 = 1 - \frac{V(r_{min})}{E_c} - \frac{b^2}{r_{min}^2}$$

Using the change in variables defined by Equations (3.10) and (3.11), we have

$$(3.14) \qquad b^2 u_{min}^2 + \frac{\alpha u_{min}}{E_c} - 1 = 0$$

where $u_{min} = 1/r_{min}$. Equation (3.14) is solved for μ_{min} using the quadratic equation, and it has the solution

$$(3.15) \qquad u_{min} = \frac{1}{r_{min}} = \frac{1}{b}\left(\frac{-\alpha}{2bE_c} \pm \left(\left(\frac{\alpha}{2bE_c}\right)^2 + 1\right)^{1/2}\right).$$

Applying Equation (3.15) to the upper limit in Equation (3.13) gives

(3.16)
$$\theta_c = \pi - 2\left\{\pm\frac{\pi}{2} - \sin^{-1}\left(\frac{\frac{\alpha}{2\,E_c\,b}}{\left(1+\left(\frac{\alpha}{2\,E_c\,b}\right)^2\right)^{1/2}}\right)\right\}$$

This can be rewritten as

(3.17)
$$\frac{\theta_c - \pi}{2} = \pm\frac{\pi}{2} + \sin^{-1}\left(\frac{\alpha}{2\,E_c\,b}\left[1+\left(\frac{\alpha}{2\,E_c\,b}\right)^2\right]^{-1/2}\right)$$

We now use Equation (3.17) to express b in terms of θ_c. Equation (3.17) can be rewritten as

$$\sin\left(\frac{\theta_c - \pi}{2}\pm\frac{\pi}{2}\right) = \pm\sin\left(\frac{\theta_c}{2}\right) = \frac{\frac{\alpha}{2\,E_c\,b}}{\left[1+\left(\frac{\alpha}{2\,E_c\,b}\right)^2\right]^{1/2}}$$

The trigonometric representation of this equation is presented in Figure 3.5, which allows us to construct the following relationship between the impact parameter b and the scattering angle θ_c:

(3.18)
$$b = \frac{\alpha}{2E_c}\cot\left(\frac{\theta_c}{2}\right) = \frac{\alpha}{2E_c}\frac{\cos(\theta_c/2)}{\sin(\theta_c/2)}.$$

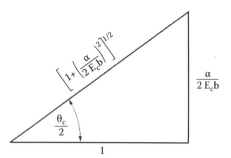

FIGURE 3.5   Trigonometric relation between the center of mass scattering angle, θ_c, and the impact parameter, b, for the Coulomb potential.

We will now use Equation (3.18) together with Equation (3.7), $d\sigma(\theta_c)/d\Omega =$ $b/\sin\theta_c \, |db/d\theta_c|$, to obtain the differential cross section for scattering into a solid angle $d\Omega$ for the Coulomb potential. Differentiating Equation (3.18) with respect to θ_c,

(3.19a)

$$\frac{db}{d\theta_c} = \frac{\alpha}{2E_c} \frac{d(\cot(\theta_c/2))}{d\theta_c} = \frac{\alpha}{2\,E_c} \frac{d}{d\theta_c}\left(\frac{\sin(\theta_c/2)}{1-\cos(\theta_c/2)}\right)$$

$$= \frac{\alpha}{4E_c \sin^2(\theta_c/2)}$$

and, multiplying by b,

(3.19b)

$$b\frac{db}{d\theta_c} = \frac{1}{2}\left(\frac{\alpha}{2E_c}\right)^2 \frac{\cot(\theta_c/2)}{\sin^2(\theta_c/2)}$$

leads to

$$\frac{d\sigma(\theta_c)}{d\Omega} = \frac{b}{\sin\theta_c}\left|\frac{db}{d\theta_c}\right| = \frac{1}{2}\left(\frac{\alpha}{2E_c}\right)^2 \frac{\cot(\theta_c/2)}{\sin\theta_c \, \sin^2(\theta_c/2)}$$

or

(3.20)

$$\frac{d\sigma(\theta_c)}{d\Omega} = \left(\frac{\alpha}{4\,E_c}\right)^2 \frac{1}{\sin^4(\theta_c/2)}$$

where the geometrical relation $\sin\theta_c = 2\sin(\theta_c/2)\cos(\theta_c/2)$ has been used.

For 1 MeV ^{4}He ions ($Z_1 = 2$) incident on silicon ($Z_2 = 14$), the value of $E_c =$ 875 keV and $\alpha = 40.3$ eV nm. For a 180° backscattering event, $\theta_c/2 = 90°$ and $\sin^4(\theta_c/2) = 1$. Then, $d\sigma(\theta_c)/d\Omega = (\alpha/E_c)^2 = 1.3 \times 10^{-10}$ nm^2, or a value of 1.3×10^{-24} cm^2. For forward scattering at $\theta_c = 2°$, the expression $\sin^4(\theta_c/2) = 1 \times 10^7$, indicates a ratio of seven orders of magnitude between forward scattering at 2° and backscattering at 180°.

The angular differential cross section is obtained from the relationship between $d\Omega$ and $d\theta_c$ defined in Equations (3.6), $d\Omega = 2\pi \sin\theta_c \, d\theta_c$, and (3.20). Using some differential algebra we have

(3.21)

$$\frac{d\sigma(\theta_c)}{d\theta_c} = \frac{d\sigma(\theta_c)}{d\Omega}\frac{d\Omega}{d\theta_c} = 2\pi\left(\frac{\alpha}{2E_c}\right)^2 \frac{\cos(\theta_c/2)}{\sin^3(\theta_c/2)}$$

Equations (3.20) and (3.21) are the Coulomb angular differential scattering cross sections, otherwise known as the *Rutherford differential cross section.* From Equations (3.20) and (3.21) we see from the $\sin(\theta_c/2)$ term in the denominator that both $d\sigma(\theta_c)/d\theta_c$ and $d\sigma(\theta_c)/d\Omega$ increase as θ_c decreases. This indicates that the Coulomb scattering process favors small-angle scattering or, in other words, that the largest cross sections are for scattering events of small angles.

For forward scattering at $\theta_c = 2°$, the expression $\sin^4(\theta_c/2) = 1 \times 10^7$, indicates a ratio of seven orders of magnitude between forward scattering at $2°$ and backscattering at $180°$.

A complete analysis of the scattering problem and transforming Equation (3.20) to the laboratory frame of reference yields (Chu, Mayer, and Nicolet 1978)

$$(3.22) \qquad \frac{d\sigma(\theta)}{d\Omega} = \left(\frac{Z_1 Z_2 e^2}{4E} \right)^2 \frac{4}{\sin^4 \theta} \frac{\{[1 - ((M_1/M_2)\sin\theta)^2]^{1/2} + \cos\theta\}^2}{[1 - ((M_1/M_2)\sin\theta)^2]^{1/2}}$$

3.4 NON-RUTHERFORD CROSS SECTIONS

The derivation of the Rutherford cross section assumes that the interaction between the particle Z_1 and the target atom Z_2 is well described by the Coulomb potential V(r). For this assumption to be correct, the particle velocity must be sufficiently large so that the particle penetrates well inside the orbitals of the atomic electrons. Under such conditions, the scattering will be due to the repulsion of the two positively charged nuclei of atomic numbers Z_1 and Z_2. However, experimental measurements indicate that the actual cross sections depart from the Rutherford at both high and low energies. The low-energy departures are caused by partial screening of the nuclear charge by the electron shells surrounding both nuclei, and the high-energy departures are caused by the interaction of the nuclei and presence of short-range nuclear forces. Typically, the real cross section, σ, is expressed in terms of the Rutherford cross section, σ_R, as

$$(3.23) \qquad\qquad\qquad \sigma = F\sigma_R$$

where F is a correction factor.

On the low-energy side we can estimate when electron screening effects become important. For the Coulomb potential to be valid, we require that a parameter called the distance of closest approach, d, be smaller than the K-shell electron radius. The distance of closest approach is given by

$$(3.24) \qquad\qquad\qquad d = \frac{Z_1 Z_2 e^2}{E}$$

where e is the charge on the electron, e^2 has the value of 1.44 eV nm, and E is the particle energy. In this analysis the K-shell electron radius can be estimated as a_0/Z_2 where $a_0 = 0.053$ nm, the Bohr radius. Using Equation (3.24) and the requirement that d be less than the K-shell electron radius sets the lower limit of the energy of the analysis beam to be

$$(3.25) \qquad\qquad E > Z_1 Z_2^2 \frac{e^2}{a_0}$$

This energy value corresponds to ~10 keV for He scattered from Si ($Z_2 = 14$) and 340 keV for He scattered from Au ($Z_2 = 79$). In practice, low-energy deviations from Rutherford occur at energies greater than the estimate given by Equation (3.25), as part of the particle trajectory is always outside the target atom's electron cloud. Results of several investigations indicate that the low-energy correction factor F in Equation (3.23) is given with adequate accuracy for light-ion analysis beams with MeV energies by

$$(3.26) \qquad\qquad F \equiv \frac{\sigma}{\sigma_R} = 1 - \frac{0.049\, Z_1 Z_2^{4/3}}{E_{CM}}$$

where E_{CM} is the center of mass energy in kiloelectron volts. In practice, replacing E_{CM} by the laboratory energy produces negligible error. For 1 MeV He ions on Au atoms, the correction factor corresponds to only ~3%.

At higher energies, departures from Rutherford scattering are due to nuclear interaction. Recent measurements and calculations regarding the onset of these high-energy departures from Rutherford backscattering are shown in Figure 3.6 for ^{4}He ions as a function of center of mass energy and target atomic number Z_2 (Bozoian 2009). The straight line in Figure 3.6 represents a rough boundary separating the region of Rutherford behavior (below the line) from the region where the cross section deviates from Rutherford by 4% (above the line). The equation resulting from a least-squares fit to the points in Figure 3.6 is

$$(3.27) \qquad\qquad \text{For } ^4\text{He: } E_{CM} \cong 0.249\, Z_2 - 0.080$$

The laboratory energy for the transition to non-Rutherford scattering for oxygen atoms ($Z_2 = 8$) can be calculated using Equation (3.27), and the relationship between center-of-mass and laboratory energies is given in Table 2.2, $E_{CM} = M_2 E_0/(M_1 + M_2)$. This calculation shows that the transition to non-Rutherford scattering should occur at $E_0 = 2.39$ MeV. Figure 3.7 shows the experimentally measured correction factor σ/σ_R for alpha particle (i.e., ^{4}He) backscattering ($\theta = 165°$) on oxygen ($Z_2 = 8$). These data show that the cross sections remain Rutherford (i.e., $\sigma/\sigma_R = 1$) up to ~2.4 MeV, with the deviation

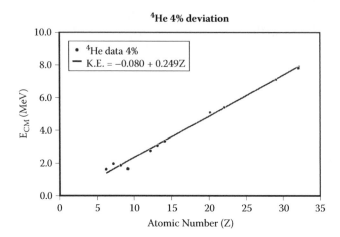

FIGURE 3.6   The center-of-mass ⁴He ion energy at which the scattering cross section deviates by 4% from its Rutherford value as a function of atomic number.

from Rutherford first resulting in the corrections factor dropping below 1 and then increasing to values greater than 1. Also visible in Figure 3.7 is the strong increase (resonance) in the scattering cross section at ~3.045 MeV. The correction factor at this energy is ~15, or the cross section is 15 times the Rutherford. This enhancement in cross section can be used to increase the sensitivity to the detection of oxygen. Indeed, many nuclear scatterings and nuclear reactions are useful for light elemental detection, as will be described in Chapters 5 and 9.

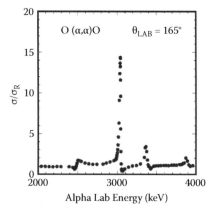

FIGURE 3.7   Experimentally measured normalized cross section for He scattering from oxygen.

PROBLEMS

3.1. A silicon surface barrier detector 2 cm in diameter is placed 4 cm away from the target. What is the solid angle of the detector?

3.2. (a) For 4 MeV He atoms incident on silicon, what is the impact parameter b? (b) Compare b to the actual size of the silicon nucleus. (c) What is the total scattering cross section?

3.3. What is the interatomic potential for a 1 MeV He ion incident at a distance of half the atomic spacing of a silicon lattice? Compare that to the value of b.

3.4. (a) What is the value of the closest approach, r_{min}, for 2 MeV He ions incident on silicon? (b) What is the value of the interatomic potential $V(r_{min})$?

3.5. For 1 MeV protons incident on germanium ($Z = 32$), what is the value of E_c and α?

3.6. Using Equation (3.22), calculate the differential scattering cross section per solid angle for 2 MeV He ions incident on Ni for laboratory scatterings of 10°, 15°, and 45°.

3.7. Transpose Equation (3.20) to the laboratory reference frame using equations in Table 2.2 in Chapter 2.

3.8. Show that, for the condition of $M_1 \ll M_2$, Equation (3.22) can be reduced to

$$\frac{d\sigma(\theta)}{d\Omega} = \left(\frac{Z_1 Z_2 e^2}{4E}\right)^2 \left[\sin-4\frac{\theta}{2} - 2\left(\frac{M_1}{M_2}\right)^2 + \ldots\right]$$

Hint: expand in a power series.

3.9. A beam of 2 MeV He ions is incident on a silver foil and undergoes Coulomb scattering in accordance with the Rutherford formula:
   a. What is the distance of closest approach?
   b. What is the impact parameter for He ions scattered through 90°?
   c. If the silver specimen is mounted as a free-standing foil of 100 nm thickness, what is the fraction of number of the incident He ions being backscattered ($\theta \geq 90°$) by the foil?

3.10. A beam of He ions is incident on a gold foil:
   a. What is the Rutherford scattering cross section at $\theta = 170°$ if the He ion energy is 2 MeV?

b. At what He ion energy does the nuclear force become not negligible and the scattering cross section of this system start to deviate by 4% from the Rutherford?

c. At what He ion energy will the electron screening reduce the Rutherford cross section by 10% for this system?

3.11. Estimate the energy ranges for He ions incident on C, O, Si, Ti, Ag, and U where their scattering cross sections do not deviate more than 4% from the Rutherford values.

REFERENCES

Bozoian, M. 2009. *Handbook of modern ion beam analysis,* 2nd ed., eds. Y. Wang and M. Nastasi, Appendix 8. Warrendale, PA: Materials Research Society.

Chu, W.-K., Mayer, J. W., and Nicolet, M.-A. 1978. *Backscattering spectrometry,* chap. 2. New York: Academic Press.

Nastasi, M., Mayer, J. W., and Hirvonen, J. 1996. *Ion–solid interactions: Fundamentals and applications.* Cambridge: Cambridge University Press.

SUGGESTED READING

Alford, T. L., Feldman, L. C., and Mayer, J. W. 2007. *Fundamentals of nanoscale film analyses.* New York: Springer.

French, A. P. 1971. *Newtonian mechanics.* New York: W. W. Norton and Co.

Goldstein, H. 1959. *Classical mechanics* Reading, MA: Addison–Wesley Publishing Co.

Johnson, R. E. 1982. *Introduction to atomic and molecular collisions.* New York: Plenum Press.

Symon, K. R. 1953. *Mechanics.* Reading, MA: Addison–Wesley Publishing Co.

Weidner, R. T., and Sells, R. L. *Elementary modern physics,* 3rd ed. Boston: Allyn and Bacon Inc.

Ziegler, J. F., Biersack, J. P., and Ziegler, M. D. 2008. *The stopping and range of ions in matter.* Chester, MD: SRIM Co.

Ion Stopping

4.1 INTRODUCTION

When an energetic ion penetrates a solid, it undergoes a series of collisions with the atoms and electrons in the target. In these collisions the incident particle loses energy at a rate of dE/dx of a few to a hundred electron volts per nanometer, depending on the energy and mass of the ion as well as on the substrate material. It is these energy loss processes that allows depth information to be obtained in ion beam analysis.

The main parameters governing the energy-loss rate are the energy, E_0, and atomic number, Z_1, of the ion and atomic number, Z_2, of the substrate if we exclude the effect of the orientation of the crystal lattice. As the incident ion penetrates the solid undergoing collisions with atoms and electrons, the distance traveled between collisions and the amount of energy lost per collision are random processes. Hence, all ions of a given type and incident energy do not have the same distance traveled for the same energy loss; conversely, for the same distance traveled by an ion, there will be a spread in the energy lost. This distribution in energy loss is referred to as energy straggling.

4.2 THE ENERGY-LOSS PROCESS

The energy-loss rate dE/dx of an energetic ion moving through a solid is determined by ion interactions with the substrate atoms and electrons. It is customary to distinguish two different mechanisms of energy loss: (1) nuclear collisions, in which energy is transmitted as translatory motion to a target atom as a whole; and (2) electronic collisions, in which the moving particle loses its kinetic energy by exciting or ejecting atomic electrons. For most purposes, this separation into elastic (nuclear) and inelastic (electronic) collisions is a convenient one and, although not strictly true, it is a good approximation. The energy-loss rate dE/dx can thus be expressed as

$$
(4.1) \qquad \frac{dE}{dx} = \left.\frac{dE}{dx}\right|_n + \left.\frac{dE}{dx}\right|_e
$$

where the subscripts n and e denote nuclear and electronic collisions, respectively.

Nuclear collisions can involve large, discrete energy losses and significant angular deflection of the trajectory of the ion (Figure 4.1). This process is responsible for the production of lattice disorder by the displacement of atoms from their positions in the lattice as well as direct backcsattering events in Rutherford backscattering spectrometry. Electronic collisions involve much smaller energy losses per collision, negligible deflection of the ion trajectory, and negligible lattice disorder. The relative importance of the two energy-loss mechanisms changes rapidly with the velocity and atomic number Z_1 of the projectile: Nuclear stopping predominates for low velocity and high Z_1, whereas electronic stopping takes over for high velocity and low Z_1. A comparison of the nuclear and electronic energy loss rates expressed in reduced energy notation ε is shown in Figure 4.2. The reduced energy ε is expressed as

$$
(4.2) \qquad \varepsilon = \frac{E\, a_{TF}\, M_2}{Z_1 Z_2\, e^2\, (M_1 + M_2)}
$$

where $e^2 = 1.44$ eV nm and a_{TF} is the Thomas–Fermi screening length, given by

$$
(4.3) \qquad a_{TF} = \frac{0.885\, a_0}{\left(Z_1^{1/2} + Z_2^{1/2}\right)^{2/3}}
$$

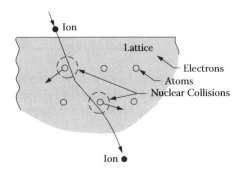

FIGURE 4.1   An ion incident on a crystal lattice is deflected in nuclear collisions with the lattice atoms and also loses energy in collisions with electrons. After passing through the sample it has lost DE in energy.

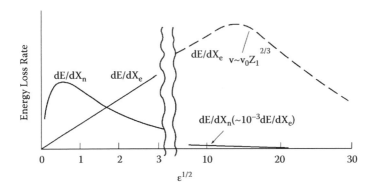

FIGURE 4.2   The reduced nuclear and electronic stopping as a function of $\varepsilon^{1/2}$.

With a_0, the Bohr radius is 0.053 nm. For the condition of 1 MeV He ions on a Si target, the Thomas–Fermi screening length $a_{TF} = 1.57 \times 10^{-2}$ nm and the corresponding reduced energy $\varepsilon = 341$.

Given that ε is proportional to ion energy E, the x axis of Figure 4.2, given in units of $\varepsilon^{1/2}$, is proportional to ion velocity. For 1 MeV He on Si, the value of $\varepsilon^{1/2}$ is ~18.5, which, according to Figure 4.2, is in the regime where the nuclear energy loss rate is approximately three orders of magnitude less than the electronic energy loss rate. This will be the case for most ion beam analysis conditions. As a result we will only treat high-energy electronic energy loss in this chapter. The reader is referred to other textbooks for treatments of low-energy electronic energy loss and nuclear energy loss (Nastasi, Mayer, and Hirvonen 1996; Ziegler, Biersack, and Ziegler 2008).

4.3   STOPPING CROSS SECTION

In addition to the energy-loss rate, it is also customary to speak of ε^A, the stopping cross section of element A, which is defined as

(4.4)
$$\varepsilon^A \equiv \frac{dE/dx}{N}$$

where N is the atomic density. The stopping cross section can be thought of as the energy-loss rate per scattering center. The stopping cross section has typical units of

$$\frac{dE/dx}{N} \left(\text{units:} \ \frac{(eV/cm)}{(\text{atoms/cm}^3)} = \left(\frac{eV \ cm^2}{\text{atom}} \right) \right)$$

The nomenclature of stopping cross sections comes from the unit of area in the numerator.

4.4 ELECTRONIC STOPPING

As we discussed in Section 4.2, the energy-loss rate of ions in solids is divided into two different mechanisms of energy loss: the energy transferred by the ion to the target nuclei (called nuclear stopping) and the energy transferred by the ion to the target electrons (called electronic stopping). The relative importance of the various interaction processes between the ion and the target medium depends mostly on the ion velocity and on the charges of the ion and target atoms.

At ion velocities, v, significantly lower than the Bohr velocity of the atomic electrons v_0 (2.188×10^8 cm/s), the ion carries its electrons and tends to neutralize by electron capture. At these velocities, elastic collisions with the target nuclei, the nuclear energy loss, dominate. However, as the ion velocity is increased, the nuclear energy loss diminishes as $1/E_0$. The electronic energy loss (i.e., collisions with the atomic electrons) soon becomes the main interaction. The total energy loss is obtained as a sum of the nuclear and electronic contributions. In the velocity range $v \sim 0.1v_0$ to $Z_1^{2/3}v_0$ the electronic energy loss is approximately proportional to velocity v or $E^{1/2}$.

At higher velocities, the charge state of the ion increases and ultimately becomes fully stripped of all its electrons at $v \geq v_0 Z_1^{2/3}$. At this point, the ion can be viewed as a positive point charge, Z_1, moving with a velocity greater than the mean orbital velocity of the atomic electrons in the shells or subshells of the target atom. When the projectile velocity v is much greater than that of an orbital electron (fast-collision case), the influence of the incident particle on an atom may be regarded as a sudden, small external perturbation. This picture leads to Bohr's theory of stopping power. The collision produces a sudden transfer of energy from the projectile to the target electron. The energy loss from a fast particle to a stationary nucleus or electron can be calculated from scattering in a central force field. The stopping cross section decreases with increasing velocity because the particle spends less time in the vicinity of the atom. In this high-energy, fast-collision regime, the values of electronic stopping are proportional to $(Z_1/v)^2$.

4.4.1 EFFECTIVE CHARGE OF MOVING IONS

As shown in Figure 4.2, the two regimes of electronic stopping are determined by the projectile's state of ionization or its effective charge. Bohr suggested that energetic ions would lose electrons whose orbital velocities were less than the ion velocity. Based on the Thomas–Fermi picture of the atom,

Bohr suggested that the ion charge fraction, or the effective ion charge, should be given by

(4.5)
$$\frac{Z^*}{Z} = \left(\frac{v}{v_0\,Z_1^{2/3}}\right)$$

where

Z is the total number of electrons that would surround the ion in its ground state (i.e., the atomic number)

Z^* is the positive charge on the ion

v is the ion velocity

v_0 is the Bohr velocity of an electron in the innermost orbit of a hydrogen atom (i.e., $v_0 \cong 2.2 \times 10^8$ cm/s).

The difference, $Z - Z^*$, is the number of electrons remaining on the ion. From Figure 4.2 and Equation (4.5), we have two extreme states for an energetic ion:

$$v < v_0\,Z_1^{2/3}$$

which implies that $Z^*/Z < 1$, that the ion is not fully stripped, and that

$$v > v_0\,Z_1^{2/3}$$

implying that $Z^*/Z \cong 1$ and that the ion is fully stripped to a bare nucleus.

Experimentally, it has been found that the ion charge fraction for heavy ions (i.e., $Z > Z_{He}$) more closely follows the form

(4.6)
$$\frac{Z^*}{Z} = 1 - \exp\left[-0.92\,v\,\Big/\left(v_0\,Z_1^{2/3}\right)\right]$$

which expands approximately to the Bohr relation given in Equation (4.5).

In the sections that follow, we will derive electronic energy loss expression in the high-velocity regimes where the ion is fully stripped.

4.4.2   HIGH-ENERGY ELECTRONIC ENERGY LOSS

In this section we will consider the case where the ion velocity is greater than $v_0\,Z_1^{2/3}$. For this condition, the ion is a bare nuclei, and its interactions with target electrons can be accurately described by a pure Coulomb interaction potential.

In 1913, Bohr derived an expression for the rate of energy loss of a charged particle on the basis of classical considerations. He considered a heavy particle, such as an α particle or a proton, of charge Z_1e, mass M, and velocity v passing a target-atom electron of mass m_e at a distance b (Figures 4.3 and 4.4). As the heavy particle passes, the Coulomb force acting on the electron changes direction continuously. If the electron moves negligibly during the passage of the heavy particle, the impulse parallel to the path, $\int F dt$ is zero by symmetry,

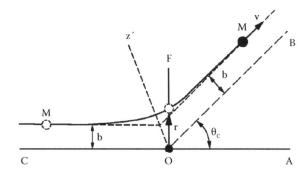

FIGURE 4.3   The nucleus is assumed to be a point charge at the origin O. At any distance r, the particle experiences a repulsive force. The particle travels along a path that is initially parallel to line OA, a distance b from it, and finally parallel to line OB, which makes an angle θ_c with OA.

since for each position of the incident particle in the $-x$ direction there is a corresponding position in the $+x$ direction that makes an equal and opposite contribution to the x component of the momentum. However, throughout the passage, there is a force in the y direction, and momentum Δp is transferred to the electron. This problem is identical to the momentum approximation introduced in Appendix 4.1 at the end of the chapter, where the momentum transferred to the target-atom electron is given by Equation (4.29) as

(4.7)
$$\Delta p = -\frac{1}{v}\frac{d}{db}\int_{-\infty}^{\infty} V\left(\left(x^2 + b^2\right)^{1/2}\right) dx$$

where
v is the ion velocity
b is the impact parameter
x is the distance along the ion's trajectory to the point r_{min}
$(x^2 + b^2)^{1/2}$ is the separation distance r between the ion and the electron (see Figure 4.4).
V is the interaction potential.

The interaction potential for this collision is purely Coulombic:

$$V(r) = \frac{Z_1 Z_2 e^2}{r}$$

or, in terms of $r = (x^2 + b^2)^{1/2}$,

(4.8)
$$V\left((x^2 + b^2)^{1/2}\right) = \frac{Z_1 Z_2 e^2}{(x^2 + b^2)^{1/2}}$$

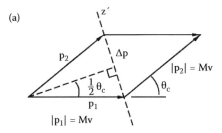

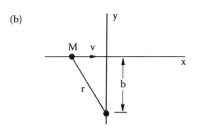

FIGURE 4.4   (a) Momentum diagram for impulse scattering (see Figure 4.3). Note that $|\mathbf{p_1}| = |\mathbf{p_2}|$ (i.e., for elastic scattering, the energy and speed of the projectile are the same before and after the collision). (b) Change-of-variable diagram for momentum (impulse) approximation.

Taking the derivative of the potential with respect to b,

(4.9)
$$-\frac{d}{db}\left(\frac{Z_1 Z_2 e^2}{\left(x^2 + b^2\right)^{1/2}}\right) = b\frac{Z_1 Z_2 e^2}{\left(x^2 + b^2\right)^{3/2}}$$

This allows us to write the momentum transferred to the electron during full passage of the ion as

(4.10)
$$\Delta p = \frac{Z_1 Z_2 e^2}{vb}\int_{-\infty}^{\infty}\frac{b^2\,dx}{\left(x^2 + b^2\right)^{3/2}} = \frac{2 Z_1 Z_2 e^2}{vb}$$

which is appropriate for glancing collisions (i.e., $\theta \cong 0$). If the electron has not achieved a relativistic velocity, and noting that $Z_2 = 1$, the electron kinetic energy following the collision will be

(4.11)
$$T = \frac{\Delta p^2}{2m_e} = \frac{2 Z_1^2 e^4}{b^2 m_e v^2}$$

where m_e is the electron mass and T is the energy transferred to the electron and lost by the ion in the collision.

The energy loss per unit path length, dE/dx, is

$$
(4.12) \qquad -\left.\frac{dE}{dx}\right|_e = n_e \int_{T_{min}}^{T_M} T\,\frac{d\sigma(E)}{dT}\,dT
$$

where n_e is the number of electrons per unit volume. The differential cross section, $d\sigma(T)$, for an energy transfer between T and dT is

$$
(4.13) \qquad d\sigma(T) = -2\pi b\,db
$$

This allows us to rewrite Equation (4.12) in terms of the impact parameter b as

$$
(4.14) \qquad -\left.\frac{dE}{dx}\right|_e = n_e \int_{b_{min}}^{b_{max}} T\,2\pi b\,db
$$

Substituting Equation (4.11) into Equation (4.14) and carrying out the integration,

$$
(4.15) \qquad -\left.\frac{dE}{dx}\right|_e = \frac{4\pi Z_1^2\,e^4\,n_e}{m_e\,v^2}\ln\frac{b_{max}}{b_{min}}
$$

To choose a meaningful value for b_{min}, we observe that if the heavy projectile collided head on with the electron, the maximum velocity transferred to a stationary electron would be 2v. The corresponding maximum kinetic energy transferred (for a nonrelativistic v) is $T_{max} = 2m_e v^2$. If this value of T_{max} is inserted into Equation (4.11), the corresponding b_{min} becomes

$$
(4.16) \qquad b_{min} = \frac{Z_1 e^2}{m_e\,v^2}
$$

If b_{max} is allowed to become infinite, −dE/dx goes to infinity because of the contribution of an unlimited number of small energy transfers given to distant electrons. But the smallest energy an atomic electron can accept must be sufficient to raise it to an allowed excited state. If I represents the average excitation energy of an electron, we choose $T_{min} = I$ and find

$$
(4.17) \qquad b_{max} = \frac{2Z_1 e^2}{\sqrt{2m_e\,v^2\,I}}
$$

When Equations (4.15) and (4.16) are substituted into Equation (4.14), we obtain

$$
(4.18) \qquad -\left.\frac{dE}{dx}\right|_e = \frac{2\pi Z_1^2\,e^4\,n_e}{m_e\,v^2}\ln\frac{2m_e\,v^2}{I}.
$$

This calculation is based on direct collisions with electrons in the solid. There is another term of comparable magnitude due to distant resonant energy transfer. The full derivation leads to a total stopping power of twice that shown before—that is,

(4.18a)
$$ -\left.\frac{dE}{dx}\right|_e = \frac{4\pi Z_1^2 e^4 n_e}{m_e v^2} \ln \frac{2 m_e v^2}{I} $$

or

(4.18b)
$$ -\left.\frac{dE}{dx}\right|_e = \frac{2\pi Z_1^2 e^4}{E} N Z_2 \left(\frac{M_1}{m_e}\right) \ln \frac{2 m_e v^2}{I} $$

where $E = M_1 v^2/2$ and $n_e = N Z_2$, with N given by the atomic density in the stopping medium.

Thus, we can regard the electronic interactions as being composed of two contributions: (1) close collisions with large momentum transfers, where the particle approaches within the electronic orbits; and (2) distant collisions with small momentum transfers, where the particle is outside the orbits.

The average excitation energy I, in electron volts, for most elements is roughly $I \cong 10 Z_2$, where Z_2 is the atomic number of the stopping atoms. Experimental and calculated values of I are given in Figure 4.5. The description of stopping power

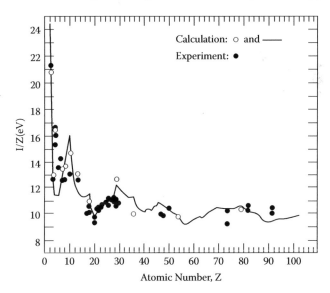

FIGURE 4.5 Calculation of mean excitation energy by Lindhard and Scharff's theory with a Hartree–Fock–Slater charge distribution. The calculation I/Z versus atomic number Z reveals structure, as was observed in many experimental measurements. (From Chu, W.- K. and Powers, D. 1972. *Physics Letters* 40A:23.)

so far ignores the shell structure of the atoms and variations in electron binding. Experimentally, these effects show up as small deviations (except for the very light elements) from the approximation given by $I \cong 10Z_2$, as shown in Figure 4.5.

The complete energy loss formula (often referred to as the Bethe formula) contains corrections that include relativistic terms at high velocities and corrections for the nonparticipation of the strongly bound inner shell electrons. For ions with $Z \geq Z_{He}$ in the energy regime of a few megaelectron volts, relativistic effects are negligible, and nearly all the target electrons participate $(I_e = NZ_2)$ in the stopping process. Consequently, Equation (4.17) can be used to estimate values of $dE/dx|_e$.

For example, the electronic energy loss of 2 MeV ^{4}He ions in Al has a value (calculated from Equation 4.17) of 315 eV/nm using values of $n_e = NZ_2 = 780/\text{nm}^3$ and $I = 10Z_2 = 130$ eV. Experiments give a value of $dE/dx|_e = 266$ eV/nm. Thus, the first-order treatment gives values to within 20% of the experimental values.

4.5   STOPPING CALCULATIONS USING SRIM

The energy loss rate, dE/dx, can be calculated using the computer program "stopping and ion ranges in matter" (SRIM) (http://www.srim.org/). SRIM is a group of programs that calculate the stopping and range of ions (up to 2 GeV/amu) into matter using a quantum mechanical treatment of ion–atom collisions (assuming a moving atom as an *ion* and all target atoms as *atoms*). This calculation is made very efficient by the use of statistical algorithms, which allow the ion to make jumps between calculated collisions and then average the collision results over the intervening gap. During the collisions, the ion and atom have a screened Coulomb collision, including exchange and correlation interactions between the overlapping electron shells. The ion has long-range interactions creating electron excitations and plasmons within the target. The charge state of the ion within the target is described using the concept of effective charge, which includes a velocity dependent charge state and long-range screening due to the collective electron sea of the target. A full description of the calculation is found in the tutorial book, *SRIM—The Stopping and Range of Ions in Solids,* by J. F. Ziegler and J. P. Biersack in 1985 (a new edition was published in 2008).

Examples of the stopping data provided by SRIM are given in Table 4.1 for He ions in Si at energies between 10 keV and 10 MeV.

4.6   ENERGY LOSS IN COMPOUNDS—BRAGG'S RULE

The process by which a particle loses energy when it moves swiftly through a medium consists of a random sequence of independent encounters between the moving projectile and the electrons attached to an atom of the solid. For a target that contains more then one element, the energy loss can be estimated by

---

TABLE 4.1   SRIM Stopping Data for He Ions in Si

Ion Energy	$[dE/dx]_e$ (eV nm^{-1})	$[dE/dx]_n$ (eV nm^{-1})
10.00 keV	53.63	10.29
25.00 keV	89.19	6.402
50.00 keV	138.0	4.155
100.00 keV	206.2	2.575
200.00 keV	283.0	1.541
300.00 keV	318.7	1.127
400.00 keV	333.4	0.898
500.00 keV	337.0	0.751
600.00 keV	334.5	0.648
700.00 keV	328.8	0.572
800.00 keV	321.3	0.513
900.00 keV	313.1	0.465
1.00 MeV	304.6	0.427
1.50 MeV	264.8	0.304
2.00 MeV	233.3	0.239
2.50 MeV	208.9	0.197
3.00 MeV	189.7	0.169
3.50 MeV	174.1	0.148
4.00 MeV	161.3	0.132
4.50 MeV	150.5	0.119
5.00 MeV	141.4	0.109
5.50 MeV	133.4	0.100
6.00 MeV	126.5	0.093
6.50 MeV	120.3	0.086
7.00 MeV	114.9	0.081
8.00 MeV	105.6	0.072
9.00 MeV	97.58	0.065
10.00 MeV	90.17	0.059

---

the sum of the losses of the constituent elements weighted by the abundance of the elements. This postulate is known as Bragg's rule and states that the stopping cross section $\varepsilon^{A_m B_n}$ of a solid of composition $A_m B_n$ is given by

$$(4.19) \qquad \varepsilon^{A_m B_n} = m\varepsilon^A + n\varepsilon^B$$

where ε^A and ε^B are the stopping cross sections of the atomic constituents A and B.

To take the specific example of SiO_2 on a molecular basis,

$$\varepsilon^{SiO_2} = \varepsilon^{Si} + 2\varepsilon^O$$

where ε^{SiO_2} is now the stopping power/molecule, so $dE/dX = N\varepsilon^{SiO_2}$ (see Equation 4.3), where N is the number of molecules/volume. Figure 4.6 shows the stopping cross section for SiO_2 on molecular basis.

4.7   ELECTRONIC ENERGY STRAGGLING

An energetic particle that moves through a medium loses energy via many individual encounters. These encounters slow the particle down and result in spreading the energy distribution of the particles. This phenomenon is called energy straggling, Ω_B. As a result, identical energetic particles, which have the same initial energy, do not have exactly the same energy after passing through a thickness Δt of a homogeneous medium.

Light particles such as H and He in the MeV energy range lose energy primarily by encounters with electrons in the target, and the dominant contribution to energy straggling is the statistical fluctuations in these electronic interactions. The straggling of the two-particle energy loss (the projectile and a stationary electron) is defined as

(4.20)
$$\Omega_B^2 = <T^2>$$

where T is the energy loss by the projectile to the electron and $<T^2>$ is the mean squared average energy transferred.

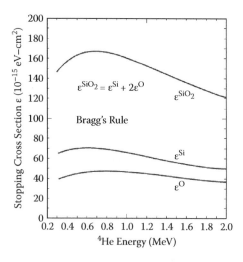

FIGURE 4.6   Stopping cross sections for He ions on Si, O, and SiO_2. The SiO_2 stopping cross section ε^{SiO_2} was determined on the molecular basis with 2.3×10^{22} molecules/cm³.

The probability of a particle with energy E undergoing a collision with an electron while traveling a distance Δt that results in an energy loss between T and $T + dT$ is given by

$$(4.21) \qquad P(T) = n_e \, \Delta t \, \frac{d\sigma}{dT}$$

where E is the energy of the moving particle, n_e is the number of electrons per unit volume, and σ is the cross section defined by Equation (3.4) in Chapter 3, $d\sigma = 2\pi b \, db$, where b is the impact parameter.

The average energy transferred by the moving particle in the distance Δt is obtained by multiplying Equation (4.21) by the transfer energy T and integrating over all possible values of T (see Equation 4.11):

$$(4.22) \qquad \langle T \rangle = \int T \, P(T) \, dT = n_e \Delta t \int_{T_{min}}^{T_M} T \, d\sigma$$

Similarly, the mean squared average energy transferred is given by

$$(4.23) \qquad \left\langle T^2 \right\rangle \equiv \Omega_B^2 = n_e \Delta t \int_{T_{min}}^{T_M} T^2 \, d\sigma = 2\pi \, NZ_2 \Delta t \int_{b_{max}}^{b_{min}} T^2 \, b \, db$$

where n_e was replaced with NZ_2, where N is the atomic density. The relationship between impact parameter and transferred energy is given by Equation (4.11). The integral thus yields

$$(4.24) \qquad \Omega_B^2 = 2\pi \, NZ_2 \Delta t \, \frac{(Z_1 e^2)^2}{m_e v^2} (T_{max} - T_{min})$$

where T_{max} and T_{min} are energies transferred corresponding to encounters with minimum and maximum impact parameters b_{min} and b_{max}, respectively. The largest energy transfer in a collision between the ion of mass M_1 and an electron of mass $m_e \ll M_1$ is $2m_e v^2$; thus, for $T_{min} \ll T_{max}$, we have

$$(4.25) \qquad \Omega_B^2 = 4\pi \, Z_1^2 e^4 NZ_2 \Delta t$$

This expression was first derived by Bohr (1915) and is often referred to as the *Bohr value of electronic energy straggling*. Described by Equation (4.25), Ω_B is usually referred to as the standard deviation of the ion energy distribution after passing through a medium of thickness Δt. The full width at half maximum

(FWHM) of the ion energy distribution, typically described by a normal or Gaussian function in most cases, would be calculated by

$$FWHM = 2\sqrt{2\ln 2}\,\Omega_B = 2.355\Omega_B$$

Bohr's theory predicts that the electronic energy straggling does not depend on the energy of the projectile and that the value of the energy variation increases with the square root of the number of electrons per unit area $NZ_2\Delta t$ in the target. A discussion of corrections to Bohr's theory of straggling can be found in Rauhala and Ziegler (2009).

The importance of straggling will become more apparent in the following chapters because energy straggling sets a fundamental limit to the depth resolution possible for ion beam energy loss techniques. For example, the amount of target material Nt (atoms/square centimeters) needed to produce 15 keV (FWHM) of energy straggling in Si with He ions is ~2.80×10^{18} atoms/cm^2, or 560 nm. In this example we have used the relationship $FWHM = 2.355\Omega_B$ and $N_{si} = 5 \times 10^{22}$ atoms/cm^3.

PROBLEMS

4.1. Calculate the velocity, v, the Thomas–Fermi screening length a_{TF}, and the reduced energy ε, for 0.1, 1, and 10 MeV He ions incident on Ge.

4.2. Show that, for a proton of velocity v in a direct head-on collision with a stationary electron, the maximum change in energy of the electron is 2 $m_e v^2$. What is the change in velocity of the target proton for a head-on collision with an incident proton?

4.3. (a) What is the effective charge for 100 keV and 10 MeV Ar in Cu? (b) What is the effective charge for 100 keV and 10 MeV He in Si?

4.4. At what velocity are the following ions fully stripped: H, He, C, Si, Ar, Xe, and Au?

4.5. (a) In the electronic stopping high-velocity regime, what is the value of dE/dx|e for 10 MeV Ar in Cu according to Equation (4.17)? (b) Is 10 MeV Ar in the high-velocity, fully stripped regime? (c) How does this compare to the stopping result from SRIM?

4.6. What is the Bohr straggling for He in 50, 100, and 1000 nm of Au?

4.7. What is the thickness of Au that can be analyzed before it becomes comparable to a detector resolution of 20 keV (FWHM)? Perform the calculation for ions of H, He, and C.

4.8. What is the interatomic potential for He ions in silicon at 1/10 the atomic spacing?

4.9. What is the value of the minimum impact parameter for 1 MeV He ions and 1 MeV protons?

APPENDIX 4.1: THE CLASSICAL IMPULSE APPROXIMATION TO THE SCATTERING INTEGRAL

In this appendix we will examine the relationship between the potential $V(r)$ and the scattering angle θ_c for the collisions where $V(r)/E_c$ remains small throughout the entire collision process. This condition is realized for collisions where b is large, which in turn leads to small-angle scattering.

The scattering cross section for central force scattering with large impact parameters can be calculated for small deflections from the impulse imparted to the particle as it passes the target nucleus. As the particle with charge Z_1e approaches the target nucleus, charge Z_2e, it will experience a repulsive force that will cause its trajectory to deviate from the incident straight line path (see Figure 4.3).

Let $\mathbf{p}_1$ and $\mathbf{p}_2$ be the initial and final momentum vectors of the particle. From Figure 4.4(a) it is evident that the total change in momentum, $\Delta\mathbf{p} = \mathbf{p}_2 - \mathbf{p}_1$, is along the z′ axis, which is the axis corresponding to the condition $r = r_{min}$. In this calculation the magnitude of the momentum does not change. From the isosceles triangle formed by $\mathbf{p}_1$, $\mathbf{p}_2$, and $\Delta\mathbf{p}$ shown in Figure 4.4(a) we have

$$\frac{\frac{1}{2}\Delta\mathbf{p}}{M\,\mathbf{v}} = \sin\frac{\theta_c}{2}$$

or, in the limit of $\theta_c \ll 1$,

(4.25) $$\frac{\Delta\mathbf{p}}{M\mathbf{v}} = \frac{\Delta(M\mathbf{v})}{M\mathbf{v}} \cong \theta_c$$

Equation (4.25) indicates that, at small deflections, θ_c can be thought of as being due to a small impulse, $\Delta\mathbf{p} = \Delta(M\mathbf{v})$, approximately perpendicular to the original direction of motion. This small-angle calculation is commonly called the *impulse* or *momentum approximation*.

The impulse approximation is appropriate for the small-angle large-impact parameter collisions that dominate the sequence of scatterings that determine the charged particle trajectory. In the impulse approximation the change in momentum is given by

(4.26) $$\Delta\mathbf{p} = \int_{-\infty}^{\infty} F_0\,dt$$

or

(4.27) $$\Delta\mathbf{p} = \frac{1}{\mathbf{v}}\int_{-\infty}^{\infty} F_0\,dx$$

where $\mathbf{F}_0$ is the component of the force acting on the ion perpendicular to its incident direction. By using the geometry of Figure 4.4(b), the force may be written with $r = (x^2 + b^2)^{1/2}$ as

$$(4.28) \qquad \mathbf{F}_0 = -\frac{dV(r)}{dy} = -\frac{dV\left(\left(x^2 + b^2\right)^{1/2}\right)}{db}$$

Then,

$$(4.29) \qquad \Delta\mathbf{p} = -\frac{1}{v}\frac{d}{db}\int_{-\infty}^{\infty} V\left(\sqrt{x^2 + b^2}\right) dx$$

or, using Equation (4.25),

$$(4.30) \qquad \theta_c = \frac{\Delta(\mathbf{Mv})}{\mathbf{Mv}} = -\frac{1}{2E}\frac{d}{db}\int_{-\infty}^{\infty} V(r)\, dx$$

for $\theta_c \ll 1$. Equation (4.30) shows that the angle θ_c is obtained from the potential $V(r)$ by one integration followed by one differentiation. Using $r = (x^2 + y^2)^{1/2}$ to change the integration variable in Equation (4.30) yields

$$(4.31) \qquad \theta_c = \frac{1}{E_c}\int_{b}^{\infty} \left(\frac{dV}{dr}\right)\frac{b}{r}\left[1 - \left(\frac{b}{r}\right)^2\right]^{-1/2} dr$$

Equation (4.31) is often referred to as the classical impulse approximation to the scattering integral.

REFERENCES

Bohr, N. 1915. Decrease of speed of electrified particles in passing through matter. *Philosophical Magazine* 30:581.

Chu, W-. K., and Powers, D. 1972. Calculation of mean excitation energy for all elements. *Physics Letters* 40A:23.

Nastasi, M., Mayer, J. W., and Hirvonen, J. 1996. *Ion–solid interactions: Fundamentals and applications.* Cambridge: Cambridge University Press.

Rauhala, E., and Ziegler, J. F. 2009. Energy loss and energy straggling. In *Handbook of modern ion beam materials analysis,* 2nd ed., eds. Y. Wang and M. Nastasi, chap. 2. Warrendale, PA: Materials Research Society.

Ziegler, J. F., Biersack, J. P., and Ziegler, M. D. 2008. *The stopping and range of ions in matter,* chap. 6. Chester, MD: SRIM Co.

SUGGESTED READING

Alford, T. L., Feldman, L. C., and Mayer, J. W. 2007. *Fundamentals of nanoscale film analyses,* chap. 3. New York: Springer.

Chu, W.-K, Mayer, J. W., and Nicolet, M.-A. 1978. *Backscattering spectrometry,* chap. 2. New York: Academic Press.

Gotz, G., and Gartner, K. eds. 1988. In *High energy ion beam analysis of solids,* chap. 1. Berlin: Academie Verlag.

Ziegler, J. F., Biersack, J. P., and Littmark, U. 1985. *The stopping and range of ions in solids,* vol. 1. New York: Pergamon Press.

Backscattering Spectrometry

5.1 INTRODUCTION

Backscattering spectrometry is a method that provides accurate depth information (typically, accuracies of a few percent, with 10–30 nm in depth resolution) about the stoichiometry, elemental area density, and impurity distributions in the near surface region of bulk materials and in thin films. Detection limits range from about a few parts per million for heavy elements to a few percent for light elements. Analysis depths using He ions are typically a few thousand nanometers. Depth profile information is obtained in a nondestructive manner. The results obtained typically do not require the use of standards and are insensitive to the sample chemical bonding. In addition, it is a quick and easy experiment to perform, typically with data acquisition times of a few tens of minutes.

The majority of backscattering analyses have been performed with ^{4}He ions in the 1–2 MeV range. The reasons for this include: (1) the backscattering cross section of ^{4}He incident on all elements with mass greater than Be is nearly Rutherford in this range (see Figure 3.6 in Chapter 3), (2) experimental and semiempirical data for energy loss are reasonably well known, and (3) accelerators that produce ions at these energies are more abundant.

5.2 EXPERIMENTAL SETUP

Backscattering spectrometry is based on collisions between an energetic incident ion and a target nucleus. The probability of a collision resulting in a backscattered event is related to the collision cross section discussed in Chapter 3. If the interaction between the incident particle and target nuclei can be described by Coulombic forces, the scattering is defined as *Rutherford,* which derives its name from Lord Ernest Rutherford, who first presented the concept of atoms having nuclei. For Rutherford backscattering, the scattering cross section is defined by Equation (3.22) in Chapter 3. The scattering cross section along with the kinematics (see Chapter 2) of the collision is independent of chemical bonding, making the backscattering measurement insensitive to matrix effects.

The experimental setup for a backscattering experiment is shown in Figure 5.1(a). A collimated beam of monoenergetic He ions is incident on a planar sample. Particles (He) backscattered to an angle θ are detected by a detector of solid angle Ω (Equation 3.1). In the most typical applications, all of the apparatus are under vacuum. Figure 5.1(b) shows a schematic representation of the elastic collision between the incident projectile of mass M_1 and energy E_0 and a target of mass M_2, which is initially at rest. After the collision, the projectile and target mass have energies of E_1 and E_2 and have been scattered to the laboratory angles θ and ϕ, respectively.

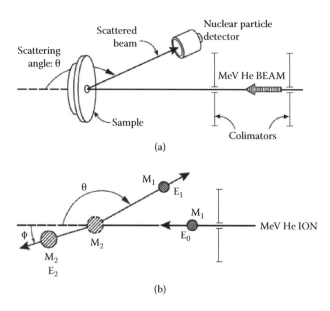

(a)

(b)

FIGURE 5.1 (a) Schematic of the experimental setup of a backscattering experiment. (b) Schematic representation of the collision process that takes place in the sample depicted in the experimental setup shown in (a).

5.3 ENERGY LOSS AND DEPTH SCALE

The energy loss by the incident projectile during the collision is defined by the kinematic factor K given in Equation (2.9) in Chapter 2. As discussed in that chapter, it is the loss of energy during the collision process that allows back-scattering spectrometry to detect different masses in the target. The energy loss dE/dx by the projectile as it traverses the specimen will allow us to extract depth information from the sample.

In addition to energy losses during collisions, the incident particles will also lose energy as they travel into the target before the collision and out of the target after the collision. This process is schematically shown in Figure 5.2. To a good approximation, the total energy loss ΔE_{in} into a depth, t, before the collision is proportional to t. That is,

$$(5.1) \qquad \Delta E_{in} = \int_0^{t/\cos\theta_1} \left.\frac{dE}{dx}\right|_{in} dx \cong \frac{t}{\cos\theta_1}\left.\frac{dE}{dx}\right|_{in}$$

where $dE/dx|_{in}$ is evaluated at some average energy between the incident energy E_0 and $E_0 - \Delta E_{in}$. This is the energy loss component that arises from the projectile incident at an angle θ_1 degrees away from the surface normal after traversing a thickness t. Thus, the energy of the incident particle just prior to the collision at depth t is

$$(5.2) \qquad E_t = E_0 - \Delta E_{in}$$

After scattering, the projectile has a reduced energy of KE_t. Therefore, the energy loss by the projectile during the scattering process at depth t is

$$(5.3) \qquad \Delta E_s = E_t - KE_t = (1-K)E_t$$

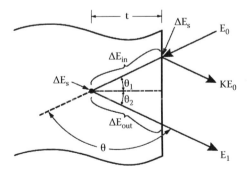

FIGURE 5.2   The energy loss components for a projectile scattered from a depth, t, in a single element target.

The energy loss on the outward path for a particle scattered θ_2 degrees away from the surface normal of the sample is given by

(5.4)
$$\Delta E_{out} = \int_0^{t/\cos\theta_2} \left.\frac{dE}{dx}\right|_{out} dx \cong \frac{t}{\cos\theta_2} \left.\frac{dE}{dx}\right|_{out}$$

The energy E_1 is the energy measure at the detector and is given by

(5.5)
$$E_1 = E_0 - (\Delta E_{in} + \Delta E_s + \Delta E_{out}) = KE_t - \Delta E_{out}$$

Alternately, if the projectile is scattered from the surface without penetrating the solid, the only energy loss will be due to kinematics and the detected projectile energy will be KE_0.

The total energy difference ΔE between projectiles scattered at the surface and at some depth t is thus given by

(5.6)
$$\Delta E = KE_0 - E_1 = K\Delta E_{in} - \Delta E_{out} \equiv [S]\,t = N[\varepsilon]\,t$$

where [S] is the energy loss factor, $[\varepsilon]$ is the stopping cross-section factor, and N is the atomic density. The energy loss factor can be defined from Equations (5.1)–(5.6) as

(5.7)
$$[S] \equiv \left[\frac{K}{\cos\theta_1} \left.\frac{dE}{dx}\right|_{in} + \frac{1}{\cos\theta_2} \left.\frac{dE}{dx}\right|_{out} \right]$$

where the subscripts "in" and "out" refer to energies at which dE/dx is evaluated, and the stopping cross-section factor is defined as

(5.8)
$$[\varepsilon] \equiv \left[\frac{K}{\cos\theta_1} \varepsilon_{in} + \frac{1}{\cos\theta_2} \varepsilon_{out} \right]$$

where ε is the stopping cross section defined in Equation (4.4) in Chapter 4 as

(5.9)
$$\varepsilon \equiv \frac{1}{N}\frac{dE}{dx}$$

Equations (5.7) and (5.8) are derived assuming that dE/dx or ε is constant along the inward and outward paths. This assumption leads to a linear relation between ΔE and the depth, t, at which scattering occurs. One can therefore assign a linear depth to the energy axis, as indicated in Figure 5.3.

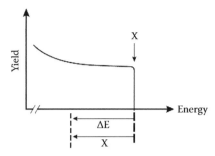

FIGURE 5.3   If the energy loss is assumed to be constant along the inward and outward paths, then the energy ΔE can be linearly related to the depth x through $\Delta E = [S]x$, as indicated in the abscissa of the backscattering spectrum.

5.3.1   SURFACE ENERGY AND MEAN ENERGY APPROXIMATION

For thin films with a thickness ≤ 100 nm, the relative change in energy along the paths is small. For these conditions, the evaluation of dE/dx can be carried out using the "surface energy approximation" in which $dE/dx|_{in}$ is evaluated at E_0 and $dE/dx|_{out}$ is evaluated at KE_0. In this approximation we have

(5.10)
$$[S_0] \equiv \left[\frac{K}{\cos\theta_1} \frac{dE}{dx}\bigg|_{E_0} + \frac{1}{\cos\theta_2} \frac{dE}{dx}\bigg|_{KE_0} \right]$$

and

(5.11)
$$[\varepsilon_0] \equiv \left[\frac{K}{\cos\theta_1} \varepsilon(E_0) + \frac{1}{\cos\theta_2} \varepsilon(KE_0) \right]$$

When the path length becomes appreciable, the surface energy approximation breaks down. A better approximation can be obtained by selecting constant values of dE/dx at a mean energy $\bar{E}$ intermediate between the energy the particle has at the end points of each track. For these conditions we have

(5.12)
$$[\bar{S}] \equiv \left[\frac{K}{\cos\theta_1} \frac{dE}{dx}\bigg|_{\bar{E}_{in}} + \frac{1}{\cos\theta_2} \frac{dE}{dx}\bigg|_{\bar{E}_{out}} \right]$$

and

(5.13)
$$[\bar{\varepsilon}] \equiv \left[\frac{K}{\cos\theta_1} \varepsilon(\bar{E}_{in}) + \frac{1}{\cos\theta_2} \varepsilon(\bar{E}_{out}) \right]$$

The mean energy $\bar{E}$ can be estimated several ways. For the inward track, the particle enters at E_0 and has energy E_t (Equation 5.2) before scattering at depth t so that $\bar{E}_{in} = \frac{E_t + E_0}{2}$. After scattering, the particle has energy KE_t, so that $\bar{E}_{out} = \frac{E_1 + KE_t}{2}$. The value of E_t is unknown, but can be estimated if the energy ΔE (Equation 5.6) can be measured. For a quick estimate it is assumed that ΔE is subdivided symmetrically between the inward and outward paths, so that E_t is approximately $E_0 - \frac{1}{2}\Delta E$. The values of $\bar{E}_{in}$ and $\bar{E}_{out}$ are then given by

(5.14)
$$\bar{E}_{in} \cong E_0 - \frac{1}{4}\Delta E$$

and

(5.15)
$$\bar{E}_{out} \cong E_1 + \frac{1}{4}\Delta E$$

5.3.2 COMPOUND TARGETS

For the situation where the target is composed of more than one element, the energy loss is assumed to be equal to the sum of the loss to the constituent elements, weighted by their abundance in the compound. This postulate is known as Bragg's rule, as discussed in Chapter 4. Thus, the stopping cross section for a mixture with the composition $A_m B_n$ is given by Equation (4.19), $\varepsilon^{A_m B_n} = m\varepsilon^A + n\varepsilon^B$, where ε^A and ε^B are the stopping cross sections for elements A and B.

For compound targets similar energy losses will occur as those depicted in Figure 5.2, with the added complexity of kinematic losses from two elements. A schematic representation of the backscattering process from an idealized free-standing compound film is presented in Figure 5.4(a). The energy loss components that arise from a projectile traversing a thickness t and then being backscattered out for the A component are

(5.16)
$$\Delta E_{in}^{A_m B_n} = \frac{t}{\cos\theta_1} \left.\frac{dE}{dx}\right|_{in}^{A_m B_n}$$

(5.17)
$$E_t = E_0 - \Delta E_{in}^{A_m B_n}$$

(5.18)
$$\Delta E_s(A) = E_t - K_A E_t$$

(5.19a)
$$\Delta E_{out}^{A_m B_n}(A) = \frac{t}{\cos\theta_2} \left.\frac{dE}{dx}\right|_{out,A}^{A_m B_n}$$

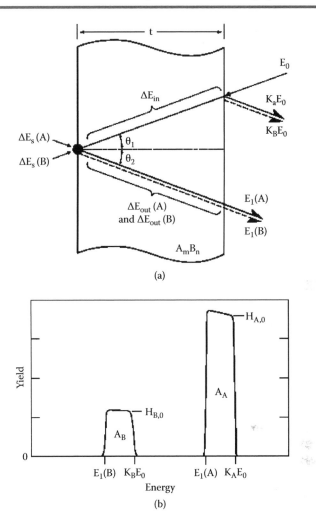

FIGURE 5.4 A schematic representation of the backscattering process from a free-standing compound film with composition A_mB_n and thickness t. (a) Energy loss relations, and (b) predicted backscattering spectrum.

Similar expressions exist for the B component of the compound target as

(5.19b)
$$\Delta E_{out}^{A_mB_n}(B) = \frac{t}{\cos\theta_2}\left.\frac{dE}{dx}\right|_{out,B}^{A_mB_n}$$

The relationship between the energy width, $K_AE_0 - E_1(A)$, and the scattering depth t is

(5.20a)
$$\Delta E_A = K_AE_0 - E_1(A) \equiv t\ N^{A_mB_n}[\varepsilon]_A^{A_mB_n}$$

and

(5.20b)
$$\Delta E_B = K_B E_0 - E_1(B) \equiv t\ N^{A_m B_n} [\varepsilon]_B^{A_m B_n}$$

where $N^{A_m B_n}$ is the number of molecules of $A_m B_n$ per unit volume and $[\varepsilon]_A^{A_m B_n}$ $([\varepsilon]_B^{A_m B_n})$ is the stopping cross-section factor for a projectile scattered from element A (B) while traversing the medium $A_m B_n$, and has the form

(5.21a)
$$[\varepsilon]_A^{A_m B_n} = \left[\frac{K_A}{\cos\theta_1} \varepsilon_{in}^{A_m B_n} + \frac{1}{\cos\theta_2} \varepsilon_{out,A}^{A_m B_n} \right]$$

and

(5.21b)
$$[\varepsilon]_B^{A_m B_n} = \left[\frac{K_B}{\cos\theta_1} \varepsilon_{in}^{A_m B_n} + \frac{1}{\cos\theta_2} \varepsilon_{out,B}^{A_m B_n} \right]$$

5.4   SCATTERING CROSS SECTION AND THE SHAPE OF THE BACKSCATTERING SPECTRUM

In the preceding discussion we assumed that collisions would occur between the projectile and a target atom that would result in a backscattering event. The likelihood of such an occurrence leads to the concept of the scattering cross section (Chapter 3) and the ability of performing quantitative composition analysis. The energy spectrum from an infinitely thick target is schematically shown in Figure 5.5. The shape of the spectrum can be understood from the relationships developed between depth and energy loss in Section 5.3 and the energy dependence of the Rutherford cross sections, Equation (3.22). For convenience, the Rutherford cross-section formula is relisted here:

(5.22)
$$\frac{d\sigma(\theta)}{d\Omega} = \left(\frac{Z_1 Z_2 e^2}{4E} \right) \frac{4}{\sin^4\theta} \frac{\{[1-((M_1/M_2)\sin\theta)^2]^{1/2} + \cos\theta\}^2}{[1-((M_1/M_2)\sin\theta)^2]^{1/2}}$$

For the experimental conditions shown in Figure 5.5, where a uniform beam of projectiles impinges at normal incidence on a uniform target, the spectrum height or yield of backscattered particles detected from a thin layer of atoms (Δt) is

(5.23)
$$Y = \sigma(\theta)\ \Omega\ Q\ N\Delta t / \cos\theta_1$$

where
$\sigma(\theta)$ is the scattering cross section at angle θ
Ω is the detector solid angle

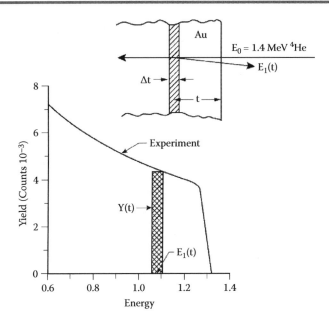

FIGURE 5.5   Backscattering spectrum for 1.4 MeV He ion incident on a thick Au sample.

Q is the measured number of incident particles

N is the atomic density, which makes $N\Delta t$ the number of target atoms per unit area in the layer Δt thick

θ_1 is the angle between the incident beam and the surface normal of the target

For thick targets, projectiles scatter from any depth t, resulting in a continuous energy spectrum starting at KE_0 and going down to low energy, as shown in Figure 5.5. The yield from a slice of materials Δt wide at depth t is given by (for $\theta = 180°$)

(5.24) $$Y(t) = \sigma(E(t)) \, \Omega \, Q \, N\Delta t/\cos\theta_1$$

where $\sigma(E(t))$ is the scattering cross section when the particle has energy E(t), at depth t. Note: In Figure 5.5, θ_1 is set to 0.

For Rutherford scattering, using Equation (5.22), we can rewrite Equation (5.24) as

(5.25) $$Y(t) \cong \left(\frac{Z_1 Z_2 e^2}{4E(t)}\right)^2 \Omega Q N\Delta t \propto \frac{1}{E(t)^2}$$

Therefore, under Rutherford scattering conditions the shape of the energy spectrum should vary as $1/E^2$. This causes an increase in signal height toward decreasing energy or deeper into the sample, as shown in Figure 5.5.

The shape of backscattering spectra and depth profiles can be obtained from computer programs used in both simulation and analysis of backscattering data. This will be discussed in more detail in Chapter 15.

5.5   COMPOSITION AND DEPTH PROFILES

Equations (5.23) and (5.24) show that when Ω, Q, and σ are known, the number of atoms per unit area, Nt, can be calculated. If the interaction between projectile and target atom is Coulombic, the cross section is easily calculated using the Rutherford formula, Equation (5.22). At higher energies, the probability that the interaction between the projectile and target atom is non-Rutherford increases (Chapter 3) and experimentally determined values of σ are required.

In the hypothetical situation of a free standing film with composition $A_m B_n$, as depicted in Figure 5.4, the total number of counts from element A is area A_A in Figure 5.4(b). Assuming that only small changes in the projectile energy (e.g., very thin films) occur on the inward and outward paths (surface energy approximation), the area A_A can be described by (Chu, Mayer, and Nicolet, 1978):

(5.26a)
$$A_A = \Omega \, Q \, \sigma_A(E_0) \, mN^{A_m B_n} \, t/\cos\theta_1$$

where $\sigma_A(E_0)$ is the scattering cross section for A atoms under the surface energy approximation and $mN^{A_m B_n}$ is the atomic density of A atoms in the compound $A_m B_n$. A similar expression can be written for A_B using the scattering cross section for B atoms, $\sigma_B(E_0)$ and the atomic density of B atoms, $nN^{A_m B_n}$:

(5.26b)
$$A_B = \Omega \, Q \, \sigma_B(E_0) \, nN^{A_m B_n} \, t/\cos\theta_1$$

By combining Equations (5.26a) and (5.26b), the ratio of the atomic densities of A and B atoms can be found to be

(5.27)
$$\frac{m}{n} \equiv \frac{N_A}{N_B} = \frac{A_A}{A_B} \frac{\sigma_B(E_0)}{\sigma_A(E_0)}$$

In some situations it may not be possible to resolve the full peak of a particular element in a backscattering spectrum. In this case Equation (5.27), which utilizes the ratio of peak areas, cannot be directly used for composition analysis. However, a composition analysis may be possible by comparing surface heights of the backscattering yield. In an analogous expression to Equation (5.26), the backscattering yield at the surface for elements A and B in Figure 5.4 is separately given by

(5.28a)
$$H_{A,0} = \Omega Q \, \sigma_A(E_0) \, mN^{A_m B_n} \, \tau_{A,0}/\cos\theta_1$$

and

(5.28b) $$H_{B,0} = \Omega Q \ \sigma_B(E_0) \ nN^{A_mB_n} \ \tau_{B,0}/\cos\theta_1$$

where $\tau_{A,0}$ and $\tau_{B,0}$ are the corresponding thicknesses of a slab of the target at the surface for elements A and B and defined by the energy width ξ of a channel in the detecting system (typically a few kiloelectron volts per channel). Projectiles scattered from within $\tau_{A,0}$ and $\tau_{B,0}$ will have a depth scale at the surface given by

(5.29a) $$\xi = \tau_{A,0} N^{A_mB_n} [\varepsilon]_A^{A_mB_n}$$

and

(5.29b) $$\xi = \tau_{B,0} N^{A_mB_n} [\varepsilon]_B^{A_mB_n}$$

where $N^{A_mB_n}$ is the number of molecules of A_mB_n per unit volume and $[\varepsilon]_A^{A_mB_n}$ ($[\varepsilon]_B^{A_mB_n}$) is the stopping cross-section factor for a projectile scattered from element A (B) while traversing the medium A_mB_n. Combining Equations (5.28) and (5.29) gives

(5.30a) $$H_{A,0} = \frac{\Omega Q \ \sigma_A(E_0) \ m\xi}{[\varepsilon]_A^{A_mB_n} \ \cos\theta_1}$$

and

(5.30b) $$H_{B,0} = \frac{\Omega Q \ \sigma_B(E_0) \ n\xi}{[\varepsilon]_B^{A_mB_n} \ \cos\theta_1}$$

Combining these equations, the ratio of atomic densities for atoms A to B for a thick target can be written as

(5.31) $$\frac{m}{n} \equiv \frac{N_A}{N_B} = \frac{H_{A,0} \ \sigma_B(E_0)[\varepsilon]_A^{A_mB_n}}{H_{B,0} \ \sigma_A(E_0)[\varepsilon]_B^{A_mB_n}}$$

5.6   EXAMPLES

In this section we will provide two examples of the types of analysis that can be carried out with backscattering spectrometry.

5.6.1   THIN FILM REACTION ANALYSIS

Figure 5.6 shows schematic Rutherford backscattering spectra for ^{4}He ions incident on a 100 nm thick Ni film on a Si substrate (a) in its as-deposited state and (b) after interdiffusion and thermal reaction. In the reaction between Ni and Si,

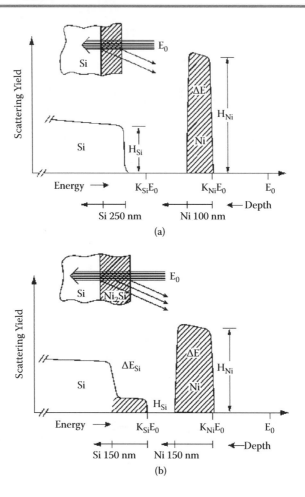

FIGURE 5.6   Schematic backscattering spectra for MeV ^{4}He ions incident on 100 nm Ni film on Si. (a) Before reaction, and (b) after reaction to form Ni_2Si.

the compound Ni_2Si is formed. Alpha particles scattered from the surface have an energy given by the kinematic equation $E_1 = KE_0$, where the kinematic factor for ^{4}He backscattered at a laboratory angle of 170° is 0.7624 for Ni and 0.5657 for Si.

As the alpha particles traverse the Ni film target, they lose energy at a rate of approximately 640 eV/nm along the incident path. A reasonable assumption in the limit of the surface energy approximation is that energy loss is linear with thickness. Thus, a 2 MeV ^{4}He ion will lose 64 keV penetrating the 100 nm Ni film before reaching the Si substrate. Immediately after scattering from a Ni atom at the interface, the ^{4}He will have an energy of 1.476 MeV. On the outward path, the scattered ^{4}He will have a slightly different energy loss due to the energy dependence of the energy loss process—approximately 690 eV/nm. The scattered ^{4}He will also have a slightly longer path length $(t/\cos\theta_2)$ reemerging to the surface where $\theta_2 = 10°$. Thus, the ^{4}He will emerge from the Ni surface with 1.406 MeV. The total energy difference between particles scattered form Ni atoms at the target surface and Ni atoms at the Ni/Si interface will be 119 keV, which can be derived from Equation (5.6).

The reaction between Ni and Si is shown in Figure 5.6(b). After the reaction the Ni energy signal ΔE_{Ni} has spread slightly due to the presence of Si atoms contributing to the energy loss. The Si has a step in its leading edge due to its reaction with Ni and the formation of Ni_2Si. As discussed for Equations (5.28)–(5.31), the ratio of the heights of Ni to Si, H_{Ni}/H_{Si}, from the Ni and Si peaks after silicide layer gives the composition of the layer. This can be calculated from Equation (5.31). In a first approximation we will note that $[\varepsilon_{Ni}^{Ni_2Si}] \approx [\varepsilon_{Si}^{Ni_2Si}]$, which allows us to obtain to within an accuracy of 5%–10%:

$$(5.32) \qquad \frac{N_{Ni}}{N_{Si}} \cong \frac{H_{Ni}}{H_{Si}} \frac{\sigma_{Si}(E_0)}{\sigma_{Ni}(E_0)}$$

If the scattering is Rutherford we can make use of the fact that $\sigma_{Si} \propto (Z_{Si})^2$ and $\sigma_{Ni} \propto (Z_{Ni})^2$, which allows us to write Equation (5.32) as

$$(5.33) \qquad \frac{N_{Ni}}{N_{Si}} \cong \frac{H_{Ni}}{H_{Si}} \left(\frac{Z_{Si}}{Z_{Ni}}\right)^2 = \frac{H_{Ni}}{4H_{Si}}$$

A better approximation can be obtained by measuring the energy width of the Ni_2Si reaction layer in the Ni and Si peaks (hashed area in Figure 5.6b) and noting from Equation (5.6):

$$(5.34a) \qquad \Delta E_{Ni}^{Ni_2Si} = \Delta t\, N^{Ni_2Si}[\varepsilon]_{Ni}^{Ni_2Si}$$

and

$$(5.34b) \qquad \Delta E_{Si}^{Ni_2Si} = \Delta t\, N^{Ni_2Si}[\varepsilon]_{Si}^{Ni_2Si}$$

Substituting the stopping cross-section factors given in Equations (5.34a) and (5.34b) into Equation (5.31) leads to

$$(5.35) \qquad \frac{N_{Ni}}{N_{Si}} = \frac{H_{Ni}}{H_{Si}} \frac{\sigma_{Si}}{\sigma_{Ni}} \frac{\Delta E_{Ni}^{Ni_2Si}}{\Delta E_{Si}^{Ni_2Si}}$$

In this case of Ni_2Si, the difference between application of Equations (5.32) and (5.35) corresponds to a 5% difference in the determination of the stoichiometry of the silicide.

5.6.2 ION IMPLANTATION

Figure 5.7 shows the RBS (Rutherford backscattering spectrometry) energy spectrum from 2.0 MeV ^{4}He ions backscattered from a silicon target implanted with ^{75}As at 250 keV to a fluence of 1.2×10^{15} As/cm^2. The Si signal gives a step with leading edge at 1.13 MeV, and the As signal (plotted on an amplified scale) has a Gaussian distribution with a peak at 1.55 MeV and a full width at half maximum (FWHM) of 60 keV. The As peak is shifted by $\Delta E_{As} = 68$ keV below the energy edge $K_{As}E_0 = 1.618$ MeV of the As at the surface. The data from Figure 5.7 are given in Table 5.1 (Chu et al. 1978)

Using Figure 5.7 and Table 5.1, we now proceed to calculate the implantation fluence, range, and range distribution for As in Si. We will assume that As is sufficiently shallow in Si that the surface energy approximation can be used to calculate the stopping cross section and Rutherford cross section. In calculating the As fluence, we use the backscattered Si signal as an internal calibration

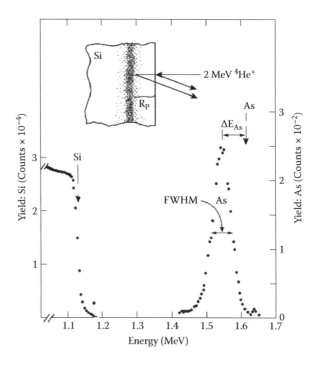

FIGURE 5.7 Energy spectrum of 2 MeV ^{4}He ions backscatterd from a silicon wafer implanted with 250 keV As ions to a nominal fluence of 1.2×10^{15} ions/cm^2. The vertical arrows indicate the energies of ^{4}He backscattered from surface atoms of ^{28}Si and ^{75}As.

TABLE 5.1   Data and Parameters Associated with Figure 5.7[a]

Data	Parameters
$H_{Si,0} = 27{,}000$ counts	$[\varepsilon_0]^{Si}_{Si} = 92.6 \times 10^{-15}$ eVcm^2
$H^{Si}_{As} = 250$ counts (at peak)	$[\varepsilon_0]^{Si}_{As} = 95.3 \times 10^{-15}$ eVcm^2
$A_{As} = 3{,}350$ counts	$\sigma_{As} = 1.425 \times 10^{-24}$ cm^2
$\Delta E_{As} = 68$ keV	$\sigma_{Si} = 0.248 \times 10^{-24}$ cm^2
$(FWHM)_{As} = 60$ keV	$K_{As} = 0.809;\; K_{Si} = 0.566$

[a] Based on the surface energy approximation. Values are $E_0 = 2.0$ MeV ^{4}He$^+$ beam at normal incidence with $\theta = 170°$ and $\xi = 5.0$ keV.

---

and treat the implanted As as a dilute impurity in Si. That is, we will combine Equations (5.26) and (5.28) to yield

$$(5.36) \qquad (Nt)_{As} = \frac{A_{As}}{H_{Si}}\frac{\sigma_{Si}(E_0)}{\sigma_{As}(E_0)}\frac{\xi}{[\varepsilon_0]_{Si}} = 1.2 \times 10^{15}\ As/cm^2$$

This is in agreement with the nominal value of the implanted fluence.

The maximum concentration of As in Si can be estimated from the peak height of the As signal. Using Equation (5.31) and the data in Table 5.1 yields

$$(5.37) \qquad \frac{N_{As}}{N_{Si}} = \frac{H_{As}}{H_{Si}}\frac{\sigma_{Si}(E_0)}{\sigma_{As}(E_0)}\frac{[\varepsilon_0]^{Si}_{As}}{[\varepsilon_0]^{Si}_{Si}} = 0.166\ at.\%$$

Using $N_{Si} = 5.00 \times 10^{22}$ atoms/cm^3 gives a value of $N_{As} = 8.30 \times 10^{19}$ atoms/cm^3.

To obtain the As concentration profile we first determine the As projected range, R_p. To calculate R_p we use Equation (5.6) and the stopping cross-section factor $[\varepsilon_0]^{Si}_{As}$, which gives the energy-to-depth conversion for scattering from As in a Si matrix. The peak position of arsenic is shifted by $\Delta E_{As} = 68$ keV below the As surface edge, and

$$(5.38) \qquad R_p = \Delta E_{As}/(N_{Si}[\varepsilon_0]^{Si}_{As}) = 143\ nm$$

When the implanted distribution is Gaussian, the depth profile can be described by a projected range R_p and a range straggling ΔR_p, which is the standard deviation of the Gaussian distribution in depth. The standard deviation is related to the FWHM of a Gaussian distribution by FWHM = $2.355\ \Delta R_p$.

As a first approximation, ignoring any detector energy resolution and energy straggling issues, the range straggling ΔR_p is given by

$$(5.39) \qquad \Delta R_p = FWHM/(2.355\,N[\varepsilon_0]^{Si}_{As}) = 53.6\ nm$$

5.7   HIGH-ENERGY BACKSCATTERING AND THE ELASTIC RESONANCE OF 8.8 MEV HE WITH ^{16}O

Up to this point we have primarily concerned ourselves with backscattering experiments at relatively low energies. As discussed in Chapter 2, going to higher ion energies offers the advantage of producing greater mass separation in the backscattering spectra. However, it is well known that the likelihood for scattering to deviate from a Rutherford nature increases with increasing projectile energy (Figure 3.6). Deviations are expected when the projectile velocity is high enough to allow it to penetrate deep into the orbitals of the atomic electrons and interact with the nucleus of the target atom. Thus, if any sense is to be made of the information obtained from high-energy backscattering experiments, it will first be necessary to know the behavior for the scattering cross sections for all the elements involved.

A good example of where high-energy backscattering has been widely applied is in the compositional analysis of the thin film high-temperature superconductors $YBa_2Cu_3O_7$. A simulated 2 MeV $^4He^+$ RBS spectrum from a 400 nm thick $YBa_2Cu_3O_7$ film deposited on $SrTiO_3$ substrate is shown in Figure 5.8. Also shown in Figure 5.8 are simulated component backscattering signals from the individual elements existing in both film and substrate (Ba, Y, Cu, Sr, Ti, and O).

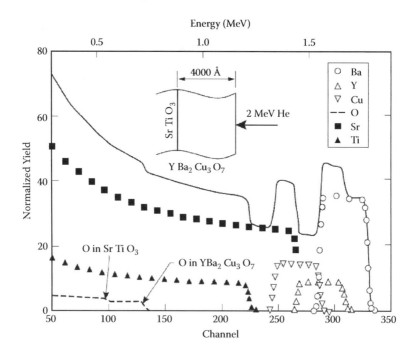

FIGURE 5.8   A simulated RBS spectrum of 2 MeV ^4He ions on a 400 nm thick $YBa_2Cu_3O_7$ film on a $SrTiO_3$ substrate. Also shown are the individual signals from Ba, Y, Cu, and O in the $YBa_2Cu_3O_7$ film and the signals from Sr, Ti, and O in the substrate.

Even without the substrate, there are considerable overlaps among Ba, Y, and Cu signals from the film. The complexity of adding the substrate greatly increases the difficulty of analyzing the oxygen concentration, because the oxygen signal will reside on the background of almost any substrate used in the synthesis of the high-temperature superconductors.

One way of increasing the separation between the heavy mass elements (Ba, Y, and Cu) and increasing the sensitivity to the oxygen is to perform experiments using the elastic resonance of 8.8 MeV He with ^{16}O. A typical backscattering spectrum for 8.8 MeV He incident on a 770 nm thick film of $YBa_2Cu_3O_7$ sitting on a $SrTiO_3$ substrate is shown in Figure 5.9. Indicated in the figure are the surface energies (KE_0) for Ba, Y, Cu, and O. The unlabeled step edges that appear at 2.80, 5.96, and 7.00 MeV correspond respectively to subsurface O, Ti, and Sr from the substrate. Clearly evident in these data is the good mass separation between Ba, Y, and Cu, as well as the large O yield from the superconductor film. However, the absolute accuracy for determining the film composition from such data depends on how well the scattering cross sections are known.

Thin film standards can be used to measure the ^{4}He scattering cross section as a function of energy. The first standard consisted of electron beam coevaporating a Y–Ba–Cu film onto a SiO_2 substrate that was capped with a Ti layer, which was used to determine the scattering cross-section ratios among Y, Ba, and Cu. The second standard consisted of evaporating Ba in the presence of O_2 onto a Ti-coated graphite substrate followed by another Ti cap layer, which was used to determine the scattering cross-section ratios between O and Ba. The substrates were chosen in each case because of their low kinematic factors.

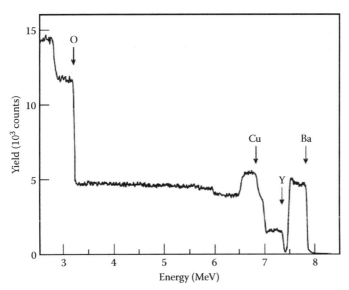

FIGURE 5.9   Typical backscattering spectrum for 8.8 MeV He ions incident on 770 nm thick film of $YBa_2Cu_3O_7$ on a $SrTiO_3$ substrate.

The Ti layers were employed to minimize environmental contamination and to aid in the adhesion to the graphite. For each standard, the film thickness was chosen to ensure elemental peak separation in a 2 MeV ^4He RBS spectrum. A schematic of the standard's structure and corresponding RBS spectra is shown in Figure 5.10.

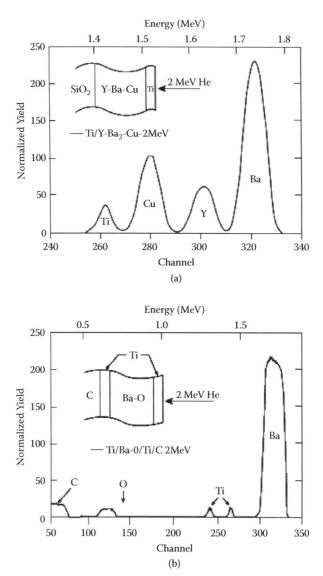

FIGURE 5.10   RBS data at 2 MeV from two reference film standards that were used to measure the relative cross sections of Cu, Y, and O relative to Ba as a function of He energy.

Performing peak integration on the data presented in Figure 5.10, applying Equation (5.27), and using the Rutherford cross section, the composition of Y and Cu relative to Ba (top spectrum) and O relative to Ba (bottom spectrum) was determined. Once the ratios of m/n are known (i.e., N_Y/N_{Ba}, N_{Cu}/N_{Ba}, N_O/N_{Ba}) at 2 MeV, where all the elements are Rutherford, Equation (5.27) can be inverted to measure the values of σ_A/σ_B at various energies by measuring the peak areas. For example, for σ_O/σ_{Ba} we have

(5.40)
$$\frac{\sigma_O(E)}{\sigma_{Ba}(E)} = \frac{N_O}{N_{Ba}} \frac{A_{Ba}}{A_O}$$

The results of this procedure for ^{4}He scattering from Ba and O in the energy range 8.2 to 9.1 MeV at $\theta = 166°$ are presented in Figure 5.11. Over this energy range, Ba is expected to have a Rutherford interaction with ^{4}He (see Figure 3.6). The dashed line at the bottom of the figure represents the ratio of cross sections assuming Rutherford scattering. The increase in sensitivity to ^{16}O at 8.8 MeV is about 32 times greater than for Rutherford scattering. The measured cross-section ratios for Y/Ba and Cu/Ba over the energy range 2 to 9 MeV are shown in Figure 5.12. The dashed line represents the Rutherford cross-section ratios. Assuming the scattering cross section for Ba remains Rutherford in this energy range, these data show that the Cu and Y cross sections start to deviate from Rutherford near 6.5 and 8.0 MeV, respectively.

With accurate cross-section data in hand, it is now possible to analyze the composition of a Y–Ba–Cu–O film from a backscattering spectrum taken with 8.8 MeV ^{4}He. A sample spectrum is presented in Figure 5.9. The analysis of these data will be left as a homework exercise.

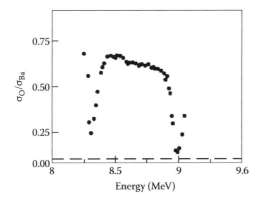

FIGURE 5.11   Ratio of scattering cross section for He ions scattering from ^{16}O and Ba as a function of He energy. The dashed line shows the ratio of the Rutherford scattering cross section.

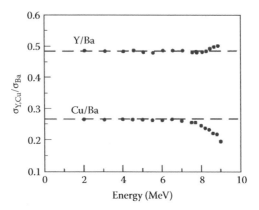

FIGURE 5.12   Ratios of scattering cross sections for He ions scattering from Y and Cu relative to Ba as a function of He energy. The dashed lines show the ratios of the Rutherford scattering cross sections.

PROBLEMS

5.1.   Consider a 200 nm free-standing film of Pt with 2 MeV ^{4}He at normal incidence and the backscattering detector located at $\theta_2 = 10°$. For the backside of Pt at t = 200 nm, determine the following: (a) the value of E_1, (b) the value of ΔE_{in}, (c) the value of E_t, (d) the value of ΔE_S, (e) ΔE_{out}, and (f) the value of ΔE. Hint: Use the surface and mean energy approximations.

5.2.   For the conditions listed in Problem 5.1, calculate [S] and [ε] using the surface and mean energy approximations. What is the percentage difference between these two approximations?

5.3.   Consider a 200 nm film of Pt on a Si substrate with 2 MeV ^{4}He at an incidence of $\theta_1 = 10°$ and the backscattering detector located at $\theta_2 = 10°$. For the He scattering off of Si atoms at the interface, determine the following: (a) the value of E_1, (b) the value of ΔE_{in}, (c) the value of E_t, (d) the value of ΔE_S, (e) ΔE_{out}, and (f) the value of ΔE. Hint: Use the surface and mean energy approximations.

5.4.   For the conditions listed in Problem 5.3, calculate [S] and [ε] using the surface and mean energy approximations. What is the percentage difference between these two approximations?

5.5.   Consider a 100 nm free-standing film of Ni_2Si (mass density of 7.23 g/cm^3) with 2 MeV ^{4}He at normal incidence and the backscattering detector located at $\theta_2 = 10°$. (a) What is the value of N^{Ni_2Si}? (b) What is the value of $\varepsilon_{in}^{Ni_2Si}$? (c) What is the value of $[\varepsilon]_{Ni}^{Ni_2Si}$? (d) What is the value of $[\varepsilon]_{Si}^{Ni_2Si}$ using the surface energy approximation?

5.6. Consider the following experimental conditions to measure Pt thin film on Si substrate: 3 MeV $^4He^{++}$ $\theta_1 = 60°$, $\theta = 170°$, $\xi = 2.5$ keV, $\Omega = 4$ msr, and $Q = 20$ μC. (a) How many He ions are there bombarding the sample? (b) If the average particle current during the experiment is 20 nA, how long did it take to acquire the spectrum? (c) How many counts would you expect in a Pt peak from a 10 nm thick Pt film with this measurement? (d) What is the Si surface height?

5.7. Using a ruler, measure the relative surface heights of Ba, Y, Cu, and O in the Y–Ba–Cu–O backscattering spectrum shown in Figure 5.9. Using these values, together with the cross section data in Figures 5.11 and 5.12, estimate N_Y/N_{Ba}, N_{Cu}/N_{Ba}, and N_O/N_{Ba}. Compare them to scattering cross section values obtained from data in Figure 5.8.

5.8. 2 MeV ^4He particles are scattered off a thin foil of an elemental material with atomic number Z_1, mass density ρ_1, mass number A_1, and thickness t_1. The backscattering spectrum was collected at the scattering angle θ, solid angle Ω, and accumulated charge Q, which yield the peak area of Y_1. (a) If the foil is replaced with another thin foil of different elements (Z_2, ρ_2, A_2, and t_2), what is the expected peak area of Y_2 for the new foil, assuming Rutherford cross sections for both foils? (b) If the foil is 10 μg/cm² of Au, what fraction of the incident He particles will be backscattered at $\theta = 150°$ into a detector with cone of the solid angle 5 msr? (c) If the foil is 10 μg/cm² of Au, what fraction of the incident He particles will be backscattered forward scattered at $\theta = 30°$ into a detector with $\Omega = 5$ msr?

5.9. For Problem 5.8, what fraction of incident He particles is backscattered from the free-standing Au foil (e.g., $\theta > 90°$)?

5.10. Assume in backscattering measurements that energy straggling is given by $\Omega_{tot}^2 = (K\Omega_{B,in})^2 + \Omega_{B,out}^2$. Calculate the amount of energy straggling in an RBS signal from a ultrathin Au layer underneath an Ag film of 400 nm thick with an analysis beam of 2 MeV ^4He ions at $\theta = 180°$? What is the total Au signal width (FWHM) if the detector resolution is 15 keV? What thickness of Ag does this correspond to?

REFERENCES

Chu, W.-K., Mayer, J. W., and Nicolet, M.-A. 1978. *Backscattering spectrometry,* chap. 2. New York: Academic Press.

SUGGESTED READING

Alford, T. L., Feldman, L. C., and Mayer, J. W. 2007. *Fundamentals of nanoscale film analyses.* New York: Springer.

Wang, Y., and M. Nastasi, eds. 2009. In *Handbook of modern ion beam materials analysis,* 2nd ed., chap. 4. Warrendale, PA: Materials Research Society.

Elastic Recoil Detection Analysis

6.1 INTRODUCTION

Elastic recoil detection analysis (ERD or ERDA) is a complementary technique to Rutherford backscattering spectrometry (RBS). In RBS, the incident particles (e.g., He) backscattered from target atoms are detected, whereas in ERD, the forward recoil target atoms (e.g., H) are directly detected. In a special case, when a common RBS beam (e.g., MeV ^{4}He) is used to analyze hydrogen and its isotopes, ERD is also referred to as forward recoil spectrometry (FReS). ERD is a very useful surface analytical technique for easy depth profiling of multiple light elements ($1 < Z < 9$). Before the introduction of ERD, light element analyses were primarily carried out with nuclear reaction analysis (NRA), which will be discussed in the next chapter. One of the advantages that ERD has over NRA is its ability of analyzing multielements or isotopes simultaneously. While ERD in the transmission mode has showed an excellent depth resolution, the majority of ERD analysis has been carried out in the reflection mode, which overcomes the ultrathin sample requirements for the transmission ERD. Interpretation of ERD data is generally more complex than that of RBS due to the use of a tilt recoil geometry (i.e., increasing multiple scattering energy spread and geometric broadening) and a range foil needed to filter out the scattering particles in the ERD measurement. However, with the advancement of data analysis software, one can now quickly and reliably analyze both ERD and RBS spectra with single computer software packages. Thus, the use of ERD and RBS as complementary tools can truly provide a complete elemental composition and the depth profiles.

The following lists a brief summary of some major developments in ERD techniques over the years:

- In 1976, the idea of ERD was first introduced by L'Ecuyer et al. to analyze light elements in a thin target (LiF or LiOH) with 25-40 MeV ^{35}Cl ions. [L'Ecuyer et al. 1976]
- In 1979, Doyle and Peercy first proposed using a conventional RBS beam (MeV ^{4}He ions) to analyze H-depth profiles distributed in Si_3N_4 layers. [Doyle and Peeray, 1979]
- In 1983, the time-of-flight (TOF) particle detection method was first applied to ERD by Groleau, Gujrathi, and Martin (TOF-ERD), where both 30 MeV ^{35}Cl$^+$ and 4 MeV ^{4}He$^+$ ions were used to investigate a Mylar film. [Groleau et al. 1983]
- In 1984, Ross and co-workers first presented an inexpensive and novel method of profiling hydrogen isotopes in beryllium where a 350 keV ^{4}He$^+$ beam and crossed electrical and magnetic fields (ExB filter) were used for particle detection (ExB-ERD). [Ross et al. 1984]
- In 1990, Tirira, Trocellier, and Frontier first described the analytical capabilities of ERD in transmission geometry using 1.8–3 MeV ^{4}He$^+$ ions in which an analyzing depth of 6 μm and a depth resolution of 35 nm at the target surface were obtained. [Tirira et al. 1990]
- In 1990, Hofsass and co-workers first introduced depth profiling of light elements using elastic recoil coincidence spectroscopy (ERCS) based on simultaneous measurement of scattered and recoiled particle energies. [Hofsass et al. 1990]
- In 1994, ΔE-E solid-state telescopes were first developed in ERD by Arnoldbik, de Laat, and Habraken to discriminate hydrogen isotopes. [Arnoldbik et al. 1994]
- In 1996, a dedicated ERD book was published by Tirira, Serruys, and Trocellier to give an excellent review of various elastic recoil detection techniques and their applications in various fields. [Tirira et al. 1996]
- In 2005, channeling ERD under a conventional range foil configuration was introduced by Shao and colleagues to study lattice location of hydrogen in silicon. [Shao et al. 2005]

6.2   ERD KINEMATICS

ERD kinematics is identical to that of RBS. Figure 6.1 shows a schematic diagram of an elastic recoil process. ERD involves measurement of the number and energy distribution of target atoms forward-recoiled by the incident heavier ions. From the physics point of view, ERD deals with a two-body elastic collision process in a central force field (Coulomb repulsion), an identical process where RBS occurs. Based on the conservation of energy and momentum equations in

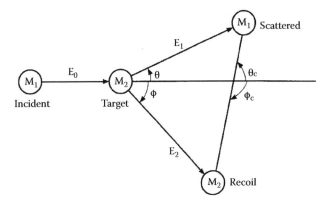

FIGURE 6.1   Schematic representation of an elastic collision between a projectile of M_1, atomic number Z_1, and energy E_0 and a target of mass M_2 (atomic number Z_2), which is at rest before the collision. After the collision, the target mass is recoiled forward at an angle ϕ and an energy E_2 and the projectile is scattered at an angle of θ with an energy E_1. The scattering and recoil angles in the center of the mass frame are designated as θ_c and ϕ_c, respectively.

Chapter 2, one can obtain the recoil kinematic factor at a recoil angle of ϕ as described in Equation (6.1):

(6.1)
$$K_R = \frac{E_2}{E_0} = \left[\frac{4\,M_1 M_2}{(M_1 + M_2)^2} \right] \cos^2 \phi$$

where M_1 and M_2 are the masses of the incident ion and recoil target atom. It is important to realize that at the same time, the forward scattering of M_1 from M_2 is possible at the same angle ϕ that is governed by the forward scattering kinematic factor:

(6.2)
$$K_S = \left[\frac{(M_1 / M_2) \cos \phi \pm [1 - (M_1 / M_2)^2 \sin^2 \phi]^{1/2}}{1 + (M_1 / M_2)} \right]^2$$

In RBS, where $M_1 < M_2$, the sign of the radical term in Equation (6.2) is always positive for all possible angles. However, in ERD, where $M_1 > M_2$, the only way to satisfy the positive sign for the radical term is for the elastically scattered projectiles to be limited within a cone of half-angle ϕ_{max} that is defined by:

(6.3)
$$\phi_{max} = \arcsin \frac{M_2}{M_1}$$

For example, to observe scattered $^4\mathrm{He}$ events from $^1\mathrm{H}$, $^2\mathrm{H}$ (D), or $^3\mathrm{H}$ (T) target atoms, the maximum detector angle will be limited to $\phi_{max} = 14.47°$, $30°$, or 48.59°, respectively. Equation (6.3) can be quite handy in choosing a favorable recoil angle geometry (i.e., the recoil angle is chosen greater than ϕ_{max}) in order to prevent excessive dead time and energy overlap due to the large elastic yield coming from the scattered beam. However, the benefit is often offset by the fact that the recoil kinematic factor and thus the recoil energy decrease with increasing ϕ. Also, if the target contains elements heavier than incident M_1, which is almost always true for most samples, the scattering events from those elements can be at any angle.

6.3   RECOIL PARTICLE IDENTIFICATION

Unlike in RBS—where only one type of particle (e.g., ^{4}He) is present in the backward angle—in ERD, both the recoil particle (e.g., ^{1}H, ^{2}H, or ^{3}H) and the scattering particle (e.g., ^{4}He) can be present at the forward detecting angle. This is called *the recoil-scatter particle ambiguity*, which is, in addition to *the mass-depth ambiguity*, already existing in both RBS and ERD. Since particle mass selection is obviously needed in ERD, there is a large difference in particle detection methods between RBS and ERD. In fact, the chronicle development of ERD techniques outlined early mainly reflects the development of various methods that were used to solve this *recoil-scatter particle ambiguity* issue in ERD. These particle detection techniques are briefly described in Figure 6.2:

- A conventional range-foil ERD as shown in Figure 6.2(a). This method is based on the fact that heavier incident particles can lose more energy per unit distance than lighter target particles, even though the former may have more kinetic energy than the latter. The detail on this characteristic nature of particle–matter interactions is described in Chapter 4. Typically, the optimum foil thickness is selected to be just thick enough to stop all scattering particles while permitting the desired recoil particles from the target to lose minimum energy when penetrating through the foil. Commercial aluminum, Mylar, and Kapton° foils (several to several tens of micrometers in thickness) are often used in conventional ERD due to their excellent thickness, uniformity, and durability. The conventional ERD method is simple to implement, but we will see later that the energy straggling from the range foil can significantly degrade energy resolution and thus reduces depth resolution of the ERD technique.
- A ΔE–E telescope for *z*- and *m*-separation as shown in Figure 6.2(b). A ΔE–E telescope is commonly used in nuclear physics to identify new particles. This method is based on principles that first use a ΔE-detector to obtain the stopping force and then use an E-detector (silicon detector or

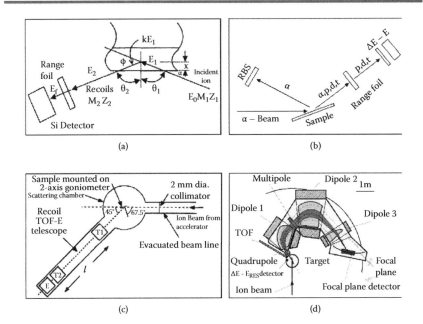

FIGURE 6.2 Schematic representation of particle identification techniques used in ERD: (a) A simple range-foil mass filter; (b) A ΔE-E telescope at Los Alamos National Laboratory in the United States; (c) A time-of-flight spectrometer at the Royal Institute of Technology in Sweden; and (d) A Q3D magnetic spectrometer at Universitat der Bundeswehr Munchen in Germany.

gas ionization chamber, etc.) to measure the remaining energy of the particles. Through coincidence detection electronics, ΔE versus E plots sort out signal counts for particles with different z and m values. The stopping force (S) is related to the recoil particle signature (Z, m, and E) by

$$S \propto Z^2 M \frac{\ln E}{E}$$

(6.4)

- A time-of-flight mass spectrometer (TOF-ERD) as shown in Figure 6.2(c). This method identifies a recoil particle of interest through the determination of its mass. The mass (m) is determined through two separate measurements: a time-of-flight spectrometer that obtains the velocity of the particle (v) and an ordinary silicon particle detector that obtains the energy of the particle (E), where $m = 2E/v^2$. The velocity (v) is determined by measuring the difference in timing signals (t) for a given particle flight path length (l), $v = l/t$. The timing signals (start and stop pulses) are normally produced by inserting two ultrathin carbon foils (~2–5 μg/cm^2) into the recoil particle beam path where the secondary electrons produced by recoil particles on each carbon foil are detected with microchannel plates to generate start and stop voltage pulses. The mass of the recoil particle is then determined by

$$(6.5) \qquad m = \frac{2E}{v^2} = 2E \cdot \left(\frac{t}{l}\right)^2$$

■ A magnetic spectrometer to identify particles based on different q/m ratios as shown in Figure 6.2(d). Charged particles with different charge/mass ratios (q/m) under a given magnetic field (B) will be deflected to different radial positions (r): $mv^2/r = qvB$, or $r = mv/qB$. By placing a silicon particle detector at different positions, the energies of the particles corresponding to different q/m ratios are detected: $E = mv^2/2$. The relationship between the radius of the recoil ion and its energy that allows for mass determination of the recoil ion is given in Equation (6.6):

$$(6.6) \qquad r = \frac{mv}{q \cdot B} = \frac{\sqrt{2mE}}{q \cdot B}$$

Sometimes, both electric and magnetic fields (ExB) are used to differentiate recoil particles. Each of the particle identification techniques has its own advantages and disadvantages. Some examples will be given in Section 6.6 on the application of these techniques.

6.4   RECOIL CROSS SECTIONS

In Chapter 3, we derived the Rutherford scattering cross section at a scattering angle of θ under a Coulomb central field as

$$(6.7) \qquad \frac{d\sigma(\theta)}{d\Omega} = \left(\frac{Z_1 Z_2 e^2}{2E}\right)^2 \frac{1}{\sin^4 \theta} \frac{\{[1-((M_1/M_2)\sin\theta)^2]^{1/2} + \cos\theta\}^2}{[1-((M_1/M_2)\sin\theta)^2]^{1/2}}$$

A similar process can be used to derive the Rutherford recoil cross section at a recoil angle of ϕ as

$$(6.8) \qquad \frac{d\sigma(\phi)}{d\Omega} = \left(\frac{Z_1 Z_2 e^2}{2E}\right)^2 \left(\frac{M_1 + M_2}{M_2}\right)^2 \frac{1}{\cos^3 \phi}$$

where E is given in MeV, $e^2 = 1.44 \times 10^{-13}$ MeV·cm, and $(d\sigma/d\Omega)_\phi$ is given in square centimeters.

Important features regarding the Rutherford recoil cross section include:

■ $(d\sigma/d\Omega)_\phi$ exhibits a minimum value for $\phi = 0°$ ($\propto 1/\cos^3\phi$).
■ The yield of recoil particles rises rapidly with decreasing incident energy ($\propto 1/E^2$).

- If $M_2 \gg M_1$, $(d\sigma/d\Omega)_\phi$ is approximately independent of the mass ratio.
- The higher incident particle Z_1 gives higher $(d\sigma/d\Omega)_\phi$ value.

6.4.1   DEVIATIONS FROM THE RUTHERFORD RECOIL CROSS SECTIONS

While recoil cross section is accurately predicted by Equation (6.8) for the majority of ion–target combinations, there are two extreme cases where the real recoil cross section may become non-Rutherford. For example, MeV energy ^{4}He-induced ERD for hydrogen isotope analysis is most likely a non-Rutherford recoil process since the Coulomb barrier between ^{4}He and ^{1}H particles is relatively low such that nuclear force can be involved in the ion–atom collisions. Another case is when heavy ions (HI) are used in HI-ERD, where the ion velocity may be so slow that the impact parameter is larger than the radius of inner-shell electron orbits (i.e., the nuclear charge is partially screened by inner shell electrons). Thus, similarly to RBS, there is an energy window for incident ions in ERD within which the recoil cross sections can be calculated by the Rutherford formula Equation (6.8). The lower limit of the energy window is determined by the screening factor K. Multiplying K in front of the Rutherford formula Equation (6.8) yields the screened recoil cross sections. For example, the screening factor of $K = 0.90$ means that the Rutherford formula Equation (6.8) would overestimate the real cross section by 10% if it were used. As long as screening is moderate (i.e., $K > 0.90$), the factor K can be accurately approximated by

(6.9)
$$K = \frac{\left(1 + \dfrac{E_S}{2E}\right)^2}{\left[1 + \dfrac{E_S}{E} + \left(\dfrac{E_S}{2E\cos\phi}\right)^2\right]^2}$$

where E_S in electron volts is given by

(6.10)
$$E_S = 48.73 Z_1 Z_2 \sqrt{Z_1^{2/3} + Z_2^{2/3}}\, \frac{M_1 + M_2}{M_1}$$

The accuracy to which the correction factor K can be approximated through Equation (6.9) for $K < 0.90$ is unclear.

On the other hand, the higher limit of the energy window is safely approximated by the so-called Coulomb barrier, defined as

(6.11)
$$E_{CB} = \left(1 + \frac{M_1}{M_2}\right)\frac{Z_1 Z_2 e^2}{4\pi\varepsilon_0 R_0}$$

The cross section can deviate by <1% from the Rutherford values when a safe minimum distance of closest approach of $R_0 = [1.25(A_1^{1/3} + A_2^{1/3}) + 5] \times 10^{-15}$ m is used. At larger incident energies, one must rely on experimental or theoretical values for the elastic scattering cross sections. Figure 6.3 shows the limits of specific energies for different projectiles on C or Si targets where the Rutherford cross sections with low screening corrections (K > 0.9 or K > 0.99) can be applied as a function of the nuclear charge of the projectiles (Dollinger and Bergmaier 2009). In addition, the limits from which the scattering cross section becomes non-Rutherford through nuclear interactions (Equation 6.11) are also plotted.

When a He ion beam is used to analyze hydrogen isotopes, the Coulomb barrier is easily overcome even at moderate energies. A collection of references for experimental data of He–H elastic cross sections can be found on the Web site of the Ion Beam Analysis Nuclear Data Library (IBANDL) (http://www.nds. iaea.org/ibandl/). For high energies or recoil angles where experimental cross-section data are unavailable, one can use the SigmaCalc tool to calculate elastic cross sections. The SigmaCalc tool is also available at the International Atomic Energy Agency Web site (http//www.nds.iaea.org/sigmacalc/). Figure 6.4 shows elastic recoil cross sections for He–p scattering calculated by SigmaCalc compared to the Rutherford cross section at various scattering angles (Dollinger and Bergmaier 2009). For the elastic scattering of He–d, there is no theory available to describe cross sections reliably, but one can refer to measured differential cross sections available in established scientific literature such as the

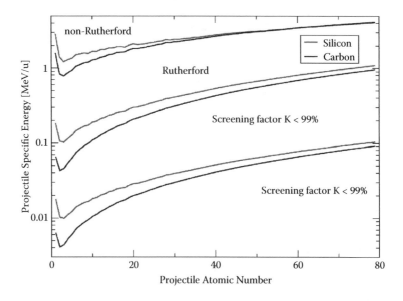

FIGURE 6.3   Limits of specific energies for projectiles where Rutherford recoil cross sections with low screening corrections (K > 0.9 or K > 0.99) can be applied as a function of the nuclear charge of the projectiles. (Dollinger, G. and Bergmaier, A. 2009. In *Handbook of Modern Ion Beam Materials Analysis,* 2nd ed., eds. Y. Q. Wang and M. Nastasi, 81–123. Warrendale, PA: Materials Research Society Press.)

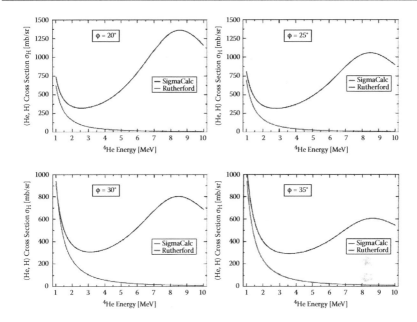

FIGURE 6.4   Elastic recoil cross sections for ^{1}H + ^{4}He – ^{4}He + p as calculated using SigmaCalc tool at various recoil angles. Rutherford cross sections are also shown for comparison. (Dollinger, G. and Bergmaier, A. 2009. In *Handbook of Modern Ion Beam Materials Analysis*, 2nd ed., eds. Y. Q. Wang and M. Nastasi, 81–123. Warrendale, PA: Materials Research Society Press.)

IBANDL database. Figure 6.5 shows the experimental elastic scattering data and an empirical fit of them for He–d scattering at various scattering angles (Kellock and Baglin 1993). The Rutherford recoil cross sections are also shown for comparison.

6.5   CONVENTIONAL RANGE-FOIL ERD

We now take a close look at the conventional range-foil ERD method since it shares similar detectors and electronics with RBS and is readily available in most ion beam accelerator laboratories. A common use of this method is to measure hydrogen depth profiles in solids by using a 3 MeV ^{4}He$^+$ ion beam, but a higher energy and heavier ion beam, such as a 28 MeV Si beam, has also been used to measure other light elements heavier than H, such as He, C, O, etc. A well-accepted measurement geometry in such an ERD method is $\alpha = 15°$ (i.e., $\theta_1 = 75°$) and $\phi = 30°$ (i.e., $\theta_2 = 75°$), as shown in Figure 6.2(a). The range foil thickness is determined by the combination of ion beam energy and target matrix elements. For example, based on an SRIM estimate, for a 3 MeV ^{4}He$^+$

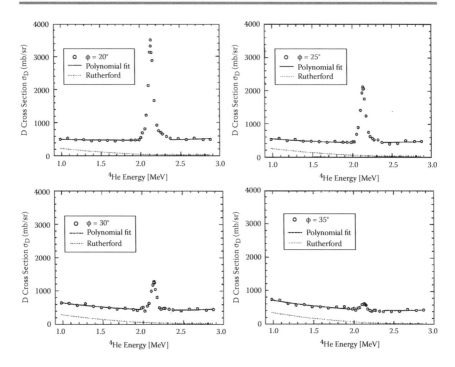

FIGURE 6.5 Measured elastic recoil cross-section data and an empirical fit of them for $^2D + {}^4He - {}^4He + d$ at various scattering angles. Rutherford cross sections are also shown for comparison. (Kellock, A. J. and Baglin, J. E. E. 1993. *Nuclear Instruments and Methods B* 79:493.)

incident ion beam, 13 μm thick Mylar [$(H_8C_{10}O_4)_n$, 1.397 g/cm³] is thick enough to stop all of the scattered ^4He particles from a Si target.

6.5.1 MASS RESOLUTION

The mass resolution defines the capability of recoil spectrometry to separate two signals arising from two neighboring elements in the target. With the use of a range foil in conventional ERD, mass resolution is the ability to separate the signal from different recoiled atoms in the spectrum along the energy axis. The larger the energy separation is, the larger the mass separation will be. If two types of atoms differ in their masses by a quantity δM_2, their corresponding differences in recoil energies after the collision, as defined by their kinematic factors, are expressed by

$$(6.12) \qquad \delta E_2 = E_0 \delta k = 4E_0 \frac{M_1(M_1 - M_2)\cos^2 \phi}{(M_1 + M_2)^2} \delta M_2$$

When δE_2 becomes the energy resolution of a particular ERD system under consideration, the mass resolution is usually defined as $\delta M_R = \delta E_2 / \delta M_2$.

Equation (6.12) shows that the mass resolution is optimized when the recoil angle goes to zero and the incident beam energy increases.

To figure out what incident particle mass M_1 produces the best mass resolution for a given system, one needs to optimize the relationship of δM_R versus M_1 through the following equation:

$$(6.13) \qquad \frac{d(\delta M_R)}{dM_1} = \frac{4E_0(M_1^2 - 4M_1M_2 + M_2^2)\cos^2\phi}{(M_1 + M_2)^4}$$

Making Equation (6.13) equal to zero results in $M_1 = 3.73\ M_2$, in which the maximum of the mass separation is obtained. This is why ^4He is effectively used to analyze ^1H in conventional range-foil ERD.

Figure 6.6 shows the detected energy of surface recoil atoms as a function of recoil mass for 1 MeV/amu projectiles of ^4He, ^{12}C, ^{28}Si, and ^{35}Cl, where a 10 μm Al range foil was used for stopping the scattered incident beams and the recoil atoms were collected at a scattering angle of 30° (Barbour and Doyle 1995). The measured surface energy positions for the different recoil masses are shown as open and closed circles, and the lines are drawn to guide the eye. Figure 6.6 indicates that

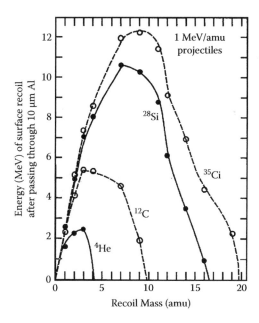

FIGURE 6.6   Detected energy of surface recoil atoms as a function of recoil mass for 1 MeV/amu projectiles of ^4He, ^{12}C, ^{28}Si, and ^{35}Cl, where a 10 μm Al range foil was used for stopping the scattered incident beams and the recoil atoms were collected at an angle of 30°. (Barbour, J. C. and Doyle, B. L. 1995. In *Handbook of Modern Ion Beam Materials Analysis*, eds. J. R. Tesmer and M. Nastasi, 83–138. Pittsburgh, PA: Materials Research Society Press.)

higher mass, higher energy projectiles yield higher energy recoils, which can then penetrate the stopping foil, often with greater mass resolution. For example, the mass resolution between H and D corresponds to an energy difference of ~0.8 MeV when using a 4 MeV ^{4}He projectile, but the mass resolution increases, giving an energy difference of ~2.1 MeV, when using a 12 MeV ^{12}C projectile or ~2.5 MeV when using a 28 MeV ^{28}Si projectile. The same figure shows that ^{3}He and ^{4}He can be separated with an energy difference of ~1.5 MeV using a 35 MeV ^{35}Cl beam but cannot be separated at all if a 12 MeV ^{12}C beam is used. Finally, Figure 6.6 shows that in some instances the mass resolution may actually be increased by decreasing M_1 because of the effect of the range foil. The energy separation between Li and Be, which are both near the peak of the curves for the Si and Cl incident beams, is greater on the 12 MeV ^{12}C curve than on the 28 MeV ^{28}Si or 35 MeV ^{35}Cl curves.

It is important to note that when the detection system itself has the ability to separate masses before their energy measurements (e.g., magnetic spectrometer), the projectile selectivity based on the kinematics as defined in Equation (6.13) is no longer a criterion. If the mass identification depends not only on the energy measurement but also on other quantities (e.g., the time resolution on the time of flight and the length resolution on the flight path in TOF-ERD), the mass resolution and selectivity need to include those factors in addition to the kinematic contribution.

6.5.2   ENERGY RESOLUTION

As stated previously, the mass resolution is related to the energy resolution of the ERD setup. At the same time, the depth resolution also depends on the energy resolution. Accurate determination of the total energy resolution of a particular ERD experimental setup is not easy because of multiple contributions to the degradation of the measured energy spectrum. In general, the evaluation of the total energy resolution is made by separately calculating individual energy spreads due to various contributions and then combining these energy spreads together. Although the individual contributions that influence the resolution are not necessarily Gaussian and do not occur independently, nevertheless, as a first approximation, one generally adds these contributions in quadrature to obtain the total energy spread as the full width at half maximum (FWHM):

$$\delta E_t^2 = \delta E_d^2 + \delta E_g^2 + \delta E_s^2 + \delta E_m^2 + \delta E_a^2 \tag{6.14}$$

where
δE_d is the energy resolution of the silicon particle detector (~15 keV)
δE_g is the geometrical broadening due to the finite detector acceptance angle and the finite beam size on the target
δE_s is the energy straggling in the target (from both ingoing and outgoing particles)

δE_{ms} is the multiple-scattering contribution in the target (from both ingoing and outgoing particles)

δE_a is the degradation in resolution due to energy straggling in the range-foil absorber

The degradations due to the surface roughness of the target and the thickness nonuniformity of the range foil are not considered in Equation (6.14). Here, we have also neglected the incident beam energy spread (~3 keV), the incident beam angular spread (the parallel beam), and the target atom (e.g., hydrogen) Doppler energy shift (<1 keV), since their contributions are marginal compared to the intrinsic energy resolution of the silicon detector. However, these contributions may not be neglected when other high-resolution ERD spectrometers (TOF, magnetic, etc.) are used.

6.5.2.1   ENERGY STRAGGLING

The energy straggling in the target can be estimated from the Bohr formula:

$$(6.15) \qquad \Omega_B^2 (keV^2) = 0.26 Z_1^2 Z_2 Nt(10^{18} \, atoms/cm^2)$$

where
Ω_B is the standard deviation of the average energy loss fluctuation
N is the target atomic density
$t = x/\sin \alpha$ for the ingoing particles
$t = x/\sin (\phi - \alpha)$ for the outgoing particles

The contributions of the energy straggling to the total energy spread for ingoing and outgoing particles are given by $\delta E_s^i = 2.355k \, \Omega_B^i$ and $\delta E_s^o = 2.355 \Omega_B^o$, respectively.

6.5.2.2   GEOMETRICAL ENERGY SPREAD (OR BROADENING)

The geometrical energy spread arises from the finite beam size on the target and the detector acceptance angle and is given by

$$\delta E_g(x) = \frac{dE_2(x)}{d\phi} \delta\phi_{det} = \left(\frac{\partial E_2(x)}{\partial k} \frac{\partial k}{\partial \phi} + \frac{\partial E_2(x)}{\partial \alpha} \frac{\partial \alpha}{\partial \phi} \right) \delta\phi_{det}$$

$$(6.16)$$

$$= \left(2kE_0 \tan\phi - \left(\frac{2kS_1 \tan\phi}{\sin\alpha} + \frac{S_2 \cot(\phi - \alpha)}{\sin(\phi - \alpha)} \right) x \right) \delta\phi_{det}$$

where $\delta\phi_{det}$ is the effective detector acceptance angle, S_1 is the stopping power of the incident particles, and S_2 is the stopping power of the recoil particles. The first term in Equation (6.16) is due to the kinematic spread and the second term is due to the path-length differences. The two terms cancel out at a certain depth x. Unlike in RBS, the energy resolution in ERD is not always the best at the surface when $x = 0$.

The effective detector acceptance angle is approximated by (Williams and Moller 1978):

$$(6.17) \qquad \delta\phi_{det} = \frac{1}{D}\sqrt{w^2 + \frac{d^2 \sin^2(\phi - \alpha)}{\sin^2\alpha}}$$

where D is the detector-target distance, w is the detector entrance width, and d is the incident beam width in the collision plane.

6.5.2.3   MULTIPLE SCATTERING

The multiple scattering leads to an energy spread via scattering kinematics. When an ion beam penetrates matter, particles undergo successive scattering events by target electrons and deviate from the original direction. As a result, initially well-collimated ions in the beam do not have the same directions after passing through a slab of the target matrix. This process, often referred to as multiple scattering, causes both angular and lateral deviations from the initial directions. The angular spread of ingoing particles, lateral spread, and spread caused by different path lengths all contribute to the total energy spread. Accurate mathematical treatments of the multiple scattering are difficult. However, the following equations give a decent approximation of the multiple scattering contributions for ingoing and outgoing particles (Turos and Meyer 1984):

$$(6.18) \qquad \delta E^i_{ms} = k(E_0 - E_1)\left(\frac{\cot\alpha}{\Gamma} - 2\tan\phi\right)\Delta\phi_i$$

and

$$(6.19) \qquad \delta E^o_{ms} = \left[(E_0 - E_1)(-2k\tan\phi) + (kE_1 - E_1)\frac{\cot(\phi - \alpha)}{\Gamma}\right]\Delta\phi_o$$

Use of Equations (6.18) and (6.19) requires more explanation. The half angles $\Delta\phi_i$ and $\Delta\phi_o$ (FWHM) of multiple scattering for ingoing and outgoing particles are taken from the tabulations of Sigmund and Winterbon (1975), and the scaling functions of Γ of Marwick and Sigmund (1975) provide the possibility to calculate the lateral spread from the angular distribution.

As an example, Figure 6.7 shows the variation of the individual contributions in Equation (6.14) with the tilt angle α when a 2.5 MeV $^4He^+$ beam is used to profile hydrogen 100 nm below the silicon surface and where a 9 μm thick aluminum foil is placed in front of the silicon surface barrier detector at the recoil angle of $\phi = 30°$ (Turos and Meyer 1984). For angles between 5° and

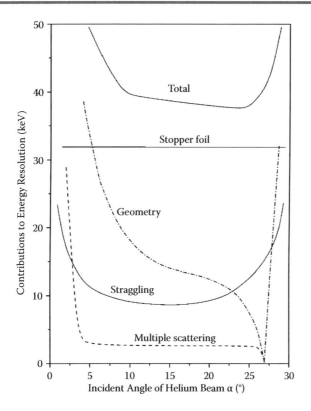

FIGURE 6.7   Magnitude of the individual contributions of the energy resolution versus tilt angle where 2.5 MeV ^{4}He$^+$ beam is used to profile hydrogen 100 nm below the silicon surface and an 8.5 µm thick aluminum foil is placed in front of the silicon surface barrier detector at the recoil angle of $\phi = 30°$. (Turos, A. and Meyer, O. 1984. *Nuclear Instruments and Methods B* 4:92.)

27°, the dominating term to the total energy spread is the energy straggling in the stopper foil. Close to the limiting angles ($\alpha = 0°$ and $\alpha = 30°$), the geometric and, eventually, the multiple scattering contributions are most pronounced. The multiple scattering contribution is mostly due to the lateral spread of the incoming beam, the other factors being negligible.

In summary, energy resolution of the conventional range-foil ERD is significantly worse than that of RBS, since

■ There is extra energy straggling in the range foil (20–40 keV).
■ Geometrical broadening is much greater due to the much tilted sample geometry in ERD.
■ Multiple-scattering energy spread is more significant again due to the ERD geometry.

6.5.3   DEPTH RESOLUTION

The slowdown of the ingoing particles on their way into the sample and of the recoil outgoing particles on their way out of the sample converts into an energy-depth relationship that is evaluated in terms of elemental depth profiles. Similarly to RBS, there is also a combined stopping factor term $\{S\}$ in ERD that relates the energy resolution, δE_t, in Equation (6.14) to the depth resolution, δx:

$$\delta x = \frac{\delta E_t}{\{S\}}$$

(6.20)

For the ERD geometry shown in Figure 6.2(a), the ingoing particle loses energy ΔE_i before reaching a depth of x,

$$\Delta E_i = E_0 - E_1 = S_1 x / \sin\alpha$$

(6.21)

where S_1 is the stopping power of the ingoing particle in the target. After the elastic collision at x, the outgoing particle has an energy of kE_1, so the energy loss of the outgoing particle (ΔE_o) when the particle comes out of the target is

$$\Delta E_o = kE_1 - E_2 = S_2 x / \sin(\phi - \alpha)$$

(6.22)

where S_2 is the stopping power of the outgoing particle in the target. Thus, the energy difference (ΔE) between the outgoing particle recoiled off the surface and that recoiled below the surface at a depth of x is obtained by using Equations (6.21) and (6.22):

(6.23)
$$\Delta E = kE_0 - E_2 = \left(\frac{kS_1}{\sin\alpha} + \frac{S_2}{\sin(\phi - \alpha)} \right) x = \{S\} x$$

where $\{S\} \equiv \frac{kS_1}{\sin\alpha} + \frac{S_2}{\sin(\phi - \alpha)}$ is called a combined stopping factor or an effective stopping factor of ERD.

Equation (6.20) suggests two approaches to increase the ERD depth resolution: decreasing the total energy spread δE_t and increasing the effective stopping power factor $\{S\}$. For a common ERD geometry ($\alpha = 15°$ and $\phi = 30°$), simply reducing the incident $^4He^+$ beam energy is an effective way to improve the depth resolution. Reducing the beam energy not only increases the effective stopping factor but also decreases the total energy spread (i.e., smaller foil straggling due to using a thinner range foil). For example, a greatly improved depth resolution ($\sim$20 nm near the Si surface) has been demonstrated by using a lower beam energy (1.3 MeV $^4He^+$) and thus a thinner range foil (5 μm Mylar),

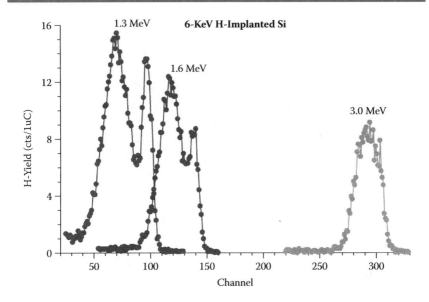

FIGURE 6.8   Comparison of recoil proton spectra of a 6 keV $^1H^+$ implant in Si (fluence ~ 5 × 10^16 ions/cm² and range ~ 110 nm below the surface) measured with different incident $^4He^+$ beam energies and Mylar range foil thicknesses: 3 MeV $^4He^+$ beam with 14.5 μm Mylar, 1.6 MeV $^4He^+$ beam with 6 μm Mylar, and 1.3 MeV $^4He^+$ beam with 5 μm Mylar.

as compared with the depth resolution of ~65 nm near the Si surface when using a standard 3 MeV $^4He^+$ beam with a range foil of 14.5 μm Mylar. This depth resolution improvement is clearly demonstrated in Figure 6.8, where 6 keV $^1H^+$ implanted Si (fluence ~ 5 × 10^16 ions/cm² and range ~ 110 nm in Si) is analyzed with different $^4He^+$ beam energies. One drawback of the reduced beam energy approach is that the maximum probing depth must be reduced (e.g., from ~2 to ~0.2 μm for measuring hydrogen in silicon when the He beam energy is reduced from 3 to 1.3 MeV).

6.5.4   IDEAL RECOIL ENERGY SPECTRUM

Converting a measured energy spectrum (counts versus recoil energy) into a desired depth profile (concentration vs. depth) requires considerable efforts. The good news is that, as for RBS, data reduction software is available for carrying out this process on ERD data. This data analysis software will be described in detail in Chapter 15. Here, we want to describe the physical process by which the depth profile is related to energy spectrum in an ERD process. Let us first consider the simplest case where the energy resolution is not included in this process. Suppose that the sample is divided into many layers (or slabs) of equal thickness δx and consider the slab at a depth of x. The incident particle energy just before the collision at depth x is $E_1 = E_0 - S_1 x/\sin\alpha$. The recoil particle has an energy of kE_1 right after the collision. This energy is decreased to $E_2 = kE_1 - S_2 x/\sin(\phi - \alpha)$ after the recoil particle leaves the target. In the end, the recoil particle has an energy of $E_f = E_2 - E_a$ before reaching the detector where E_a is the energy loss in the range foil.

The recoil particle counts at a given detected energy E_f and energy width δE_f are related to concentration $C(x)$ and the recoil cross section $\sigma(E_1)$ at the depth x through the following formula:

$$(6.24) \qquad N(E_f)\delta E_f = Q\Omega\sigma(E_1)C(x)\frac{\delta x}{\delta E_f}\frac{\delta E_f}{\sin\alpha}$$

where Q is the accumulated beam charge and Ω is the solid angle subtended by the detector.

In most cases, the following relationships are very good approximations:

$$(6.25) \qquad \delta x = \frac{\delta(kE_1)}{\{S\}}; \frac{\delta(kE_1)}{\delta E_2} = \frac{S_2(kE_1)}{S_2(E_2)}; \frac{\delta(E_2)}{\delta E_f} = \frac{S_a(E_2)}{S_a(E_f)}$$

where S_a is the stopping power of the recoil particle in the range foil.

Combining Equation (6.24) and Equation (6.25), we obtain a general expression for the ideal recoil energy spectrum:

$$(6.26) \qquad N(E_f)\delta E_f = Q\Omega\sigma(E_1)C(x)\frac{1}{\{S\}}\frac{S_2(kE_1)}{S_2(E_2)}\frac{S_a(E_2)}{S_a(E_f)}\frac{\delta E_f}{\sin\alpha}$$

6.5.4.1   SURFACE SPECTRUM HEIGHT OF A THICK TARGET

If E_f is the recoil energy from the surface atoms ($E_{f,0}$) and δE_f is the channel width (ξ) of the multichannel analyzer (MCA), the $N(E_f)\delta E_f$ above will become the surface spectrum height $H_{f,0}$ of a thick target:

$$(6.27) \qquad H_{f,0} = N(E_{f,0})\xi = Q\Omega\sigma(E_0)C(x)\frac{\xi}{\{S_0\}}\frac{S_a(kE_0)}{S_a(E_{f,0})}\frac{1}{\sin\alpha}$$

where the imagined thickness of the sample surface layer that corresponds to a channel width ξ is given by $\tau_0 = \xi/\{S_0\}$ and the surface effective stopping factor $\{S_0\} = kS_1(E_0)/\sin\alpha + S_2(kE_0)/\sin(\phi - \alpha)$. This surface recoil spectrum height is similar to the surface RBS spectrum height we discussed earlier in Chapter 5 but with an additional scaling term $S_a(kE_0)/S_a(E_{f,0})$ due to the use of a range foil. We will show an example about the importance of this scaling term in ERD spectrum quantification later.

6.5.4.2   RECOIL YIELD FROM A THIN TARGET

For a thin film sample (δx) where the energy loss in the film becomes negligible, the recoil yield in the film is obtained by a simple integral of Equation (6.26):

$$(6.28) \qquad A(E_0) = \int_0^{\delta x} N(E_f)\delta E_f = Q\Omega\sigma(E_0)(C_0\delta x)\frac{1}{\sin\alpha}$$

where the total amount of interested elements in the film is estimated as the product of the concentration and the thickness ($C_0\, \delta x$) (e.g., the implant fluence (at./cm^2) for an ion implanted target). This equation is identical to that used for thin film analysis in RBS.

6.5.5   ACTUAL RECOIL ENERGY SPECTRUM

An actual recoil energy spectrum must include a contribution of the energy resolution of the given ERD setup during the measurement. For a practical estimate, the actual energy spectrum is considered a convolution of an ideal recoil energy spectrum shown in Equation (6.26), $N(E_f)\delta E_f$, and the total energy resolution of the ERD experimental setup, $G(E_f - E'_f)$:

$$(6.29) \qquad N'(E'_f) = \int_{-\infty}^{+\infty} G(E_f - E'_f)N(E_f)dE_f$$

The energy resolution function, $G(E_f - E'_f)$, is normally approximated with a Gaussian function:

$$(6.30) \qquad G(E_f - E'_f) = \frac{1}{\sqrt{2\pi\delta_t^2}}e^{-\frac{(E_f-E'_f)^2}{2\delta_t^2}}$$

where δ_t is the variance of the Gaussian distribution (standard deviation) and is related to the FWHM of the total energy resolution δE_t in Equation (6.14) by $\delta_t = \delta E_t/2.355$.

6.5.6   SHAPE OF RECOIL ENERGY SPECTRUM

For a thick uniform concentration target, its RBS spectrum of yield versus energy collected from a silicon particle detector trends with a negative slope, following the energy dependence of Rutherford scattering cross section ($\sigma \propto 1/E^2$). However, the ERD spectrum of a thick uniform target appears to have a positive slope. Figure 6.9 shows (a) a measured forward recoil ^{1}H spectrum

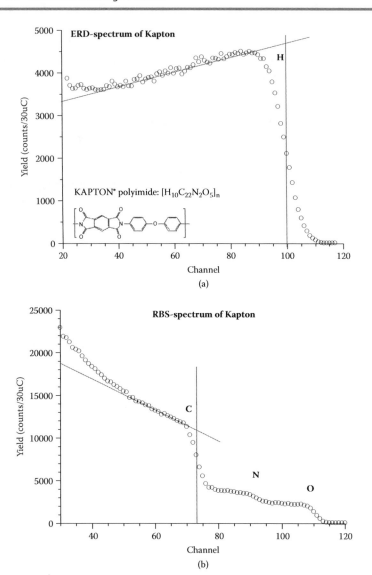

FIGURE 6.9 Comparison between ERD spectrum shape and RBS spectrum shape from a uniform Kapton® polyimide target, $[H_{10}C_{22}N_2O_5]_n$: (a) forward recoiled ^1H spectrum and (b) backscattered ^4He spectrum by C, N, and O elements. The ERD and RBS spectra were collected concurrently with a 1.3 MeV ^4He beam under the measurement geometry of $\alpha = 15°$ (target tilt), $\theta = 165°$ (RBS detector), $\phi = 30°$ (ERD detector), and a range foil of 4.5 µm Mylar in front of the ERD detector. The solid lines in the spectra are drawn to help determine the surface spectrum height of two major elements in the Kapton (H and C).

and (b) a backscattered ^{4}He spectrum off C, N, and O elements from a uniform Kapton® polyimide target, $[H_{10}C_{22}N_2O_5]_n$. The ERD and RBS spectra were collected concurrently with a 1.3 MeV ^{4}He beam under the measurement geometry of $\alpha = 15°$ (target tilt), $\theta = 165°$ (RBS detector), $\phi = 30°$ (ERD detector), and a range foil of 5 μm thick Mylar in front of the ERD detector. The solid lines in the spectra are drawn to help determine the surface spectrum height of two major elements in the Kapton (H and C). The positive-sloped ERD spectrum in Figure 6.9(a) is contributed by relative importance to four factors that define the ideal recoil spectrum as described in Equation (6.26): recoil cross section $[\sigma(E_1)]$, reciprocal effective stopping power $[1/\{S\}]$, energy loss ratio in the matrix $[S_2(kE_1)/S_2(E_2)]$, and energy loss ratio in the absorber $[S_a(E_2)/S_a(E_f)]$. Figure 6.10 shows these factors as a function of depth in a silicon target for a typical ERD experimental setup, where a 3 MeV ^{4}He beam is used for a measurement geometry of $\alpha = 15°$ and $\phi = 30°$ with a 13.5 μm Mylar as the range foil (Wang et al. 1990). Even though the depth dependence of the ERD cross section contributes to a negative slope trend (similar to the RBS), as shown in Figure 6.10, the other three factors—especially the energy loss ratio factor in the absorber $[S_a(E_2)/ S_a(E_f)]$—have contributed in an opposite way so that they have resulted in an overall positive-sloped spectrum shape for ERD from a uniform thick target. Although the effect of the depth dependence of the system energy resolution

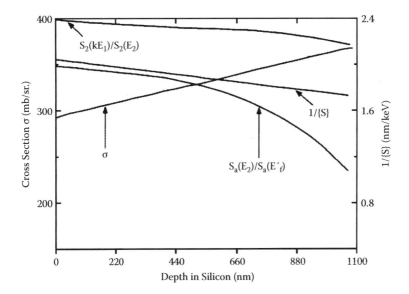

FIGURE 6.10   Depth dependence of recoil cross section, $\sigma(E_1)$, reciprocal effective stopping power, $1/\{S\}$, energy loss ratio in the target Si, $S_2(kE_1)/S_2(E_2)$, and in the absorber Mylar, $S_a(E_2)/S_a(E_f)$, where a 3 MeV ^{4}He beam is used for a geometry of $\alpha = 15°$, $\phi = 30°$ and a 13.5 μm Mylar as the range foil. (Wang, Y. Q. et al. 1990. *Nuclear Instruments and Methods B* 47:427.)

on the shape of the ideal recoil spectrum is not included in this discussion so far, it is reasonable to argue that the energy resolution does broaden the front and back edges of the spectra of a thick uniform target but will not change the overall depth dependence of the recoil yield.

6.5.7   ERD SENSITIVITY

Sensitivity is an important parameter for any analytical technique with trace element analysis. In theory, the element detection sensitivity of ERD is determined by two factors: recoil cross section of the specific element and the ion radiation damage (elemental diffusion, surface desorption, sputtering, etc.) to the target. The latter will ultimately limit accessible beam current and recoil signal counts. In reality, other factors, including the background level in the region of interest in the detector signals, will also limit sensitivity. This background may have different origins, such as limited elemental separation, dark counts from electronic noise, scattered atoms from somewhere other than the area or depth of interest in the target, wrong signals in the detector (i.e., pile up effects), plural or multiple scattering effects, secondary reactions, etc. In addition, certain elements, such as hydrogen, are so mobile during the analysis that data correction to the zero accumulated charge is often required to obtain the true content. The detailed discussions can be found in (Dollinger and Bergmaier, 2009). The consensus is that a simple range-foil ERD by 1–3 MeV ^{4}He$^+$ beam exhibits a hydrogen detection sensitivity of as low as 0.1 at.%, while a high-energy and heavy-ion (e.g., 28 MeV Si) ERD equipped with a sophisticated magnetic spectrometer may have a parts per million sensitivity for certain light elements such as C and O.

6.5.8   QUANTIFICATION OF ERD

As a prime example of the conventional range-foil ERD, hydrogen analysis with a MeV ^{4}He$^+$ ion beam is chosen to demonstrate how quantitative hydrogen analysis can be done using ERD. Compared to RBS, ERD analysis is more complicated because of the extra factors involved: the less known recoil cross-section data for (^{4}He, p), the less direct energy-to-depth conversion as a result of the mass-filtering range foil, and the poor energy resolution due to both the range foil and the ERD glancing incidence-exit geometry. The glancing incidence-exit geometry also makes it more difficult to measure the total charge on the target accurately and directly. To simplify these uncertainties, a known hydrogen standard is often used in quantitative hydrogen analysis.

Two types of H standards are commonly used in ERD: thick target standards with the known H atoms/cm^3 and thin film standards with the known H atoms/cm^2 (Wang 2004). Both the standards have uniform lateral H distribution and smooth surfaces, which are important to the glancing incidence-exit ERD configuration. Kapton polyimide and H-implanted Si are the most frequently used primary H standards in each category. Kapton polyimide $[H_{10}C_{22}N_2O_5]_n$ has a well-known and high H concentration (25.64 at.%, 2.25×10^{22} H atoms/cm^3) and is easily obtainable (DuPont). The surface spectrum height of the Kapton standard is determined by Equation (6.27). On the other hand, H-implanted silicon usually has a lower but also well-known total H content or H implant fluence ($2 \sim 5 \times 10^{16}$ H at./cm^2) and is radiation stable and reusable. The peak area of the H implant Si standard is determined by Equation (6.28).

To determine the surface spectrum height for Kapton accurately, several precautions must be carefully considered in the experiment. As shown in Figure 6.11(a), the surface spectrum height, which should be read from the

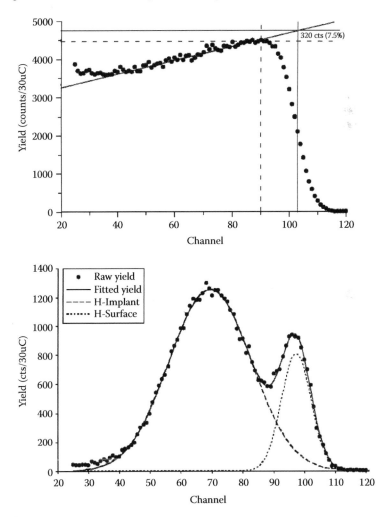

FIGURE 6.11   Measured ERD spectra with 1.3 MeV ^{4}He$^+$ beam from (a) Kapton thick target with 1 nm Pt-coating layer; and (b) 6 keV H-implant in Si. (Wang, Y. Q. 2004. *Nuclear Instruments and Methods B* 219–220:115.)

intersection of the solid vertical and sloped lines, can be easily underestimated by 7%–8% if one neglects the surface spectrum broadening due to the poor energy resolution or careless reading of the Kapton spectrum. On the other hand, the surface spectrum height can be overestimated by a noticeable amount (5%–6%) because of the spectral shift (to a higher channel) due to beam charging on the insulating Kapton surface. Beam charging also introduces more noise in the detection system and affects the charge integration accuracy. The precoating of 1–2 nm Pt on Kapton has adequately eliminated these charging-related problems and, at the same time, the concurrently measured Pt signal by an RBS detector provides a more reliable reference for the charge measurement in ERD, as compared with the RBS oxygen or carbon spectrum height collected from the bare Kapton. Lastly, hydrogen loss (H mobility) in Kapton during the ERD measurements, as a result of the beam irradiation and/or thermal heating, can underestimate the surface spectrum height if not properly corrected. Normally, one needs to collect a hydrogen mobility curve of H spectrum height versus accumulated beam charge first and then extrapolate to zero beam charge to obtain the true spectrum height. Precoating of 1–2 nm Pt on Kapton was found also to reduce significantly the hydrogen loss during the ERD measurement, perhaps by reducing the surface thermal spikes. The details can be found in Wang, 2004.

To determine the peak area of the implanted H in Si accurately, one must carefully separate the surface H contamination. H contamination, typically 5–10×10^{15} H at./cm^2, is often neglected in determining the surface spectrum height for Kapton due to the very high H content in Kapton, but can significantly influence the accurate determination of the implanted H content because these two H signals are often comparable in signal intensity and in very close depth proximity. As shown in Figure 6.8 and again in Figure 6.11(b), with the proper beam energy and range-foil selection (e.g., 1.3 MeV ^{4}He$^+$ and 5 µm Mylar), one can separate a 6 keV ^{1}H$^+$ implanted signal (nominally 5×10^{16} at./cm^2) from the surface hydrogen even though the implanted-H peak concentration is only ~110 nm below the Si surface.

When using a known uniform standard (e.g., Kapton) to determine the fluence of a H-implanted target or using a known H implant standard (e.g., 6 keV ^{1}H$^+$ in Si) to determine the H content of a thick hydrogenated film, Equations (6.27) and (6.28) need to be used together. For example, supposing that we do not know the exact fluence in our 6 keV ^{1}H$^+$ implanted Si target but would like to use the Kapton standard to determine it; the following formula will represent the implanted fluence in Si:

$$(6.31) \qquad D = (C_0 \delta x) = \frac{A}{H(E_{f,0})} \frac{\sigma(E_0)}{\sigma(E)} \frac{\xi}{\{S_0\}} \frac{S_a(kE_0)}{S_a(E_{f,0})}$$

Here we assume that the same beam charge Ω and same detector solid angle Ω are used to collect both spectra, so these quantities are cancelled out in Equation (6.31). Two important theoretical correction terms in Equation (6.31), $\sigma(E_0)/\sigma(E)$ and $S_a(kE_0)/S_a(kE_{f,0})$, can sometimes be overlooked since they are either insignificant (the first ratio) or absent (the second ratio) in the more familiar RBS relative analysis (Wang 2004). In fact, these corrections can be very important in ERD relative analysis.

Let us put some numbers in Equation (6.31) to get perspective. For the given recoil geometry of $\phi = 30°$, the recoil kinematics factor for ^{4}He on ^{1}H is $k = 0.48$ based on Equation (6.1). Using SRIM-2000 stopping data (Ziegler, Biersack, and Littmark 1985), the surface recoil proton energy $kE_0 = 0.624$ MeV decreases to $kE_{f,0} = 0.336$ MeV after passing through 5 μm Mylar foil, and thus the stopping ratio $S_a(kE_0)/S_a(kE_{f,0})$ equals 0.679. On the other hand, seen by the incoming beam of $E_0 = 1.3$ MeV, the H implant peak is "apparently" located ~425 nm below the Si surface in the current ERD setup. The mean beam energy decreases to $E = 1.175$ MeV when the beam reaches to such a depth. Considering Rutherford recoil cross sections for such a low ^{4}He$^+$ beam energy region, as defined in Equation (6.8), the term $\sigma(E_0)/\sigma(E)$ becomes 0.817, which is significantly smaller than the unity. In other words, one could simply overestimate the implant fluence by nearly 18% if the term $\sigma(E_0)/\sigma(E)$ is neglected, and in a worse case could even inflate the implant fluence by as much as 55% if both terms were simply treated as the unity. In summary, after taking careful consideration of both measurement and theoretical corrections, the implant fluence of 6 keV H-implanted Si is determined to be $(5.75 \pm 0.30) \times 10^{16}$ H atoms/cm^2 depending on which database of He stopping power in Si is used. The ~5% uncertainty in the analysis is mainly due to the uncertainty of He stopping data in Kapton, where Bragg's additive rule is used to figure out the stopping power for Kapton from the stopping powers of individual elements (H, C, N, O) in the Kapton compound. As a comparison, the average H-implant fluence for this Round Robin H-analysis provided by all other participating ion beam labs is $(5.71 \pm 0.13) \times 10^{16}$ atoms/cm^2 (Boudreault et al. 2004).

6.6   SELECTED APPLICATIONS

6.6.1   DIFFUSION OF HYDROGEN IN POLYSTYRENE

The first example deals with measurements of hydrogen isotopes (H and D) and their diffusion in polymers such as polystyrene with MeV He ion beams under a conventional range-foil configuration. For the purpose of demonstration, an example ERD analysis will be given only for the sample before annealing treatments were performed. Specifically, the pristine sample contained a 400 nm thick polystyrene (PS) layer on a thin layer of deuterated PS-PMMA (polymethyl methacrylate) on a 400 nm thick poly vinylchloride (PVC) layer, as schematically shown in the insert of Figure 6.12(a) (Barbour and Doyle 1995). The ERD measurement on the sample was done with 2.8 MeV He$^+$ beam under a standard recoil geometry where the beam was incident at 75° from the sample normal and the ERD detector was placed at a recoil angle of 30°. A 12 μm aluminized-Mylar range foil was placed in front of the surface barrier silicon detector. The ERD spectrum in Figure 6.12(a) from the sample was collected with the detector solid angle of 3.6 msr and the total charge of 3 μC.

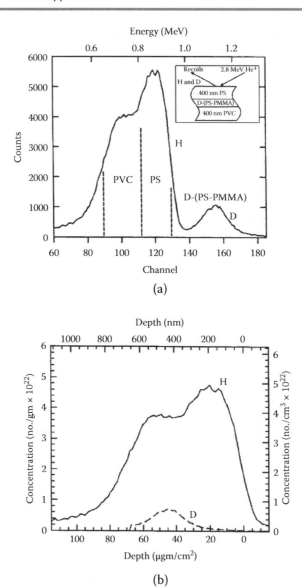

FIGURE 6.12   ERD measurement of hydrogen and deuterium distribution in a multilayer polymer film with a 2.8 MeV ^{4}He$^+$ beam and a 12 μm aluminized-Mylar range foil: (a) ERD spectrum collected with the detector solid angle of 3.6 msr and the total charge of 3 μC (see insert for the sample structure); and (b) the converted depth profiles of H and D in polystyrene based on the slab analysis, where the concentration and depth scales are shown in density-exclusive units along the left and bottom axes, respectively, while the right and top axes were determined by assuming the density of the sample is 1.04 g/cm^3. (Barbour, J. C. and Doyle, B. L. 1995.)

Figure 6.12(a) shows that the hydrogen signal exists from approximately channel 70 to channel 135, while the deuterium signal is shown from channel 135 to above channel 180. No background exists to subtract from the H or D peaks in this spectrum, although the back edge of the D signal will be abruptly terminated in this analysis in order to eliminate the appearance of the H signal. To convert the ERD spectrum into depth profiles of H in PS and D in D-(PS-PMMA), one may directly use a data analysis software package such as RUMP, SIMNRA, WinNDF, etc. as described in Chapter 15 or one could also use the traditional slab analysis based on Equations (6.23) and (6.26). Figure 6.12(b) shows the converted depth profiles of H and D in polystyrene based on the slab analysis of Figure 6.12(a). The concentration and depth scales are shown in density-exclusive units along the left and bottom axes, respectively, in Figure 6.12(b), while the right and top axes were determined by assuming that the density of the sample is 1.04 g/cm^3.

6.6.2   MEASUREMENT OF HYDROGEN ISOTOPES BY ΔE-E ERD

Figure 6.13 shows an example of an ERD spectrum measured with a ΔE-E telescope detection system (Prozesky et al. 1994). The analyzing beam was 4 MeV ^{4}He with the recoil angle at 30°. The angles of the target surface to beam and target surface to detector were both selected as 15°. The entrance collimator of the beam was circular with a diameter of 0.7 mm, and the collimator defining the solid angle in front of the ΔE detector was rectangular with a height of 10 mm and slit width of 0.9 mm, defining a solid angle of 2 msr. A Ni stopper foil with a thickness of 6.25 μm was put in front of the rectangular slit to prevent scattered ^{4}He from reaching the ΔE detector. The E detector was a 100 μm thick Si surface barrier detector with a quoted energy resolution of 16 keV for 5.486 keV ^{4}He ions, and the ΔE detector was a 13.6 μm thick Si detector with a quoted noise resolution of 35 keV. With beam currents of 15 to 30 nA, the system gave rise to count rates of between 1 and 2 kHz in the two detectors.

Figure 6.13(a) shows a three-dimensional plot of hydrogen isotopes in titanium hydride after the total measurement charge of 30 μC. The X- and Y-axes represent the total energies (the sum of E and ΔE) of the recoils, and the ΔE signals. The counts at these energies are represented in the Z-axis, with each bin respectively corresponding to 40 keV in the total energy spectrum and 20 keV in the ΔE spectrum. The loci corresponding to H, D, and T are indicated in the figure. The H signals are completely separated from the other isotopes, mainly due to the lower kinematic factor and the relatively high mass difference with respect to the heavier isotopes. The D and T are closer in both ΔE and E, due to the smaller difference in kinematic factor and mass. The slight curvature of the loci with decreasing energy is a result of the change of the stopping powers of the hydrogen isotopes as a function of energy.

Figure 6.13(b) is a two-dimensional plot of ΔE at a constant E value (in the SSB detector) between 470 and 510 keV, indicating the separation of the signals in the

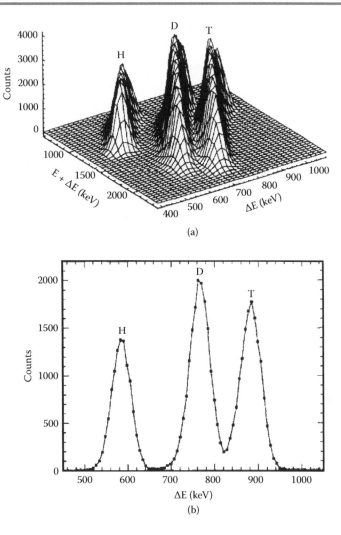

(a)

(b)

FIGURE 6.13   Hydrogen isotope measurement using a 4 MeV ^{4}He$^+$ beam and a ΔE-E telescope detector. (a) A three-dimensional plot of hydrogen isotopes in titanium hydride after the total measurement charge of 30 mC where the X, Y, and Z axes represent the total energies (the sum of E and ΔE) of the recoils, the ΔE signals, and the counts at these energies with each bin corresponding to 40 keV in the total energy spectrum and 20 keV in the ΔE spectrum, respectively. (b) A two-dimensional plot of ΔE at a constant E value (in the SSB detector) between 470 and 510 keV, indicating decent separation of each isotope signal in the DE detector. (Prozesky, V. M. et al. 1994. *Nuclear Instruments and Methods B* 84:373.)

ΔE detector. The separation of the two loci of D and T is very important in the subsequent quantitative analysis of the D and T content of targets. The quality of separation is seen where the valley between the D and T peaks corresponds to about 1/10 of the peak heights of D and T. This corresponds to an effective ΔE detector resolution for these ions of 55 keV at around 1.2 MeV, which compares favorably with the noise resolution of 35 keV and the Bohr straggling of 16 keV, yielding an expected theoretical resolution of 39 keV. The separation of the D and T peaks allows extraction of more than 95% of the peak contents of the two isotopes, which is adequate for analytical processing of the data. Count rates were kept low enough to ensure that pileup events in the ΔE detector, which in the case of D would interfere with events from T, were kept negligible. The maximum depth information obtainable for single isotopes in this experimental setup is about 4.1×10^{18} Ti at./cm^2 for H, 5.3×10^{18} Ti at./cm^2 for D, and 4.8×10^{18} Ti at./cm^2 for T, calculated according to the energy where recoils are fully stopped in the ΔE detector. Deeper analyzing depths can be obtained through the use of higher bombarding energies, with the disadvantages of worse separation in the ΔE detector and generally lower ERD cross sections.

6.6.3   MULTIELEMENT ANALYSIS WITH A HEAVY ION TOF-ERD

Time-of-flight ERD involves the simultaneous measurement of both the velocity and total energy of the atoms recoiled from the target. As with traditional range-foil ERD, the recoil atom energy is collected in a semiconductor detector such as a surface barrier Si detector (labeled E), but only after the atom traverses the time-of-flight detection telescope. A TOF telescope usually contains two ultra-thin carbon foils (2–5 μg/cm^2) in the recoil beam path separated by a known flight path length. The timing signals are produced by recoil-induced secondary electrons from the carbon foils that are then detected by channel electron multiplier plates or other electron detectors. An example of a three-dimensional TOF-ERD spectrum is shown in Figure 6.14(a) for 12 MeV Au projectiles incident on a sample containing a C-B-C-B-C-B... multilayer on a Si wafer substrate and the recoiled B, C, O,...; particles are detected at a recoil angle of 30° with a time-of-flight mass spectrometer followed by a silicon energy detector (Knapp, Barbour, and Doyle 1992). In the spectrum of Figure 6.14(a), the time axis has been reinverted so that increasing time channels reflect lower velocities. The "front" edges of the two boron isotopes, the carbon, the silicon substrate, and even an oxygen contaminant are easily identified in the three-dimensional spectrum as islands, loci of E versus t data points. Notice that the locus of counts in this energy-time space follows the $E \sim 1/t^2$ dependence. The following summarizes a few key points to understanding the three-dimensional spectrum:

- Recoils from the surface of the sample (B + C + O) have the highest energies and lowest flight times and are therefore positioned furthest away from the reader.

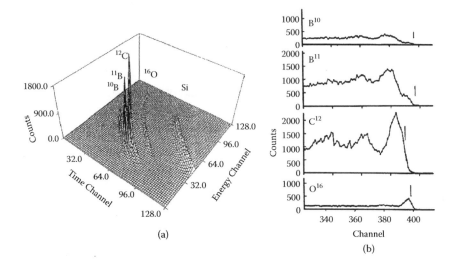

FIGURE 6.14   Light element measurement using a 12 MeV Au beam and a TOF-ERD spectrometer. (a) Yield versus energy channel (from a surface barrier detector) and flight time (from a time to amplifier converter) for atoms recoiled from a multilayered BC/Si target. (b) The converted energy-domain spectra of four recoils (^{10}B, ^{11}B, ^{12}C, and ^{16}O) from the corresponding time-domain spectra shown in (a). The surface edge position for each mass is indicated by the vertical lines. (Knapp, J. A., Barbour, J. C., and Doyle, B. L. 1992. *Journal of Vacuum Science Technology A* 10:2685.)

■ Recoils that occur at the deepest region of the BCO layer, −100 nm thick, have lower energies and therefore higher flight times.

■ The B, C, and O signals terminate with increasing flight time and decreasing energy, corresponding to the thickness of the BCO layer.

■ The deepest region of the BCO layer also represents the uppermost region of the Si substrate, and Si recoil atoms from this interface region have greater energy and lower flight times than the Si deep in the substrate.

■ Since the Si wafer is much thicker than the range of analysis for the Au beam, the Si recoil counts continue to lower energies and higher flight times beyond the limits set for acceptance by the detection system.

The manipulation of this three-dimensional data requires the definition of regions of interest around each mass island and then the projection of this region of interest onto either the energy or flight-time axis. Since the recoil ions of different masses separate themselves so nicely, these projections eliminate the mass-energy ambiguity present in most traditional ERD spectra. If the data are projected onto the energy axis, they can be directly converted to concentration profiles using the similar methods described previously for the conventional range foil system. This approach is quite sufficient for many TOF-ERD experiments where the mass resolution is more important than the depth resolution. However, if better depth resolution is required, the data must be projected onto the time axis because the energy resolution obtained from the good timing resolution is greater than that achievable from the surface barrier detector.

The result of transforming the data plotted in Figure 6.14(a), first to a time-domain spectrum and then to an energy-domain spectrum, is shown in Figure 6.14(b). While the data are still in the form of counts versus energy, details of the depth profiles of ^{10}B, ^{11}B, ^{12}C, and ^{16}O are quite apparent. The topmost layer is a BC-oxide, followed by a C-rich layer that contains a small amount of B and O, followed by a B-rich layer containing small amounts of C and O, and so on. Although the energy resolution deteriorates very quickly with Au-beam TOF-ERD as a result of energy straggling in the sample, the front edge of the oxygen (at the sample surface) suggests an excellent depth resolution of several nanometers as compared to several tens of nanometers in the typical range foil ERD. Using the slab analysis or computer software packages mentioned earlier, one can then convert the energy spectrum in Figure 6.14(b) into depth profiles of ^{10}B, ^{11}B, ^{12}C, and ^{16}O in the sample.

6.7 SUMMARY

Elastic recoil detection (ERD) is a simple and fast method for quantitative light-element depth profiling in thin films (<1 μm thick) and complements the heavy-element depth profiling obtained from RBS analysis and thus should be a part of every MeV ion beam analysis system. ERD shares many attributes of RBS analysis; that is, similarly to RBS, conventional ERD uses a simple detector scheme with a surface barrier detector and a stopper foil to obtain spectra. However, unlike RBS, where only one type of particle (e.g., ^{4}He) is detected, in ERD, both recoiled (e.g., ^{1}H, ^{2}H, or ^{3}H) and scattered (e.g., ^{4}He) particles can be present at the forward detecting angle. This so-called *recoil-scatter particle ambiguity,* which is, in addition to *mass-depth ambiguity,* already existing in both RBS and ERD, requires further particle differentiation or discrimination in recoil particle detection. The recoil particle detection scheme ranges from a simple range foil method, time of flight, ΔE-E telescope to a more sophisticated magnetic spectrometer, etc.

When it is employed on small electrostatic accelerators, ERD can be used for determining concentration depth profiles or areal densities of hydrogen isotopes (H, D, and T) with MeV He projectiles. When ERD is employed on larger accelerators with the ability to accelerate heavy ions to high energies (several tens of megaelectron volts), this technique can be used for concentration depth profiling for even heavier elements out to fluorine. The analysis of the ERD data can be fully analytical, which may at first appear to be difficult because of the requirement for slab analyses, but with modern computers and analysis codes such as SIMNRA, WiNDF, etc. (see examples in Chapter 15), the spectrum can be turned into a concentration profile within minutes. Finally, conventional ERD gives adequate depth resolution (tens of nanometers) and sensitivity (~0.1 at.%) for many types of thin film analyses performed in materials science. If greater depth resolution or greater sensitivity is required, more exotic detection schemes can be used—such as those used with TOF-ERD, ΔE-E telescope, magnetic spectrometer, etc., which have a depth resolution an order of magnitude greater than that for conventional ERD.

PROBLEMS

6.1. A simple range-foil ERD in Figure 6.2(a) is used to measure 50 keV ^{4}He (~1 × 10^{17} at./cm^2) implant depth profile in Si. Suppose that the ERD detector (SSD with ~15 keV resolution and 100 μm depletion depth) is fixed at $\phi = 30°$, the sample surface is tilted at $\alpha = 15°$ from the incident beam, and 2 μm Al foils are available for you to stack up using them as range foils. If available beams for you to choose for this analysis include ^{12}C, ^{16}O, and ^{19}F, estimate the optimum beam energies you would use to do the measurement for each of the three ion species so that the beam energy is high enough to obtain the entire implanted depth profile and yet low enough to still have a reasonable depth resolution.

6.2. Kevin used the same ERD setup as in Problem 6.1 to determine hydrogen and deuterium ratio (H/D) in a single-layer polymer thin film coated on Si substrate. He chose 1.6 MeV ^{4}He$^+$ beam and three stacks of 2 μm Al foil as a range foil for the measurement. The ERD spectrum of his first specimen shows two well-separated Gaussian type peaks: one from D and one from H. He obtained the integrated counts of D-peak and H-peak as 1100 and 5200, respectively. (a) Assuming the Rutherford recoil cross sections at this energy, determine the H/D ratio in the specimen. (b) Using the correct cross sections listed in Figures 6.4 and 6.5, redetermine the H/D ratio in this specimen.

6.3. To determine the absolute amount of H and D in the polymer specimen in Problem 6.2, Kevin also collected a spectrum from a piece of Mylar sheet under the same measurement conditions and obtained the surface spectrum height of 480 for the recoil hydrogen. What are the H and D contents (at/cm^2) in the polymer specimen? What is the thickest polymer film the current ERD setup can measure without H and D signal overlap?

6.4. In the ERD measurement in Problem 6.3, the collimating slit for the incident ^{4}He$^+$ beam is set as 1 mm in width and 3 mm in height, the aperture slit for the recoil atoms is set as 2 mm in width and 3 mm in height, and the distance between the target and the detector is set as 50 mm. Estimate the energy resolution and depth resolution of the

ERD setup when used to detect the surface hydrogen in silicon. At which depth in Si is the geometric energy spread minimized? Estimate the energy and depth resolutions of the ERD at such a depth in Si.

REFERENCES

Arnoldbik, W. M., de Laat, C. T. A. M., and Habraken, F. H. P. M. 1994. On the use of a ΔE-E telescope in elastic recoil detection. *Nuclear Instruments and Methods B* 64:832.

Barbour, J. C., and Doyle, B. L. 1995. Elastic recoil detection: ERD. In *Handbook of modern ion beam materials analysis,* eds. J. R. Tesmer and M. Nastasi, 83–138. Pittsburgh, PA: Materials Research Society Press.

Boudreault, G et al. 2004. Round Robin: Measurement of H implants in Si by ERDA. *Nuclear Instruments and Methods* B222:547.

Dollinger, G., and Bergmaier, A. 2009. Elastic recoil detection: ERD. In *Handbook of modern ion beam materials analysis,* 2nd ed., eds. Y. Q. Wang and M. Nastasi, 81–123. Warrendale, PA: Materials Research Society Press.

Doyle, B. L., and Peercy, P. S. 1979. Techniques for profiling ^{1}H with 2.5 meV Van de Graaff accelerators. *Applied Physics Letters* 34:811.

Groleau, R., Gujrathi, S. C., and Martin, J. P. 1983. Time of flight system for profiling recoiled light elements. *Nuclear Instruments and Methods* 218:125.

Hofsass, H. C., Parikh, N. R., Swanson, M. L., and Chu, W.- K. 1990. Depth profiling of light elements using elastic recoil coincidence spectroscopy (ERCS). *Nuclear Instruments and Methods B* 45:151.

Kellock, A. J., and Baglin, J. E. E. 1993. Absolute cross section for D(^{4}He,D)^{4}He forward scattering. *Nuclear Instruments and Methods B* 79:493.

Knapp, J. A., Barbour, J. C., and Doyle, B. L. 1992. Ion beam analysis for depth profiling. *Journal of Vacuum Science Technology A* 10:2685.

L'Ecuyer, J., Brassard, C., Cardinal, C., Chabbal, J., Deschenes, L., Labrie, J. P., Terrault, B., Martel, J. G., and St.-Jacques, R. 1976. An accurate and sensitive method for the determination of the depth distribution of light elements in heavy materials. *Journal of Applied Physics* 47:881.

Marwick, A. D. 1975. Small-angle multiple scattering of ions in the screened Coulomb region: 2 Lateral spread. *Nuclear Instruments and methods* 126:317.

Prozesky, V. M., Churms, C. L., Pilcher, J. V., and Springhorn, K. A. 1994. ERDA measurement of hydrogen isotopes with a ΔE-E telescope. *Nuclear Instruments and Methods B* 84:373.

Ross, G. G., Terreault, B., Gobeil, G., Abel, G., Boucher, C., and Veilleux, G. 1984. Inexpensive, quantitative hydrogen depth profiling for surface probes. *Journal of Nuclear Materials* 128/129:730.

Sigmund, P., and Winterbon, K. B. 1974. Small-angle multiple scattering of ions in the screened coulomb region: I. Angular distributions. *Nuclear Instruments and Methods* 119:541.

Shao, L., Wang, Y. Q., Lee, J. K., Nastasi, M., Thompson, P. E., Theodore, D., and Mayer, J. W. 2005. A technique to study the lattice location of hydrogen atoms in silicon by channeling elastic recoil detection analysis. *Applied Physics Letters* 87:131901.

Tirira, J., Serruys, Y., and Trocellier, P. 1996. *Forward recoil spectrometry.* New York: Plenum Press.

Tirira, J., Trocellier, P., and Frontier, J. P. 1990. Analytical capabilities of ERDA in transmission geometry. *Nuclear Instruments and Methods B* 45:147.

Turos, A., and Meyer, O. 1984. Depth profiling of hydrogen by detection of recoiled protons. *Nuclear Instruments and Methods B* 4:92.

Wang, Y. Q. 2004. Hydrogen standards in elastic recoil detection analysis. *Nuclear Instruments and Methods B* 219–220:115.

Wang, Y. Q., Liao, C. G., Yang, S. S., and Zheng, Z. H. 1990. A convolution analysis method for hydrogen concentration profiles by elastic recoil detection. *Nuclear Instruments and Methods B* 47:427.

Williams, J. S., and Moller, W. 1978. *Nuclear Instruments and Methods* 157:213.

Ziegler, J. F., Biersack, J. P., and Littmark, U. 1985. *The stopping and range of ions in matter.* New York: Pergamon Press. (Stopping power compilation SRIM2000 at http://www.srim.org)

SUGGESTED READING

Tirira, J., Serruys, Y., and Trocellier, P. 1996. *Forward recoil spectrometry.* New York: Plenum Press.

Wang, Y., and Nastasi, M. eds. 2009. In *Handbook of modern ion beam materials analysis,* 2nd ed., chap. 5. Warrendale, PA: MRS Publisher.

Nuclear Reaction Analysis

7.1 INTRODUCTION

In the preceding chapters of this book, the majority of ion–solid interactions could be described by classical scattering laws. In this chapter we will examine the effects of using radioactivity through *projectile–nucleus* interactions to characterize materials. We will specifically consider the condition of *prompt radiation analysis*, where the radiation emitted instantaneously from a projectile–nucleus interaction is detected. The prompt radiation from a nuclear reaction is emitted within times less than 10^{-12} s after the nuclear reaction is initiated. Contrary to the prompt radiation analysis, charged particle activation analysis is another form of the reaction process, which is less frequently used by the general ion beam analysis (IBA) community and will not be included in this chapter.

If energetic particles from an accelerator are allowed to interact with bulk matter, there is the possibility of a nuclear reaction taking place. Early experiments by Rutherford in 1919 observed the following change/transmutation of nuclear species: $\alpha + {}^{14}N \rightarrow {}^{17}O + p$. In this case, energetic α-particles bombarding ${}^{14}N$ resulted in the formation of ${}^{17}O$ plus a proton. Another example of a nuclear reaction was discussed briefly in the section on non-Rutherford cross sections in Section 3.4 of Chapter 3. As the projectile's incident energy increases, the projectile can penetrate a target atom to within the regime of the attractive nuclear force, as indicated by the potential energy diagram shown in Figure 7.1. Such nuclear interactions modify the simple Rutherford scattering process, and nuclear scattering occurs. For non-Rutherford or nuclear scattering, the cross section is derived from the effects of the *strong nuclear force*. This is the same force that is responsible for nuclear fission (e.g., nuclear reactors) and nuclear fusion (e.g., stars, fusion reactors). A detailed description of strong nuclear interactions involves the quantum mechanics of the nucleus (discrete nuclear energy levels, etc.), just as a detailed description of electronic interaction involves the quantum mechanics of the atom (discrete electron energy levels, etc.).

The quantum mechanics of the nucleus is beyond the scope of this book, and we therefore proceed with a simple qualitative picture, employing empirical data from the nuclear physics community. A potential energy diagram of a particle interacting with a nucleus is presented in Figure 7.1. For an incident particle to reach the attractive nuclear regime of a nucleus, it must have sufficient energy to overcome the repulsive Coulomb barrier (electromagnetic force). The Coulomb barrier is the energy that the incident charged particle must have for the distance of closest approach to be such that the two nuclei just touch:

$$(7.1) \qquad E_b = \frac{Z_1 Z_2 (m_1 + m_2)}{m_2 (A_1^{1/3} + A_2^{1/3})}$$

where Z, A, and m are the atomic number, atomic mass, and the mass of the nucleus, and subscripts 1 and 2 denote incident particle and target nucleus, respectively. In reality, the incident particle could be found within the attractive nuclear regime even if the incident particle had energy $E_0 < E_b$ because of quantum mechanical tunneling effects.

In most nuclear reactions we have two particles or nuclei interact to form two different nuclei. A common notation used to express a nuclear reaction is

$$(7.2) \qquad a + X \rightarrow b + Y$$

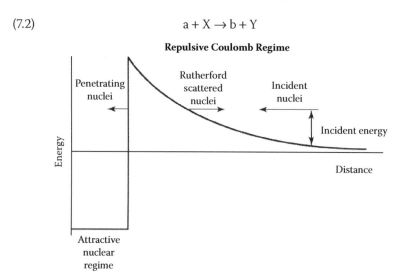

FIGURE 7.1   A potential energy diagram of a particle interacting with a nucleus.

where (a + X) are the reactants and (b + Y) are the products. We assumed that each side of the equation includes a very light nuclide, which is identified by the lower case. Any reaction must meet the requirement that the sum of the atomic numbers and the sum of the mass number of the reactants and products must balance. A common shorthand notation for this reaction is X(a,b)Y.

A complete description of the nucleus contains information about its total number of protons, Z, and number of nucleons, which we term the mass number, A, and is expressed as $^{A}_{Z}X$, where a nucleus with four neutrons and three protons would be expressed as $^{7}_{3}Li$. For example, a reaction between a neutron and ^{10}B in full notation would be written as $^{1}_{0}n + ^{10}_{5}B \rightarrow ^{4}_{2}He + ^{7}_{3}Li$ or, in shorthand, as $^{10}B(n,\alpha)^{7}Li$.

Once the incident particle (e.g., proton or α-particle) passes the Coulomb barrier, two primary nuclear reaction processes may occur depending on the energy of the incident particle.

7.1.1   COMPOUND NUCLEUS PROCESS

At relatively low energies, the incident particle may be captured by the target nucleus to form an intermediate *compound nucleus.* The compound nucleus is now in a highly excited state and the kinetic energy of the incident particle also adds to the excited energy. In the compound nucleus process, it is assumed that the excitation energy is randomly distributed among all the nucleons in the resultant nucleus and that none of them has enough energy to escape immediately. The highly excited nucleus can now deexcite in many different ways (channels) by emitting γ-rays, nucleons (protons, neutrons), and groupings of nucleons (deuterons, α-particles, etc.).

An example of a compound nucleus formed by the reaction between a proton and ^{12}C to form a ^{13}N compound nucleus is shown in Figure 7.2. As the figure shows, the highly excited compound nucleus, $^{13}N^{*}$, can deexcite in a multitude of ways.

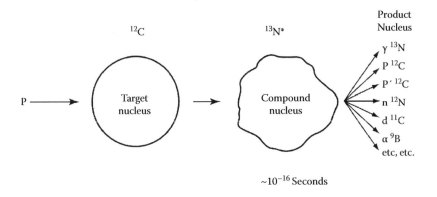

FIGURE 7.2   A compound nucleus $^{13}N^{*}$ formed by the reaction between a proton and ^{12}C and its multitude of decay channels.

7.1.2   DIRECT INTERACTION PROCESS

For higher energy light particle (e.g., proton, α-particle) bombardment, there is a greater chance to transfer sufficient energy to single nucleons (protons, neutrons) or groups of nucleons (deuterons, α-particles), so that they may be directly ejected from the nucleus. Some of the direct interactions that can occur during proton irradiations are (p,n) and (p,α), (α,p) and (α,n) for incident α-particles, etc.

7.2   ENERGETICS AND KINEMATICS

7.2.1   ENERGETICS

Nuclear reactions obey the following conservation laws:

1. Conservation of nucleons (A)
2. Conservation of charge (Z)
3. Conservation of energy (E)
4. Conservation of momentum (p)

The conservation of energy (E) in nuclear reactions means the conservation of the total energy. The total energy of a particle is the sum of its rest energy and kinetic energy. Applying the conservation of total energy to the basic reaction (Equation 7.2), we have

$$(7.3) \qquad (m_X c^2 + E_X) + (m_a c^2 + E_a) = (m_Y c^2 + E_Y) + (m_b c^2 + E_b)$$

where *E*s are the kinetic energies (for which we can use the nonrelativistic approximation $E = \frac{1}{2}mv^2$ at low energies), *m*s are the rest masses, and c is the speed of light. The collisional aspects of this reaction are shown in Figure 7.3. If the exact rest masses of the reactants and the products of a nuclear reaction are totaled, there is likely to be a difference between the two because mass and energy may be exchanged according to the Einstein equation $E = mc^2$. The conversion of mass to energy is accomplished using 1 amu = 931.4 MeV/c^2.

The mass difference will correspond to either an emission of energy, called an *exoergic* or *exothermic* reaction, or an absorption of energy, called an *endoergic* or *endothermic* reaction. Thus, the complete nuclear reaction should be written as

$$(7.4) \qquad\qquad a + X \rightarrow b + Y + Q$$

where Q is the energy balance, often referred to as the *reaction Q value*. If energy is released by the reaction, the Q value will be positive. If the Q value is

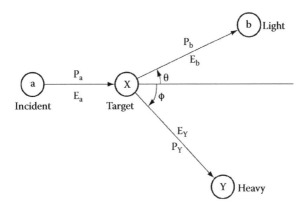

FIGURE 7.3   Collisional and kinematic schematics of a nuclear reaction.

negative, energy must be supplied and there will be a definite threshold energy below which these *endoergic* or *endothermic* reactions will not occur. The threshold energy is defined later.

The Q value is defined as the initial mass energy minus the final mass energy of the reaction:

(7.5) $$Q = m_{initial}c^2 - m_{final}c^2 = (m_X + m_a)c^2 - (m_Y + m_b)c^2$$

This is the same as the excess kinetic energy of the final products

(7.6) $$Q = E_{final} - E_{initial} = (E_Y + E_b) - (E_X + E_a)$$

For $Q > 0$, $m_{initial} > m_{final}$ or $E_{final} > E_{initial}$, and for $Q < 0$, $m_{initial} < m_{final}$ or $E_{final} < E_{initial}$. The changes in mass and energy must be related by the Einstein equation $E = mc^2$ and any changes in the kinetic energy of the system of reacting particles must be balanced by an equal change in its rest energy.

Let us consider the reaction

$$^{35}_{17}Cl + ^{1}_{0}n \rightarrow ^{32}_{15}P + ^{4}_{2}He + Q$$

The reaction is balanced with respect to nucleons and charge in that the reactants and products have the same total number of nucleons (36) and protons (17). The reaction energy Q is equivalent to the differences in mass between the reactants and products.

$$\Delta m = m(^{35}Cl) + m(n) - m(^{32}P) - m(^{4}He)$$

$$= 34.96885 + 1.00867 - 31.97391 - 4.00260$$

$$= +0.00101 \text{ amu}$$

Therefore, $Q = +0.00101 \times 931.4 = +0.94$ (MeV).

Similarly, the Q value can be calculated for the following reaction: $^{14}N(p,n)^{14}O$, with $Q = -5.931$ MeV.

7.2.2   KINEMATICS

Now let us look at kinematics of nuclear reactions. For the reaction displayed in Figure 7.3, we assume the target nuclei (X) to be at rest (i.e., $E_X = 0$). We further define the reaction plane by the direction of the incident particle (a) and one of the outgoing particles (e.g., b). Under these conditions, conserving the momentum perpendicular to the plane shows that the motion of the second outgoing particle (e.g., Y) must also lie in the plane. Therefore, the reaction depicted in Figure 7.3 shows the basic geometry in the reaction plane. Conserving linear momentum along and perpendicular to the incident particle direction (in the reaction plane) gives

(7.7a) $$p_a = p_b \cos\theta + p_Y \cos\phi$$

(7.7b) $$p_b \sin\theta = p_Y \sin\phi$$

Combining Equations (7.7a) and (7.7b) by eliminating ϕ yields

(7.8) $$p_Y^2 = p_a^2 + p_b^2 - 2p_a p_b \cos\theta$$

Replacing $p^2 = 2mE$, Equation (7.8) becomes

(7.9) $$m_Y E_Y = m_a E_a + m_b E_b - 2(m_a m_b E_a E_b)^{1/2} \cos\theta$$

Combining Equation (7.9) and Equation (7.6) where $E_X = 0$, we can obtain Q, Equation (7.10), by eliminating E_Y:

(7.10) $$Q = (\frac{m_a}{m_Y} - 1)E_a + (\frac{m_b}{m_Y} + 1)E_b - \frac{2(m_a m_b E_a E_b)^{1/2}}{m_Y} \cos\theta$$

In carrying out nuclear reaction analysis, the kinetic energy and momentum of the incident particle, E_a and p_a, are parameters that we control, and Q is a known quantity for a given reaction. The solution for the Q equation can be obtained as

(7.11)
$$E_b^{1/2} = \frac{(m_a m_b E_a)^{1/2}}{m_Y + m_b} \cos\theta \pm$$
$$\frac{\{m_a m_b E_a \cos^2\theta + (m_Y + m_b)[m_Y Q + (m_Y - m_a)E_a]\}^{1/2}}{m_Y + m_b}$$

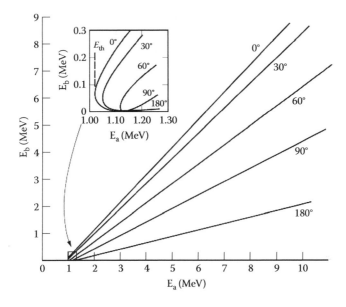

FIGURE 7.4   Emitted particle energy as a function of incident beam energy for reaction ^{3}H(p,n)^{3}He with Q = −763.75 keV. Between 1.019 and 1.147 MeV, the relationship between E_a and E_b is pretty much linear.

This is the kinematic equation for a two-body nuclear reaction. Equation (7.11) shows that E_b is characteristic for a given E_a and Q. The value Q is found experimentally from energy measurements using Equation (7.6). The residual nucleus can be left in ground or excited states, each corresponding to a different Q value for the same reaction, and hence to a different value of E_b. The energy spectrum of the emitted particle b will exhibit a series of peaks specific to the reaction and lead to the detection of a given nucleus m_X. The energy of a peak allows the identification of the reaction and hence the nucleus m_X, and the intensity of the peak measures the amount of the m_X species in the sample.

Equation (7.11) is plotted in Figure 7.4 for the reaction ^{3}H(p,n)^{3}He for which Q = −763.75 keV. Except at low energy between 1.019 and 1.147 MeV, the relationship between E_a and E_b is pretty much linear. This allows us to approximate Equation (7.11) within a wide energy range as

(7.12)
$$E_b = \alpha E_a + \beta$$

where α and β are specific to the reaction under study and depend on θ. In the linear region, the data in Figure 7.4 show that, for a fixed incident particle energy and fixed angle θ, the energy of the outgoing particle is automatically determined.

Another feature shown in Figure 7.4 is that there is an absolute minimum value of E_a, below which the reaction does not occur. This occurs for $Q < 0$ (endoergic reactions) and is called the threshold energy E_{th}:

(7.13)
$$E_{th} = -Q \frac{m_Y + m_b}{m_Y + m_b - m_a}$$

This condition is essential for the expression under the square root in Equation (7.11) to be positive. For $E_a < E_{th}$, the reaction is impossible. At the threshold, the particles emerge at $0°$. If $Q > 0$ there is no threshold condition and the reaction proceeds, barriers permitting, even for very low energies.

7.3 CLASSIFICATION OF NUCLEAR REACTIONS

The interaction between a and X can lead to a number of different outgoing channels (reactions) b and Y, as shown in Figure 7.2. The following is a list of some of the prompt nuclear reactions that can occur during proton irradiation:

(p,p) Elastic scattering (Rutherford)
(p,p) Compound nucleus elastic scattering (non-Rutherford)
(p,p') Inelastic scattering (nucleus at an excited state)
(p,γ) Prompt γ-ray emission
(p,n) Prompt neutron emission
(p,d) Prompt deuteron emission
(p,α) Prompt α emission

Occurring probabilities for each of these reaction channels largely depend on the nuclear structure of the new compound nucleus formed as well as the energy of incident projectiles. We will explore some of these reactions in more detail later.

7.3.1 ELASTIC SCATTERING

In the elastic interaction or elastic scattering, the projectile and the target are not modified by the interaction. The equation for elastic scattering is

$$a + X \rightarrow a + X \ (Q = 0)$$

This interaction is always possible and can be initiated by simple Coulomb repulsion or by more complicated nuclear interactions. When Coulomb forces are dominant, the process is called Rutherford scattering. When nuclear forces are involved (but the target nuclei remain in the ground state), the process is called non-Rutherford or elastic nuclear scattering.

7.3.2   INELASTIC SCATTERING

When there is no change in particles but the target is left in an excited state, the process is called inelastic scattering. The equation for inelastic scattering is

$$a + X \rightarrow X^* + a' + Q \ (Q < 0)$$

The asterisk indicates that X is in an excited state and the reaction is conventionally written $X(a,a')X^*$. Using Equation (7.13), and the fact that $m_a = m_b$, the theoretical threshold energy for this reaction is written as

(7.14)
$$E_{th} \cong -Q \frac{m_a + m_X}{m_X}$$

When nucleus X is left in an excited state the corresponding level usually decays by either γ-ray emission or by ejection of protons (a'). An example of this reaction is

$$^{19}F + p \rightarrow {}^{19}F^* + p' \ (Q = -0.0110 \text{ MeV})$$

This is followed by $^{19}F^* \rightarrow {}^{19}F + \gamma$ with the emitted γ-ray of 110 keV.

7.3.3   REARRANGEMENT COLLISIONS

When the products b and Y are different from the reactants a and X, we are dealing with a nuclear rearrangement collision or, more simply, a nuclear reaction:

$$a + X \rightarrow b + Y + Q$$

In this reaction the total number of protons and neutrons is conserved (Section 7.1). Examples of this reaction are

$$^{14}N + d \rightarrow {}^{12}C + \alpha + 13.575 \text{ MeV}$$

$$^{14}N + d \rightarrow {}^{12}C^* + \alpha + 9.142 \text{ MeV}$$

In the second reaction the carbon nucleus is left in its first excited state, $E_1 = 4.433$ MeV, which promptly returns to the ground state by emitting 4.433 MeV γ-rays. Both of these reactions are exoergic (energy released), leading to the emission of energetic, fast α-particles that can be detected in the depth profile determination of nitrogen in samples.

7.3.4   RADIATIVE CAPTURE

A radiative capture reaction is induced when the projectile is captured by the target to form a nucleus C in a highly excited state that subsequently emits γ-rays to return to its ground state:

$$a + X \rightarrow C^* \rightarrow C + \gamma$$

An example of this reaction is

$$^{27}Al + p \rightarrow {}^{28}Si + \gamma$$

This reaction is a proton capture reaction, usually written as (p,γ). These reactions are very intense in light nuclei, where they originate from a resonant process.

7.4   NUCLEAR ENERGY LEVELS AND NUCLEAR DECAYS

Energetic particles or γ-rays are often emitted from the disintegration of a compound nucleus formed during a nuclear reaction. This can occur in two distinct stages where (a) the incident particle is absorbed by the target nucleus to form a compound nucleus, followed by (b) the disintegration of the compound nucleus by ejecting a particle and/or a γ-ray. To understand the various pathways for the decay of an excited nucleus, energy level diagrams, as shown in Figure 7.5, are used. In the energy level diagram we designate the compound nucleus as $^A_Z X$ and assume that the nuclear reaction produces the

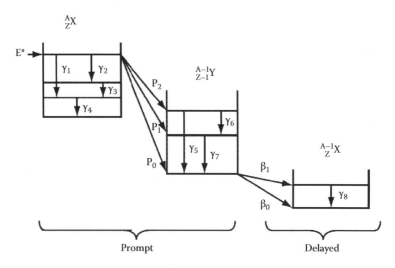

FIGURE 7.5   Nuclear energy level representation of various decay pathways for an excited nucleus $^A_Z X$.

compound nucleus in the excited state E˙. In the diagrams, the energy levels are represented by horizontal lines, the spacing between them giving the energy differences.

The excited energy level E˙ can decay either by the emission of the radiative capture γ-rays, γ_1, γ_2, γ_3, and γ_4, to reach the ground state, or by ejection of the protons p_0, p_1, and p_2, of three distinct energies. The ejected protons feed excited states of the residual nucleus $^{A-1}_{Z-1}Y$, which can also deexcite by the emission of γ-rays (γ_5, γ_6, and γ_7) to yield the ground state of $^{A-1}_{Z-1}Y$. The $^{A-1}_{Z-1}Y$ nucleus is also unstable and decays by the emission of a β-particle to the excited or ground state of $^{A-1}_{Z}X$. The transitions γ_1 through γ_7 and proton emissions are most likely to occur very rapidly, within $10^{-14} \sim 10^{-18}$ s of the formation of the compound nucleus. However, the half-life for the β-decay and the γ_8 will be very much longer: 10^{-3} s ~ years. The former is used in prompt nuclear reaction analysis as discussed in this chapter and the latter is used in charged particle activation analysis.

A generic energy diagram for the reaction $a + X \rightarrow Y_i + b_i + Q_i$ is shown in Figure 7.6. In energy diagrams, the energy is always indicated in center-of-mass units, and atomic mass units are also used. The energy relationship between laboratory (lab) and center-of-mass (c.m.) systems for a projectile, a, colliding with nucleus, X, is given by

(7.15)
$$E_a^{c.m.} = \frac{m_X}{m_a + m_X} E_a^{lab}$$

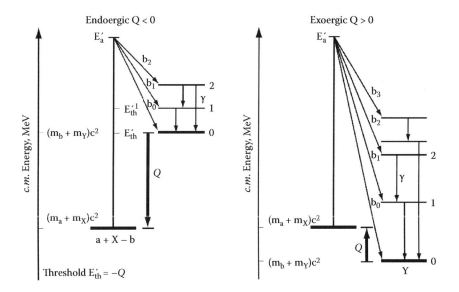

FIGURE 7.6   A generic energy diagram for the reaction $a + X \rightarrow Y_i + b_i + Q_i$, where the energy is indicated in center-of-mass units.

When the nuclear reaction Q value is indicated with its sign, the threshold energy is calculated according to Equation (7.13). Since we can easily approximate $m_a + m_x \approx m_b + m_y$, the threshold energy can be rewritten as

(7.16)
$$E^i_{th} \approx -Q_i \frac{m_a + m_X}{m_X}$$

Converting to c.m units using Equation (7.14) yields

(7.17)
$$E^{c.m.,i}_{th} \approx -Q_i$$

Equation (7.17) means that the minimum particle energy needed to initiate an endoergic nuclear reaction at the c.m. reference frame is the reaction Q value.

The energy diagram does not give the energy for the outgoing particle b, which depends on the scattering angle θ. Nuclear reaction Q values stated in reaction tables are given in the lowest energy state (Q = Q_0) or ground state of the residual nucleus. The Q value corresponding to a given excited state is obtained from the ground state value by

$$Q_i = Q_0 - E^i$$

where E^i is the energy of excitation of the residual nucleus.

As a real example, Figure 7.7 shows energy-level diagrams of different nuclei for the case of fluorine analysis with a proton beam. It shows three different reaction channels (p,n), (p,γ), and (p,αγ) through which the initial ^{19}F nuclei are converted into new nuclei, ^{19}Ne, ^{20}Ne, and ^{16}O, respectively. A plot of cross section versus proton energy is also included for $^{19}F(p,\alpha\gamma)^{16}O$ resonance reactions (Bird and Williams 1989). When the proton beam energy is high enough that the total energy of the reactants channel reaches an excited energy level of expected compound nuclei, there will be a resonant reaction rate for the reaction to occur. This narrow and sharp resonance nature of the reaction cross sections is quite useful in depth profiling of target elements.

7.5   NUCLEAR REACTION CROSS SECTION

Reaction probability between an incident particle and a target nucleus can be approximated by the geometrical cross section presented by the target nucleus to a point-sized projectile. The radius of a nucleus is given rather accurately by the empirical formula

(7.18)
$$R = R_0 A^{1/3}$$

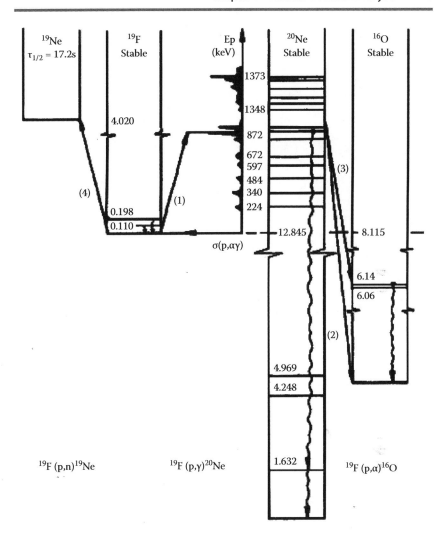

FIGURE 7.7 Example of nuclear energy level diagrams from the proton bombardment of ^{19}F nuclei.

where A is the mass number and R_0 is a constant equal to 1.4×10^{-13} cm. We can estimate the geometrical cross section by considering a medium weight nucleus such as ^{66}Zn. The area represented by this nucleus is the geometric cross section, which is

$$\sigma_{geo} = \pi \, (1.4 \times 10^{-13} \times 66^{1/3})^2 = 1.006 \times 10^{-24} cm^2$$

Since most cross sections are of the order of 10^{-24} cm^2, it has become common to associate the cross section in the unit of barn. This came about

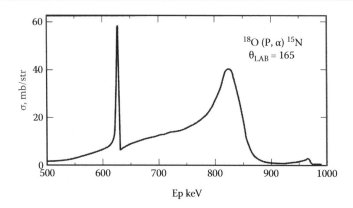

FIGURE 7.8 The ^{18}O(p, α)^{15}N reaction cross section as a function of proton energy. The cross section varies smoothly with energy below and above a resonance at 0.629 MeV.

when early nuclear physicists described the uranium nucleus as *big as a barn*, where

$$1 \text{ barn} = 10^{-24} \text{ cm}^2$$

The actual cross section will be different from the estimate given by σ_{geo}. Figure 7.8 shows that the ^{18}O(p,α)^{15}N reaction cross section varies smoothly with energy below and above a resonance at 0.629 MeV. The value of reaction cross sections can be found in Wang and Nastasi (2009b) and also in the online ion beam analysis nuclear data library (http://www-nds.iaea.org/exfor/ibandl.htm).

The Breit–Wigner treatment provides a quantitative way to calculate the cross section near resonances. In this treatment, the probability of the nuclear reaction, X(a,b)Y, may be denoted by a cross section σ(a,b). According to a two-step compound nucleus view of nuclear reactions,

$$\sigma(a,b) = \sigma_c(a) \times (\text{relative probability of emission b})$$

where $\sigma_c(a)$ is the cross section for the formation of the compound nucleus. The relative probability for the emission of b is Γ_b/Γ, where Γ_b is the transition rate for emission of b (also called the partial level width for b), and Γ is the total level width, $\Gamma = h/\tau$, where τ is the mean lifetime for a state so that

$$\sigma(a,b) = \sigma_C(a)\Gamma_b/\Gamma$$

In general, the values of the cross sections and level widths depend on the energy of the incident particle and on the charge and mass of the target nucleus. In its simplest form, the Breit–Wigner formula gives the value of the cross section in the neighborhood of a single resonance level in the compound nucleus formed by an incident particle with zero angular momentum. Under these conditions, the formula is

$$\sigma(a,b) = \frac{\lambda^2}{4\pi} \frac{\Gamma_a \Gamma_b}{(E-E_R)^2 + (\Gamma/2)^2} = \frac{\sigma_R(a,b)}{1+[(E-E_R)/(\Gamma/2)]^2}$$

(7.19)

where

λ is the de Broglie wavelength of the incident particle (λ = h/mv)

E_R is the energy at the peak of the resonance

E is the energy of the incident particle

Γ_a is the partial level width for the emission of the particle a in the inverse reaction.
The resonant cross section when E = E_R is expressed as

(7.20)

$$\sigma_R(a,b) = \frac{\lambda^2}{\pi} \frac{\Gamma_a \Gamma_b}{\Gamma^2}$$

For a given incident charged particle, the resulting product detected from the sample and corresponding cross section can be quite varied and include particles such as protons, α-particles, and neutrons as well as electromagnetic radiation such as γ-rays. An example of cross sections obtained following proton bombardment of ^{7}Li is shown in Figure 7.9. These data show the variety of products possible following the reaction between energetic protons and ^{7}Li. Visible are the cross sections as a function of proton energy for (p,p), (p,α), (p,n), and (p,γ) reactions. The energy dependence of the (p,α) cross section varies smoothly, while the other cross sections display sharp resonances, which occur

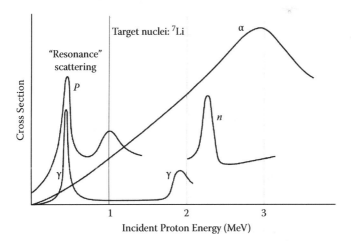

FIGURE 7.9   Various reaction cross sections as a function of proton energy, following proton bombardment of ^{7}Li nuclei.

when the proton energy is great enough for a particular (quantized) excited nucleus to form.

Based on the emitted products resulting from different types of charge particles induced by nuclear reactions (e.g., Figure 7.9), nuclear reaction analysis may also be grouped into three types:

(Charged particle, charged particle)
(Charged particle, γ-ray)
(Charged particle, n)

7.6 NUCLEAR REACTION ANALYSIS

This section will focus on prompt radiation analysis. The presence of an element in a sample is detected through the nuclear radiation emitted instantaneously from a nuclear reaction produced in the target by the irradiating beam. The field of nuclear reaction analysis is quite broad and extensive and will not be covered in depth here. Our goal is to provide some basic insight into the methods. A complete discussion of nuclear reaction analysis can be found in Wang and Nastasi (2009a).

Based on depth profiling analysis methods, nuclear reaction analysis may be grouped into two types:

1. Particle energy spectrum analysis

- Nuclear reaction cross section is fairly smooth (similar to Rutherford backscattering spectrometry [RBS]).
- This is not applicable to (charge particle, γ-ray) reactions.

2. Reaction yield distribution analysis

- Nuclear reaction cross section presents a sharp resonance.
- (Charge particle, γ-ray) reactions are most suitable.
- (Charge particle, charge particle) reactions with high Q-values are also used.

7.6.1 PARTICLE ENERGY SPECTRUM ANALYSIS

For (*charged particle, charged particle*) reactions, the energy spectrum of the emitted particle can be used to obtain the depth profile of its corresponding target nucleus if the energy dependence of the reaction cross section is nearly constant. For (charged particle, charged particle) reaction analysis, a mass separation is necessary in the detection channel since the elastic scattering process

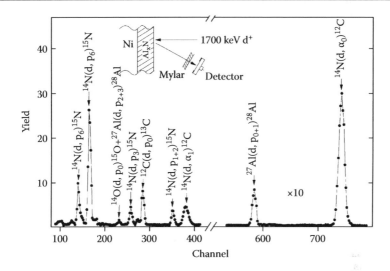

FIGURE 7.10   Energy spectra of various charged particles emitted from a thin aluminum nitride target upon a deuterium beam irradiation of 1.7 MeV.

always competes with other reaction-product channels. The absorber foil technique, ExB mass selector, ΔE-E telescope, or time-of-flight techniques that were presented in Chapter 6 will find the same use here. The actual energy spectrum is a convoluted result of the ideal energy spectrum and the energy resolution function of the nuclear reaction analysis (NRA) setup. Figure 7.10 shows energy spectra of various charged particles emitted from a thin aluminum nitride target upon a deuterium beam irradiation. The 1.7 MeV deuterium beam is high enough to trigger many reaction channels from Al and N nuclei, but not high enough to overcome Coulomb barrier repulsion from Ni nuclei in the substrate. The abundant, elastically scattered deuterium particles are stopped in a thin absorber to prevent count-rate saturation of the detector and electronic systems.

The yield of emitted particles depends on the reaction cross section. However, unlike RBS, there is no simple analytical expression for the nuclear reaction cross section and experimental values are needed. If the cross section is known, the absolute value of nuclei per square centimeter in thin layers can be determined independently of the concentration profile and other components of the target. Similarly to RBS, the yield data resulting from the emitted particles in a nuclear reaction are proportional to the number of nuclei per square centimeter in the sample through the following expression:

(7.21) $$A = Q\Omega\sigma(E_a)Nt / \cos\theta$$

where
A is the number of counts in the peak
Q is the number of incident particles

Ω is the detector solid angle

$\sigma(E_a)$ is the cross section

Nt is the number of nuclei per square centimeter

θ is the angle between the incident beam relative to the surface normal.

As the area A is dependent only on the number of nuclei per square centimeter (Nt) in a different matrix, the measurement can be made to a well-known standard—that is, $A / A_{st} = (Nt)/(Nt)_{st}$. In this analysis it is assumed that the thin layer being analyzed is on the order of hundreds of nanometers and that the reaction cross section of the nucleus changes slowly in the vicinity of the incident beam energy E_a.

For thick samples, the depth dependence information in the energy spectrum analysis is due to energy loss of the incident ions as they penetrate into the sample prior to the nuclear reaction and the energy loss suffered by the charged particle emitted by the nuclear reaction. The conversion between depth and energy loss can be obtained by an expression similar to that used in RBS:

$$(7.22) \qquad \Delta E = [S]_{nr} \, x = [\varepsilon]_{nr} \, xN$$

where ΔE is the energy difference between the detected particles originating from the surface and from a depth x, N is the atomic density, and $[S]_{nr}$ and $[\varepsilon]_{nr}$ are the nuclear reaction stopping power factor and stopping cross section factor, respectively. The nuclear reaction stopping cross section factor is defined as

$$(7.23) \qquad [\varepsilon]_{nr} = \alpha \, \varepsilon_{in}(E_a) / \cos \theta_{in} + \varepsilon_{out}(E_b) / \cos \theta_{out}$$

where E_a and E_b are the incident and emitted outgoing particle energies and θ_{in} and θ_{out} are the angles between the incident and outgoing beams relative to the surface normal, respectively. The relationship between E_a and E_b is given by Equation (7.12). The reaction factor α, defined as $\alpha \equiv \partial E_b / \partial E_a$, weights the energy loss in the inward path in the same fashion as the kinematic factor in RBS and can be expressed as the derivative $\partial E_b / \partial E_a$. For some reactions this derivative is greater than one, and the stopping power of the incident particle can be large at low incident energies. For light nuclei at backward angles, the derivative is smaller than zero. The rest of the data analysis in terms of energy resolution, depth resolution, spectrum deconvolution, etc. is similar to that for RBS/elastic recoil detection (ERD) and will not be repeated here.

Figure 7.11 shows an example of oxygen depth profiling using reaction $^{16}O(d,\alpha)^{14}N$. This reaction, at low deuterium energy, emits only a ground state α-group, α_0. For α-particles corresponding to the first excited state of the ^{14}N nucleus, the reaction has a negative Q value of 0.829 MeV (and therefore a threshold energy) and will not occur for a deuterium energy below 933 keV. At low deuterium bombarding energies, the α_0 energy at large angles is low, and the stopping power or energy loss per unit length therefore is relatively high and provides improved depth resolution. Figure 7.11

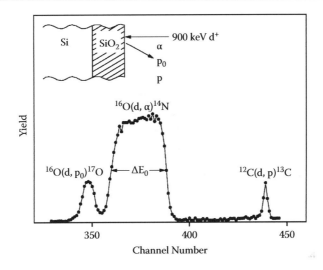

FIGURE 7.11   The energy spectrum of charged particles observed for a 600 nm thick SiO_2 layer on Si substrate with 900 keV deuterium beam.

shows the energy spectrum observed for a 600 nm thick SiO_2 layer on Si substrate with the beam at normal incidence to the target and the Si detector at 145°. In order to avoid interference from protons in $^{16}O(d,p)^{17}O$ reaction, the depletion depth of the Si detector was limited to 26 μm. While the α-particles were stopping in this Si thickness and deposited their full energy in the detector, the protons deposited only a portion of their energy and were displaced thereby to lower energies in the particle energy spectrum. This example clearly shows the advantage of making measurements with particles having high stopping power. Thus, while the proton groups from the $^{16}O(d,p)^{17}O$ are quite narrow and cannot be used for depth profile measurements, the $α_0$ group is quite wide and can be so used.

7.6.2   REACTION YIELD DISTRIBUTION METHOD

For certain nuclear isotopes, a better depth resolution can be obtained by using narrow resonant nuclear reactions. (*Charged particle*, γ) resonant reactions are often preferred since the γ-ray energy spectra from the reactions occurring at the surface and from the reactions occurring below the surface in target matrix are virtually the same since there is no energy straggling effect for outgoing γ-rays and the attenuation of γ-rays in the matrix is negligible.

Figure 7.12 shows the principle of the reaction yield distribution method for depth profiling. For resonant reactions, the reaction products (usually γ-rays) are only produced at a narrow energy window Γ (resonance energy width) centered at the resonant energy E_R. Incident ions with energy E_a, greater than the

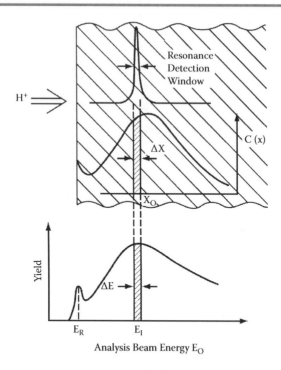

FIGURE 7.12   Principle of concentration profile measurements using resonant reaction yield distribution method.

resonance energy, E_R, are incident on the sample. The ions lose energy as they traverse the sample, ultimately losing sufficient energy at depth x where E_R is reached; at this point the nuclear reaction occurs at a rate proportional to the impurity concentration. Stepping up the incident energy above E_R, the reaction occurs at deeper and deeper depths, providing a reaction yield distribution ($Y(E_a)$ vs. E_a) that resembles the full concentration profile ($C(x)$ vs. x). The depth at which the γ-rays are emitted is calculated from

(7.24)
$$x = (E_a - E_R)\cos\theta_{in} / S_{in}$$

The energy loss of the incident particle S_{in} can be calculated using the mean energy approximation, $E_{mean} = (E_a + E_R)/2$, as the energy change is usually small.

In principle, the yield $Y(E_a)$ is related to the depth profile $C(x)$ through the following equation for an incident energy E_a impinging on the surface with θ_{in} from the normal:

(7.25) $Y(E_a) = \iiint Q\varepsilon\Omega g(E_a,E_1)f(E_1,E_2,x)\sigma(E_2)C(x)dE_1\,dE_2\,dx\,/\cos\theta_{in}$

where

$g(E_a,E_1)$ is the energy spread function of incident beam E_a from the accelerator at a possible energy E_1

$f(E_1,E_2,x)$ is the energy straggling distribution of the incident beam E_1 at a depth x

E_2 is the energy of the incident particle before the reaction

ε is the intrinsic detection efficient of the γ-detector

Ω is the solid angle subtended by the detector

Q is the number of incident particles during the data acquisition

Finding an analytical solution for depth profile $C(x)$ in Equation (7.25) is practically impossible. In reality, numerical solutions are obtained using computer simulation codes in which the thick sample is sliced into many thin slabs and, within each slab, the concentration, energy loss, energy straggling, and nuclear cross section are treated as constant values. Basically, one starts with a constant concentration depth profile and then the corresponding yield distribution is calculated and compared with the experimental yield data, adjusting the concentration depth profile until the calculated yield distribution curve matches the experimental yield data within the experimental error. Then, the concentration depth profile is found.

Since the incident beam energy spread, $g(E_a,E_1)$, and the beam energy straggling, $f(E_1,E_2,x)$, usually follow Gaussian distributions, the integral for these two terms can be approximately described by a combined Gaussian function, called the energy response function:

(7.26) $$F(E_a,E_2,x) = \int g(E_a,E_1)f(E_1,E_2,x)dE_1$$

Using the energy response function, the yield distribution function, Equation (7.25), is simplified as

(7.27) $$Y(E_a) = \iint Q\varepsilon\Omega F(E_a,E_2,x)\sigma(E_2)C(x)dE_2\,dx\,/\cos\theta_{in}$$

By introducing a depth resolution function,

(7.28) $$B(E_a,x) = \int F(E_a,E_2,x)\sigma(E_2)dE_2$$

This can be numerically calculated; the yield distribution function in Equation (7.25) is further simplified as

$$(7.29) \qquad Y(E_a) = \int Q \varepsilon \Omega B(E_a, x) C(x) dx / \cos \theta_{in}$$

Energy resolution in resonant depth profiling is determined by the total energy spread of the incident beam in the matrix (δE_{in}) and the reaction resonant width (Γ). The total energy spread of the incident beam consists of two contributions: the intrinsic beam energy spread from the accelerator (δE_b) and the beam energy straggling in the target (δE_s). *If we use Bohr straggling approximation, then both of these terms can be well described using Gaussian functions and thus the total energy spread is*

$$(7.30) \qquad \delta E_{in}^2 = \delta E_b^2 + \delta E_0^2 x$$

where δE_0^2 is the straggling prefactor in Bohr approximation.

While the beam energy spread is usually approximated as a Gaussian distribution, the resonant cross section described by the Breit–Wigner formula is a Lorenze distribution, so the combined effect for depth resolution function $B(E,X)$ in Equation (7.28) can only be obtained numerically.

Depth resolution for resonant reaction analysis is then calculated by

$$(7.31) \qquad \delta x = \delta E_t / S_{in}$$

where S_{in} is the stopping power of incident beam in the target and δE_t is the total energy resolution that includes the beam energy spread (δE_{in}) and the width of the resonance reaction (Γ).

Depending on the particular reaction and target matrix, the depth resolution of resonance NRA can vary from a few nanometers to several tens of nanometers. Obviously, the deeper the reaction takes place in the solid, the more straggling that occurs prior to reaching the analyzed depth interval. Thus, the probed depth window widens at deeper locations, and the depth resolution worsens accordingly. However, the integrated intensity and hence the sensitivity remain the same. Figure 7.13 illustrates the profiling of fluorine in glass using the 340 keV resonance of $^{19}F(p,\alpha\gamma)^{16}O$ (Deconninck and van Oystaeyen 1983). The calibrated experimental resolution function is plotted as a function of depth for the 340 keV resonance and the closest higher energy resonance of 484 keV. The background yield is very low, and fluorine profiles up to 1.5 µm in a glass can be measured without interferences. In this case, interferences from the other fluorine resonances cannot be resolved using a Ge detector, as all resonances decay through emission of the same γ-rays. This is typical for all rearrangement-reaction resonances. Figure 7.14(a) shows the measured γ-rays

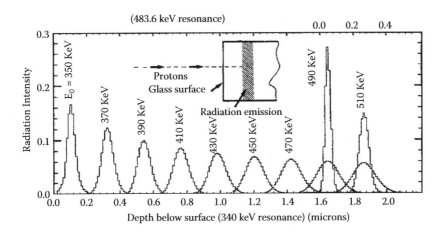

FIGURE 7.13 Calculated yields initiated by the 340 keV resonance in fluoridated glass at different beam energies. The curve corresponds to the experimental resolution. Interference from the next resonance at 483 keV is also represented. (Deconninck, G. and van Oystaeyen, B. 1983. *Nuclear Instruments and Methods B* 226:29.)

spectra obtained from this reaction and other reactions for C and N analyses when a high-purity Ge detector is used (Wang and Nastasi 2009a). These measurements are virtually background free and thus trace level detection of these light isotopes (parts per million to tens of parts per million) can be efficiently done using nuclear reaction analysis. As a reference, Figure 7.14(b) shows a typical laboratory background spectrum measured with an unshielded Ge detector where the origins of the strongest peaks are indicated. The strongest peaks are due to ^{208}Tl (2614.6 keV) and ^{40}K (1461 keV), which are often used for detector energy calibration. The natural background occurs below 3 MeV and it can be significantly reduced by properly shielding the detector and reducing data collection time using a higher beam current.

For certain applications, an analytical solution of Equation (7.25) is possible if the energy straggling in the sample and the accelerator energy spread are completely neglected. With this assumption, Equation (7.25) becomes

$$(7.32) \quad Y(E_a) = \int Q \varepsilon \Omega \sigma(E_a) C(x) \, dx \, / \cos \theta_{in} = \int Q \varepsilon \Omega \sigma(E_a) C(E_a) \, dE_a \, / \, S_{in}(E_a)$$

realizing that over energies at which the cross section is nonzero, $C(E_a)$ and $S_{in}(E_a)$ are constant and can be taken outside the integral sign, and $\sigma(E)$ is given by the Breit–Wigner formula, Equation (7.19).

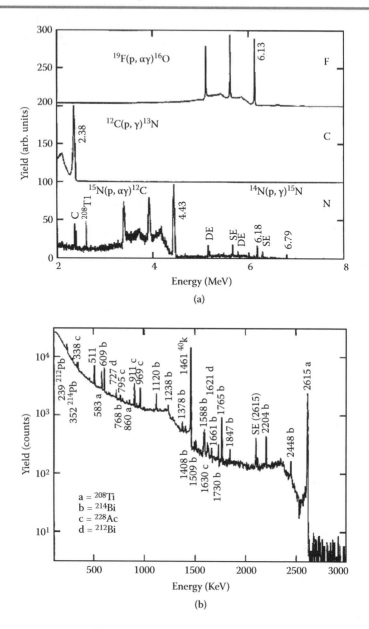

FIGURE 7.14 (a) Typical γ-rays spectra obtained from ^{19}F, ^{12}C, ^{15}N, and ^{14}N upon proton beam bombardments. (b) Typical laboratory background spectrum measured with an unshielded Ge detector where the origins of the strongest peaks are indicated.

Integrating this cross section over energy yields

(7.33)
$$Y(E_a) = \frac{\pi Q \varepsilon \Omega \sigma_R \Gamma C(x)}{2 S_{in}(E_R)}$$

Therefore, the yield to concentration conversion is done by

(7.34)
$$C(x) = k S_{in}(E_R) Y(E_a)$$

where $k = \frac{2}{\pi Q \varepsilon \Omega \sigma_R \Gamma}$, a constant that is only related to the specific reaction (cross sections' parameters) and experimental setup (the detector efficiency).

Theoretically, k can be calculated for a specific reaction and the detector. However, it is dangerous to trust the literature for absolute cross sections to the accuracy needed for analysis. In practice, the constant k is determined for an experimental chamber by making a measurement on a sample of known elemental content (e.g., fluorine content). It is important that k is independent of the material being analyzed. Two types of standard samples are commonly used:

- Known total amount: atoms per square centimeter (usually done by ion implantation)
- Known concentration: atoms per cubic centimeter (relying on knowing stopping power)

Finally, let us see an example of depth profiling of light elements using resonant nuclear reaction analysis, where $^1H(^{15}N,\alpha\gamma)^{12}C$ at the resonant energy of $E_R = 6.385$ MeV is used for hydrogen depth profiling. While there are a number of resonant reactions available for hydrogen measurement, $^1H(^{15}N,\alpha\gamma)^{12}C$ resonance at $E_R = 6.385$ MeV provides the narrowest resonance width ($\Gamma = 1.8$ keV) and the strongest resonant cross section ($\sigma_R = 1650$ mb). The gamma spectrum of $^1H(^{15}N,\alpha\gamma)^{12}C$ at $E_R = 6.385$ MeV is identical to that generated from its inverse reaction, $^{15}N(p,\alpha\gamma)^{12}C$ at $E_R = 429$ keV, as shown in Figure 7.14(a). The gamma energy is 4.43 MeV, resulting from the deexcitation of the first excited state of the ^{12}C nucleus. Figure 7.15 shows hydrogen depth profiling in two different specimens with a ^{15}N beam: (a) hydrogen was implanted into a Si wafer with 40 keV protons to 1×10^{16} ions/cm^2, and (b) a thin film of hydrogenated amorphous Si (a-Si:H) on Si substrate (Wang and Nastasi 2009a). Both raw data (γ-ray counts vs. beam energy) and final results (H concentration vs. depth) are given. ^{15}N beam energy to depth conversion was done using Equation (7.24), using the beam stopping power in Si at the resonant energy ($S_{in} = 1.45$ keV/nm). Gamma counts to H concentration conversion were done using Equation (7.34), where k is determined to be $\sim 4.5 \times 10^{18}$ nm/keV/cm^3 using the implanted H sample in Figure 7.15(a) as a standard.

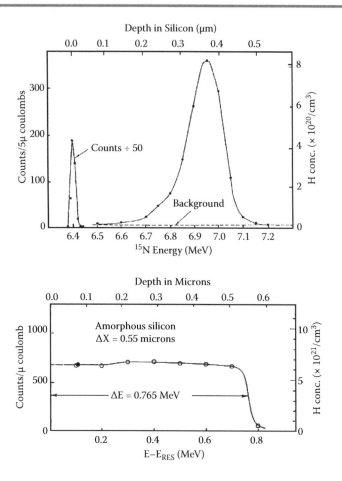

FIGURE 7.15   Hydrogen depth profiling in two different specimens with $^1H(^{15}N,\alpha\gamma)^{12}C$ resonance reaction at $E_R = 6.385$ MeV. (Top) hydrogen was implanted into Si wafer with 40 keV protons to 1×10^{16} ions/cm^2; and (b) a thin film of hydrogenated amorphous Si (a-Si:H) on Si substrate.

PROBLEMS

7.1. Estimate the Coulomb barrier for protons and alphas incident on carbon, silicon, and iron.

7.2. Write the full notation for an alpha incident on ^9B and the full notation for a proton incident on ^{12}C to form ^{13}N.

7.3. What is the rest mass of electron, proton, neutron, deuteron, triton, ^3He, α, and ^{14}Si in terms of MeV units?

7.4. Write the full notation of nuclear reactions in the Sun. Is the reaction exoergic or endoergic?

7.5. Calculate Q values for the following reactions: (1) $^2D(^3He,p)^4He$; (2) $^3T(d,n)^4He$; (3) $^{27}Al(p,n)^{27}Si$; and (4) $^{27}Al(p,\gamma)^{28}Si$.

7.6. Compare the values of the geometrical cross section ($\sigma_{geom} = \pi R^2$) with the Rutherford scattering cross section for 1 MeV p, d, and α-particles incident on ^{14}N at $\theta = 180°$.

7.7. A 1050 keV deuterium beam is used to detect C and O traces on a Ag foil through (d,p) reactions. If only proton emissions from the ground state nuclei are detected at a Si detector located at 150° from the beam direction, what are the proton energies from C and O elements, respectively?

7.8. Nuclear reaction cross sections are typically a couple of orders of magnitude smaller than elastic scattering cross sections. For the reactions in Problem 7.7, if a range foil of Al is placed in front of the detector to block the backscattered deuterons from the Ag target, estimate the minimum thickness of Al foil needed. For such a range foil, how much energy loss of corresponding protons produced from C and O elements would there be? (Hint: you may use SRIM code to obtain energy loss data; see Chapter 4.)

7.9. Compare the depth scales in electron volts per angstrom for detection of F in a Cu thin film with RBS for 2 MeV ^4He ions incident ($\theta = 180°$) and for $^{19}F(p,\alpha)^{16}O$ reaction with 1.25 MeV protons. Which technique offers better depth resolution for surface F detection?

7.10. For X(a,b)Y resonant nuclear reactions, there is an inverse reaction a(X,Y)b. If the energy of the resonance at E_a is known, what is E_x in terms of E_a and the masses of the reactants (m_a and m_x)? Evaluate your answer for two sets of reactions: (1) $^3He(d,p)^4He$ and $^2D(^3He,\alpha)^1H$ for $E_d = 400$ keV; and (2) $^{15}N(p,\alpha\gamma)^{12}C$ and $^1H(^{15}N,\alpha\gamma)^{12}C$ for $E_p = 429$ keV.

REFERENCES

Bird, J. R., and Williams, J. S. 1989. *Ion beams for materials analysis.* Sydney, Australia: Academic Press.

Deconninck, G., and van Oystaeyen, B. 1983. High resolution depth profiling of F, Ne, and Na in materials. *Nuclear Instruments and Methods in Physical Research,* 218: 165–170.

Wang, Y., and Nastasi, M. eds. 2009(a). In *Handbook of modern ion beam material analysis,* 2nd ed., chaps. 6–10. Warrendale, PA: Materials Research Society.

————, eds. 2009(b). In *Handbook of modern ion beam material analysis,* 2nd ed., appendices. Warrendale, PA: Materials Research Society.

SUGGESTED READING

Alford, T. L., Feldman, L. C. and Mayer, J. W. 2007. *Fundamentals of nanoscale film analysis,* chap. 13. New York: Springer.

Baldin, A. M., Gol'Danskii, V. I., and Rozenthal, I. L. 1961. *Kinematics of nuclear reactions,* chap. 3. New York: Pergamon Press.

Deconninck, G. 1978. *Introduction to radioanalytical physics.* Amsterdam: Elsevier.

Halliday, D. 1950. *Introductory nuclear physics,* chap. 10. New York: John Wiley & Sons.

Krane, K. S. 1988. *Introductory nuclear physics,* chap. 11. New York: John Wiley & Sons.

Particle-Induced X-Ray Emission Analysis

8

8.1 INTRODUCTION

Particle-induced x-ray emission (also often referred to as proton-induced x-ray emission [PIXE]) is based upon the ejection of inner-shell electrons from target atoms by the energetic incident particle impact, and the spectroscopy of the subsequently emitted x-rays during the electronic deexcitations. If one includes electrons among the incident particles in the context, then PIXE as an analytical tool has a rather long history, dating back to an experiment done by Moseley in 1913 during which he used a demountable x-ray tube and a flat crystal spectrometer with photographic recording. The fact that particles heavier than electrons can give rise to x-ray emission was shown as early as 1912 by Chadwick, where alpha particles from a radioactive source were used. However, it was quickly realized that with such a low-intensity excitation source, it was impossible to study any details of the x-ray emission process or to use it for analytical purposes.

The replacement of the radioactive sources with particle accelerators in connection with rapid growth in nuclear physics research during the 1950s considerably pushed the PIXE technique forward both in experimental and theoretical aspects. Great progress in nuclear detector technology during the 1960s made multielement analysis possible and greatly aided the expanded use of PIXE in the next two decades. The definitive review by Johansson and Johansson in 1976 was probably the beginning of an international PIXE explosion. Since then, the rapid development and expansion is evidenced by the development of microbeam PIXE and external (in air) PIXE techniques. These special versions of PIXE have found great applications in environmental, biomedical, geological, art, and archaeological research. The steady growth of interest in PIXE prompted the inauguration of dedicated international conference series on PIXE in Lund, Sweden, in 1976 and later establishment of the *International Journal of PIXE* by World Scientific Publisher in 1990. More details of PIXE can be found in the book by Johansson and Campbell (1988); in Chapter 5 by Cohen and Clayton in *Ion Beams for Materials Analysis* (Bird and Williams 1989); in the book by Johansson, Campbell, and Malmqvist (1995); and in Chapter 11 by Campbell in the second edition of *Handbook of Modern Ion beam Materials Analysis* (Wang and Nastasi 2009). New development and applications of PIXE can be found in various international conference series including "Particle-Induced X-Ray Emission and Its Analytical Applications," "Ion Beam Analysis," "Applications of Accelerators in Research and Industry," "European Conference on Accelerator Applications in Research and Technology," "Nuclear Microbeam Technology and Applications," etc.

8.2 FUNDAMENTALS

Figure 8.1 shows a typical PIXE spectrum of an intermediate thickness specimen with the different x-ray peaks identified (Johansson 1992). It consists mainly of the K_α and K_β peaks from the light and medium-heavy elements, but L x-rays from lead (Pb) are also present. The continuous background radiation below 10 keV is bremsstrahlung emitted by secondary electrons, and there is also a high-energy tail consisting of bremsstrahlung produced directly by the proton beam. Thus, PIXE analysis involves two basic aspects: assigning observed x-ray peaks (energies) to appropriate elements (element identification) and converting observed x-ray peaks (areas) into elemental concentrations (concentration quantification). This section describes the fundamental physics of element identification and concentration quantifications involved in the PIXE analysis.

8.2.1   ENERGIES OF X-RAY LINES: THE BASIS FOR ELEMENT IDENTIFICATION

Particle-induced x-ray emission is a two-stage process: First, the particle creates vacancies in the electron inner shells of the target atom through electronic ionizations and/or excitations; second, these vacancies are filled by out-shell electrons and the excessive energy taken away by either radiative (characteristic x-ray photons) or nonradiative (characteristic Auger electrons) processes. Fluorescence yield (ω) defines the partition between characteristic x-ray and Auger emissions and is solely determined by the atomic structure of the element. These characteristic x-rays or Auger emissions are the fingerprints of the target atoms to be analyzed. Figure 8.2 shows the most commonly occurring transitions for an initial K or L or M shell vacancy in one of any of their subshells; the upper portion of the figure indicates the relative intensities of the various lines that form the characteristic x-ray spectrum and indicates the spectral distribution for Pb (Thomas and Cachard 1978). Both the conventional (Siegbahn) and spectroscopic notations are given in Figure 8.1. The difference between these two notations arises in their naming of the target atom electron energy levels. The conventional notation uses the symbols K, L_i (i = 1 to 3), M_i (i = 1 to 5), N_i (i = 1 to 7), etc. to label the electron subshells going outward from the nucleus (decreasing in binding energy). These letters are used to label the characteristic x-ray lines according to the electron transitions that produce them.

The spectroscopic notation uses the three quantum numbers n, l, and j to label the electron subshells. The principal quantum numbers (n = 1, 2, 3,...) are analogous to the major shells K, L, M,... The symbols s, p, d, f,... are used to represent the subshells with orbital angular momentum l = 0, 1, 2, 3,... respectively, while the quantum number j = l + s where s = ½ is the usual electron spin number. Each subshell is then labeled nlj; hence, the L_2 subshell, in the spectroscopic notation, is written $2p_{1/2}$ and the $K\alpha_1$ x-ray is produced by the transition $1s_{1/2} - 2p_{3/2}$. X-Ray emission is mainly due to dipole radiation in which electron transition selection rules are obeyed ($\Delta l = \pm 1$ and $\Delta j = 0 \pm 1$) (the photon has

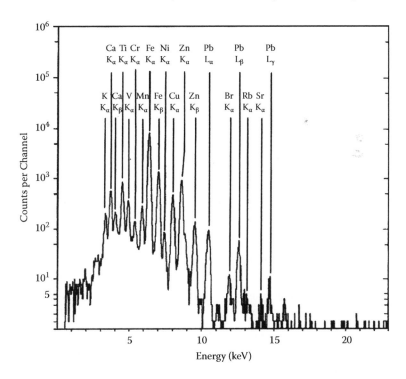

FIGURE 8.1 Typical PIXE spectrum of an intermediate thickness specimen with the different x-ray peaks identified. (Johansson, S. A. E. et al. 1995. *Particle-Induced X-Ray Emission Spectrometry (PIXE).* Chichester, UK: Wiley.)

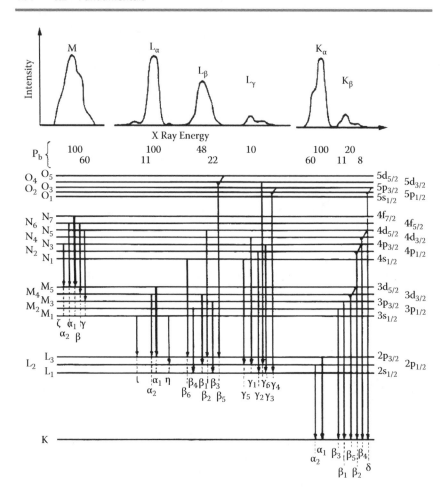

FIGURE 8.2 Characteristic x-ray lines that are generally grouped into three main subgroups—α, β, and γ—according to their x-ray energies.

a unit angular momentum); the energy is given by the difference in binding energies of two subshells involved. For example, the atomic structure of Cu determines that the binding energies for its K, L_1, L_2, L_3 subshells are 8.979, 1.096, 0.951, and 0.931 keV, respectively; thus, the $K\alpha_1$ (K–L_3) and $K\alpha_2$ (K–L_2) x-ray energies for Cu are 8.048 keV (8.979–0.931) and 8.028 keV (8.979–0.951), respectively. The K–L_1 transition ($1s_{1/2}$–$2s_{1/2}$) is radiatively forbidden since the angular momentum is zero.

Although each element has a unique set of characteristic x-rays that in theory allow it to be unequivocally identified in the periodic table, due to x-ray detection limitations in experiment, there are x-ray energy overlaps from certain elements and one should be cautious if these elements coexist in a given target. The experimental limitation includes primarily the finite energy resolution (~135 eV) and limited energy detection range (1 to 25 keV) of the Si(Li) detector. In general, $K_\alpha(Z)$ and $K_\beta(Z-1)$ x-rays overlap for elements between K and Zn; K lines of F to Ti are partially overlapping with L lines and sometimes even M lines of heavier elements. For example, K_α of S (2.307 keV) overlaps with L_α of Mo (2.293 keV) and M_α of Pb (2.346 keV), K_α of Ti (4.508 keV) overlaps with L_α of Ba (4.467 keV), etc.

Figure 8.2 shows that the x-ray lines are generally grouped into three main subgroups, α, β, and γ, according to their x-ray energies. In general, the α lines are lower in energy and more intense than the β lines, the β lines are lower in energy and more intense than γ lines, etc. Approximately 13 K lines, 37 L lines, and 39 M lines have been observed in the experiments by high-resolution x-ray detectors such as Bragg crystal spectrometer (energy resolution of ~5 eV). However, for a commonly used Si(Li) x-ray detector (energy resolution $\sim$ 135 eV) in the PIXE application, only compound lines and line groups K_α, K_β, L_α, L_β, L_γ, M_α, M_β, and so on could be resolved. Figure 8.3 shows x-ray energy (a) and relative intensity ratio of K_β/K_α (b) as a function of atomic number (Bambynek et al. 1972). $I(K_\beta)/I(K_\alpha)$ increases as increasing atomic number and gradually levels off to an average value of ~0.25 when the atomic number is above 60 (Nd). Also, as a rule of thumb, $I(L_\beta)/I(L_\alpha)$ and $I(L_\gamma)/I(L_\alpha)$ are approximately 0.8 and 0.1, respectively, for most the elements.

8.2.2 X-RAY LINE INTENSITIES: THE BASIS FOR CONCENTRATION QUANTIFICATION

As previously mentioned a vacancy in a particular subshell can be filled by various transitions from outer subshells. These transitions can be radiative (x-ray photon productions) or nonradiative (Auger electron emissions). Thus, the K-shell x-ray production cross section σ_k^x is related to the K-shell ionization cross section σ_k^I by

$$(8.1) \qquad \sigma_k^x = \omega_k \sigma_k^I$$

where the coefficient ω_k is the K-shell fluorescence yield interchangeably used for radiative transition rate and defined as the ratio of the radiative width, $\Gamma_R(R)$, for x-ray photon emission to the total width, $\Gamma_T(R)$, of the K-shell. For the K-shell there is only one subshell and the fluorescence yield lies between 0 and 1.

Radiative transition rates and thus fluorescence yields for elements heavier than boron elements have been well predicted using the relativistic Hartree–Slater potential for these atoms. As a first estimate, the radiative transition probability for x-ray production from an initial state i to a final state j is determined by

$$(8.2) \qquad W = \frac{4}{3} \frac{(\hbar\omega_{ji})^3}{(\hbar c)^3} \frac{(e^2)}{\hbar} \frac{a_0^2}{Z^2}$$

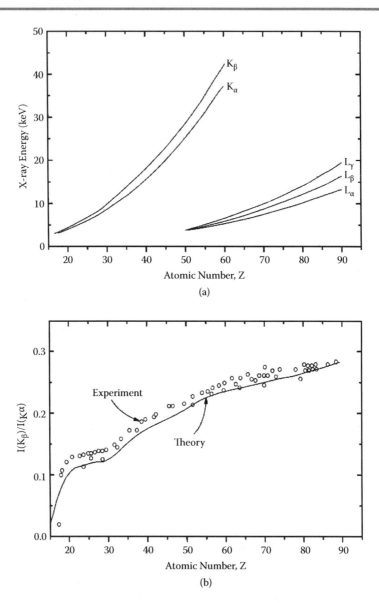

FIGURE 8.3 (a) X-Ray energy and (b) relative intensity ratio of K_β/K_α (b) as a function of atomic number. $I(K_\beta)/I(K_\alpha)$ gradually increases as increasing atomic number and gradually levels off to an average value of ~0.25 when the atomic number is above 60.

where

$\hbar\omega_{ji}$ is the energy of the emitted radiation, $\hbar\omega_{ji} = E_B^i - E_B^j$

E_B is the binding energy of the initial or final state

Z is the atomic number

a_0 is the Bohr radius ($a_0 = 0.53$ Å)

e is the electronic charge

c is the speed of light

h is the Planck's constant h divided by 2π

For a hydrogenic atom, the energy of the transition ($\hbar\omega_{ji} = E_B^i - E_B^j$) is expressed by

$$(8.3) \qquad \hbar\omega_{ji} = 13.6Z^2 \left(\frac{1}{n_i^2} - \frac{1}{n_j^2} \right) (eV)$$

For example, Ni 2p → 1s transition energy ($n_i = 1$, $n_j = 2$, and $Z = 28$) is calculated to be ~8 keV, and the transition probability is ~3.15 eV/$\hbar$. Combining Equations (8.2) and (8.3) yields the information that the transition probability is proportional to Z^4, which is consistent with the experimental observation.

The fluorescence yield generally depends on the target atomic number Z_2 and its initial charge state, but is independent of the projectile atomic number Z_1 and incident energy E_1. For light ions incident on heavy targets (i.e., $Z_1/Z_2 \ll 1$), the fluorescence yield has generally been considered to be independent of the charge state of the target atom, and the neutral atom fluorescence yields are used. Figure 8.4 shows K-shell fluorescence yields as a function of atomic number. It is clear that for a state consisting of a K-vacancy with atomic number $Z > 40$, radiative processes dominate.

The x-ray production cross section for a particular branch such as K_α is thus determined by

$$(8.4) \qquad \sigma_{k\alpha}^x = S_\alpha \sigma_k^x$$

where the branch ratio S_α is the width of the K_α transition relative to the total K-shell width.

For L-shell (or M-shell), there exist multiple subshells, so average shell fluorescence yield $\bar{\omega}_L$ (or $\bar{\omega}_M$) is used to relate L-shell (or M-shell) ionization cross sections σ_L^i (or σ_M^i) to L-shell (or M-shell) x-ray production cross sections σ_L^x (or σ_M^x) in Equation (8.1). Transitions between the subshells of an atomic shell have the same principal quantum number, n, making it possible for a primary vacancy created in one of the subshells to shift to a higher subshell before the vacancy is filled by an x-ray transition. These are called Coster–Kronig transitions and the probability of shifting a vacancy from a subshell i to a higher subshell j (both in the same shell), is denoted by f_{ij}. For the L shell, the effective

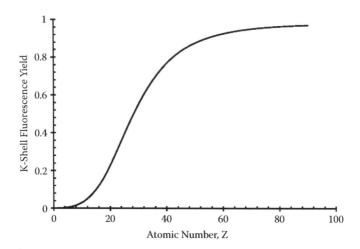

FIGURE 8.4 K-shell fluorescence yields as a function of atomic number. It is clear that for a state consisting of a K-vacancy with atomic number Z > 40, the radiative process dominates.

subshell fluorescence yields v_i and the subshell fluorescence yields ω_i and are related as follows:

$$(8.5a) \qquad v_{L1} = \omega_{L1} + f_{12}\omega_{L2} + (f_{13} + f_{12}f_{23})\omega_{L3}$$

$$(8.5b) \qquad v_{L2} = \omega_{L2} + f_{23}\omega_{L3}$$

$$(8.5c) \qquad v_{L3} = \omega_{L3}$$

Using these notations, the total L-shell and L-subshell ionization cross sections are related to their x-ray production cross sections in the following way:

$$(8.6) \qquad \sigma_L^x = \sigma_{L1}^x + \sigma_{L2}^x + \sigma_{L3}^x = v_{L1}\sigma_{L1}^I + v_{L2}\sigma_{L2}^I + v_{L3}\sigma_{L3}^I = \bar{\omega}_L\sigma_L^I$$

where $\bar{\omega}_L$ is the L-shell average fluorescence yield. Fluorescence yields (ω_i) and Coster–Kronig probabilities (f_{ij}) for different elements can be found in Appendix 10 of the *Handbook of Modern Ion Beam Materials,* second edition (Wang and Nastasi 2009).

With the known fluorescence yields and Coster–Kronig probabilities for a given target element, one only needs to know the ionization cross section to predict the x-ray production cross section, as defined in Equation (8.1) for K x-ray production and Equation (8.6) for L x-ray production. Inner shell ionization is caused by the time-dependent electric field created by the passage of a charge near an atom. In a classical sense, the field from a proton is precisely the same (aside from a sign) as that of an electron at the same velocity. The velocity

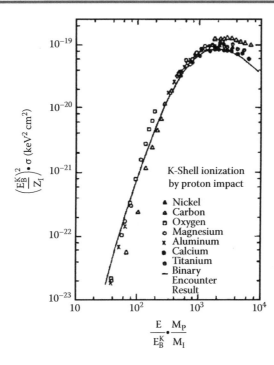

FIGURE 8.5   A reduced plot of proton-induced K-shell ionization cross sections. The horizontal axis is the product of E/E_B^K, the ratio of the ion energy to the K-shell binding energy, and M_p/M_1, the ratio of proton mass to ion mass. The vertical axis is $(E_B^K/Z_1)^2 \sigma$, the product of the square of the K-shell binding energy and the reduced ionization cross section σ/Z_1^2. This scaling allows the formulation of a universal plot for the ionization cross section.

matching criteria require proton kinetic energies of 1,836 times the electron energies, which corresponds to the MeV ions to match keV electrons. Figure 8.5 shows a reduced plot of proton-induced K-shell ionization cross sections (Garcia 1970). The horizontal axis is the product of E/E_B^K, the ratio of the ion energy to the K-shell binding energy, and M_p/M_1, the ratio of proton mass to ion mass. The vertical axis is $(E_B^K/Z_1)^2 \sigma$, the product of the square of the K-shell binding energy and the reduced ionization cross section σ/Z_1^2. This scaling allows the formulation of a universal plot for the ionization cross section.

More advanced atomic theories have been developed to calculate ionization cross sections. For example, for protons incident on the K-shell and L-shell, the tabulated energy-loss Coulomb-repulsion perturbed-stationary-state relativistic (ECPSSR)-Dirac–Hartree–Slater (DHS) theoretical ionization cross sections of Chen and Crasemann (1985, 1989) are in good agreement with experimental values and thus are included in most of the PIXE analysis codes. ECPSSR-DHS cross sections have not been computed for deuterons or helium ions, so recourse must be had for these ions to the older ECPSSR-SH

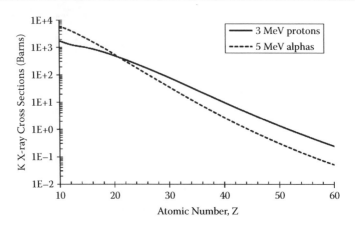

FIGURE 8.6 Calculated K x-ray production cross sections (barns) as a function of atomic number (from 10 to 60) induced by 3 MeV proton and 5 MeV alpha bombardment—two commonly used beam energies for PIXE analysis.

treatment (Liu and Cipolla 1996) based on screened hydrogenic wave functions. Figure 8.6 shows K x-ray production cross sections (barns) as a function of atomic number (from 10 to 60) induced by 3 MeV proton and 5 MeV alpha bombardment—two commonly used particles and beam energies for PIXE analysis. The K-shell ionization cross sections for other beam energies and L-subshell ionization cross sections can be found in Appendix 10 of the *Handbook of Modern Ion Beam Materials Analysis*, second edition (Wang, Y. and Nastasi, M. 2009).

8.3   QUANTITATIVE ANALYSIS

To show the power and complexity of the PIXE technique, Figure 8.7 displays a PIXE spectrum (Ryan et al. 2001) recorded with an HPGe detector, chosen to illustrate the high dynamic range that is usually encountered. The HPGe detector extends detectable x-ray energies up to 40 keV as compared with the Si(Li) detector. The specimen has intense x-ray peaks from lighter elements of high concentration (Fe ~ 10%) and from much weaker peaks at higher x-ray energies arising from the trace elements (typically tens of parts per million). Invariably, this dynamic range, combined with electronic limitations on achievable counting rate, demands the use of an absorber (200 μm aluminum in this case) between sample and detector to strongly suppress the dominant lower-energy x-rays while transmitting the majority of the higher energy x-rays. Because each element emits several x-ray lines, even a small number of elements can result in a complicated spectrum, with overlapping peaks superimposed on a bremsstrahlung background of complex shape. To convert the x-ray intensity peaks in

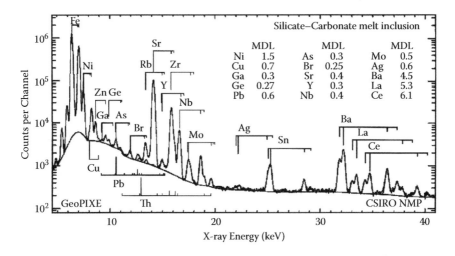

FIGURE 8.7   Typical PIXE spectrum of a geological sample showing major elements at low x-ray energy and trace elements at high energy. (Reprinted from Ryan, C. G. et al. 2001. *Nuclear Instruments and Methods B* 181:578–585. With permission from Elsevier.)

Figure 8.7 into concentrations of elements present in the specimen, one often needs to use PIXE data analysis computer codes such as Gupix, Geo-pix, or WiNDF. The basis of these analysis packages is the intensity–concentration relationship, which is described next.

8.3.1   INTENSITY–CONCENTRATION RELATIONSHIP

For a basic PIXE measurement geometry shown in Figure 8.8, with a homogeneous sample thick enough to stop the ion beam and the justifiable neglect of straggling (since the bulk of the x-ray production arises from the earliest part of the proton track), the intensity or yield Y_Z of the principal characteristic x-ray ($K\alpha_1$ or $L\alpha_1$) of element Z (atomic mass A_Z) with a concentration of C_Z is

$$(8.7) \qquad Y_Z = \frac{N_{av}\omega_z b_z t_z \varepsilon_z (\Omega/4\pi)}{A_z} N_p C_z \int_{E_0}^{E_f} \frac{\sigma_z(E)\, T_z(E)\, dE}{S_M(E)}$$

where
E_0 is the initial ion energy
E_f is the final energy on exit from the sample
$S_M(E)$ is the overall stopping power of the sample matrix
$\sigma_Z(E)$ is the ionization cross section

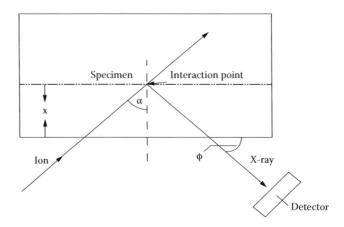

FIGURE 8.8   Generalized geometry for PIXE.

ω_Z is the fluorescence yield

b_Z is the branching fraction of the principal line within the particular x-ray series (e.g., Kα within the K-series)

t_Z is the transmission through any absorbers present

Ω is the detector solid angle

ε_Z is the detector intrinsic efficiency

N_p is the number of protons incident

N_{av} is Avogadro's number

The dominant linear dependence of Y_Z on C_Z is modified by the integral term, known as the matrix correction, which can only be evaluated if the concentrations of all elements present are known. The fact that the matrix effects are physically simple and well understood and the underlying database accurately known gives PIXE a significant advantage in contrast to other analysis techniques in which matrix effects are difficult to calculate and may not even be reproducible. Within this matrix term,

$$(8.8) \qquad T_z(E) = \exp\left[-\left(\frac{\mu}{\rho}\right)_M \frac{\cos\alpha}{\sin\phi} \int_{E_0}^{E} \frac{dE}{S_M(E)} \right]$$

where the matrix mass attenuation coefficient $(\mu/\rho)_M$ is the concentration-weighted sum of the mass attenuation coefficients of the matrix elements, α is the angle between the incident beam and the sample normal, and φ is the angle between the outgoing x-rays and the sample surface.

It should be mentioned that Equation (8.7) has neglected incident beam energy straggling and secondary fluorescence contributions. Overall stopping power of the incident proton beam in the sample matrix as a function of depth in Equation (8.7) can be reliably predicted by SRIM code (Ziegler, Biersack, and Ziegler 2008), and the omission of energy straggling of MeV protons in Equation (8.7) is believed to change the predictions of the preceding equation by less than 0.1% and can be therefore ignored. However, the secondary fluorescence effect must be included in any code that generates x-ray yields from element concentrations. The secondary fluorescence contributes to the intensity arising from the x-rays of a dominant element being absorbed within the sample and fluorescing those of another element. This contribution may be calculated accurately using the various quantities introduced before, together with the photoelectric cross sections for the various atomic shells.

When a specimen is so thin that proton energy loss and x-ray attenuation are negligible, the matrix effects drop out, and the integral in Equation (8.7) becomes $x/\cos \alpha$, where x is the specimen thickness. Thus, there is a direct linear relationship between Y_Z and C_Z, as described in the following:

$$(8.9) \qquad Y_Z = S_Z N_p C_Z x$$

Here, $S_Z = \frac{N_{av}\omega_z b_z t_z \varepsilon_z \sigma_z (\Omega/4\pi)}{A_z \cos\alpha}$ is a sensitivity factor (i.e., x-ray yield per proton per unit areal density for element Z). Although Equation (8.9) lends itself to absolute analysis without standard since S_z can be predicted theoretically, most PIXE researchers still prefer to analyze relative to standards in order to (1) minimize reliance on the database for ionization cross section and ion stopping power, (2) avoid the need for absolute calibration of beam charge measurement, and (3) avert the need to determine x-ray detector properties. Therefore, in practical use, the sensitivity factor S_z is measured for a given PIXE system using thin-film standards, such as those provided by MicroMatter Corp., and then the sensitivity curve of Kα or Lα intensity versus areal concentration as a function of Z is obtained. This measured curve subsumes all the physics and all the detector terms in Equation (8.7) and reduces the analysis to a direct comparison with standards; the accuracy of concentrations in unknown specimens is then determined by the accuracy of concentrations in standards and the accuracy of the peak area determination (spectrum fitting). The sample substrate must also be thin and free of elemental contaminants; polycarbonate foils of a few micrometers thickness are good candidates. This thin sample approach has been widely used in PIXE analysis of aerosols. Typically, thickness of less than 1 mg/cm^2 is often considered as a thin specimen in PIXE analysis with a 3 MeV proton beam.

More often, we encounter samples that are thick enough to stop the proton beam completely ($E_f = 0$) and we know the major elements (comprising >99.9%) and their concentrations in the matrix, or the major elements can be easily determined by other techniques such as Rutherford backscattering spectrometry (RBS) or electron probe microanalysis (EPMA) separately, or can even be measured by a simultaneous PIXE analysis using a second detector tuned for major elements. For specimens of this nature, it is clear that all the matrix (M) effects (due to proton slowing and x-ray attenuation) are contained in the integral I_Z in Equation (8.7). This integral is equivalent to the ZAF or $\phi(\rho Z)$ matrix correction methods of electron microprobe analysis, but it involves neither a series of approximate models (ZAF) nor Monte Carlo simulations [$\phi(\rho Z)$]. In PIXE, the integrals in Equation (8.7) can be computed directly using the database and the matrix concentrations. The problem is therefore reduced to the simple linear equation:

$$(8.10) \qquad\qquad Y_Z = HC_z Y_1(Z) N_P \varepsilon_z t_z$$

Here, $Y_1(Z)$ is the theoretical x-ray yield of the element Z per unit solid angle, per unit concentration, and per unit beam charge, and the H factor is a single instrumental constant that includes the detector solid angle and matrix integral contributions. This constant H can be determined with standards; the advantage of this approach is that, if one accepts both the database and the detector parameters as being sufficiently accurate, then these standards need not have identical matrixes to those of the specimens.

Many biological and environmental samples deposited as thin layers on thin polymer backings are considered as intermediate thickness in PIXE analysis. These samples are not thick enough to reduce the beam energy to zero, but cannot be considered in the extreme case as "thin." In this case, the sample matrix and thickness must be known a priori (or the latter measured via transmitted ion energy loss) in order for the integral in Equation (8.7) to be evaluated between the proper limits of ion energy.

8.3.2   BACKGROUND RADIATION

An examination of PIXE spectra shows that the characteristic x-ray peaks are superimposed on a continuous background, which forms a limiting factor to sensitivity. Origin of this background mainly comes from bremsstrahlung radiation, which occurs when charged particles are decelerated by Coulomb scattering in the sample. Bremsstrahlung intensity is roughly proportional to the square of the deceleration, $a^2 = (F/m)^2$, where F is the Coulomb interaction force and m is the mass of the decelerating particle. This suggests that the electron bremsstrahlung is approximately $(m_p/m_e)^2$ (or 3.4×10^6) times more intense than proton bremsstrahlung under the same electric interaction field. Since the theoretical cross section for bremsstrahlung radiation contains a term $[(Z/A)_{projectile} - (Z/A)_{target}]^2$, there are projectile–target combinations where this primary bremsstrahlung has zero intensity. For light to medium mass elements, $(Z/A)_{target} \sim 0.5$; thus, deuteron and alpha beams theoretically produce much smaller primary bremsstrahlung than a proton beam. At the same time, in general, alpha beams also produce less γ-ray background than deuteron beams.

As mentioned before, the main component in PIXE spectral background is secondary electron bremsstrahlung (SEB), which arises when an electron ejected from a bound state is decelerated by Coulomb scattering in the sample; its maximum energy, corresponding to all the energy imparted to an electron in a head-on collision with a proton, is $\frac{4m_e}{m_p} E_p$, which is 6 keV for the typically used 3 MeV proton beam. Quasi-free electron bremsstrahlung (QFEB) is emitted in cases where the target electron velocities are so low compared with those of the projectile that they are essentially free; for 3 MeV protons its upper energy limit is 1.6 keV. Atomic bremsstrahlung (AB) is emitted by an ejected electron falling back into its initial bound state; this is much less intense than SEB but extends to higher photon energies. These contributions are schematically shown in Figure 8.9 (Ishii and Morita 1987).

Despite the effort made based on the bremsstrahlung theory to develop an analytical description of background shape for use in PIXE spectrum inter-pretation, in practice, real specimens contain many elements spanning a wide range of atomic numbers, and the projectile energy varies from its initial value down to zero; therefore, the mathematical description of background is very complex. In addition, background is sometimes augmented by (p,γ) reactions of light elements (e.g., F, Na, Al) in the specimen through Compton scattering. At sufficiently high counting rate, pairs of x-ray events that fall within the resolving time of the electronic system are summed to produce an apparent single event, resulting in a continuum to the right of the characteristic peaks of a given ele-ment. This so called pulse pileup distortion on spectral peaks and background can be reduced using an electronic pileup inspector or an on-demand beam deflector, but is still difficult to remove or model. Finally, insulating targets pose special difficulties because localized high voltages on the target surface accel-erate free electrons, producing very high background up to tens of kiloelec-tron volt x-ray energy. If not properly managed, this increased background not

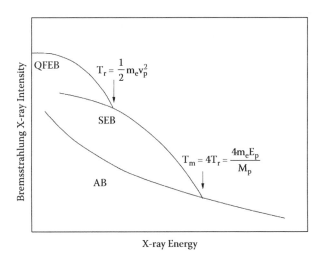

FIGURE 8.9 Components of the PIXE background as defined in the text. (Ishii, K. and Morita, S. 1987. *Nuclear Instruments and Methods B* 22:68.)

only reduces detection sensitivity but also may completely swamp smaller x-ray peaks in the spectrum. A thin carbon coating has proved to be a very effective means in reducing this charge-related background problem. Most workers therefore adopt empirical approaches to background determination and subtraction.

8.3.3   SPECTRUM FITTING

For x-ray energies examined by PIXE (1–25 keV), a typical element will contribute more than 20 characteristic x-rays; most of these peaks will have a silicon escape peak. The range of peak heights tends to be large, reflecting concentrations that vary from percent to parts per million levels. This plethora of peaks is superimposed on a continuous background, as described in the previous section, mainly due to electron bremsstrahlung but sometimes including Compton events from gamma rays and pileup peaks due to high counting rates. A common approach to the processing of such a complex spectrum is to model it by an analytic function, which includes modified Gaussians to describe the characteristic x-ray peaks and a polynomial or exponential polynomial to represent the underlying continuous background. The model function is fitted to the experimental PIXE spectrum by means of a nonlinear least-squares fitting procedure. The fitting produces values for the parameters of the model, so the required x-ray peak areas are easily calculated. Before the fit begins in any PIXE analysis code, the database relative x-ray intensities for each element must be corrected for matrix absorption, filter transmission, and detector efficiency. Additional details regarding the spectrum fitting can be found in Campbell, 2009.

In principle, PIXE could be a standard-less method, given the accurate database on ion stopping powers, ionization cross sections, fluorescence yields, etc. but the analytical community would not be convinced by such an approach. One of the concerns is the uncertainty of an x-ray detector's intrinsic efficiencies at the extremes of x-ray energy. At energies above about 20 keV, the detector's intrinsic efficiency falls below 100%, and the approach is vulnerable to errors in detector thickness and to inadequate treatment of Compton interactions in the efficiency model. At energies below about 4 keV, entrance windows, contacts, ice buildup, the escape of photo and Auger electrons, and the poorly understood effects referred to as ICC (incomplete charge collection) all conspire to render calculated intrinsic efficiency inaccurate. At the other extreme, one could rely entirely on standards, but the effort to create standards pertinent to every situation encountered would be tremendous and unrealistic.

A well-accepted pragmatic approach in most PIXE analysis codes is to combine these extremes and use both fundamental parameters and a small set of standards. Several PIXE codes also take different approaches to peak and background description and employ different databases. Some take the peaks to be Gaussian while others adopt the more rigorous Voigtian approach. Some ignore the peak asymmetry that arises from electron transport and charge collection

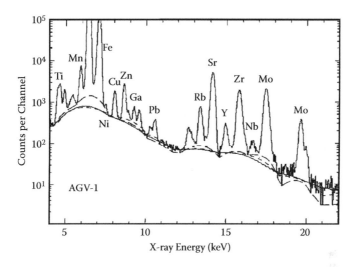

FIGURE 8.10 PIXE spectrum obtained from US Geological Survey rock standard AGV-1 in glass form, showing different background approximations using GEOPIXE code. (Ryan, C. G. et al. 1988. *Nuclear Instruments and Methods B* 34:396.)

effects at low x-ray energy and from Compton scatter at high energy, while others provide means to include this in the peak model. Continuous background can be modeled with multiparameter semiempirical functions that are included in the overall fitting function, but use of mathematical approaches to strip off the background is simpler and has been demonstrably successful in the cases of the GEOPIXE (Ryan et al. 1990) and GUPIX (Maxwell, Campbell, and Teesdale 1988) codes. It is vital to include peak pileup, and the GEOPIXE and GUPIX codes do so to second order; the GUPIX code also offers the option of peak-plus-continuum pileup (Maxwell and Campbell 2002). Figure 8.10 shows a PIXE spectrum from the US Geological Survey rock standard AGV-1 in glass form, showing different background approximations using GEOPIXE code (Ryan et al. 1988). The measurement was done with 3 MeV protons to an accumulated charge of 14 μC. Figure 8.11 shows a PIXE spectrum obtained from a garnet containing approximately 120 ppm nickel, where the fitting residue is shown in the bottom of the figure using a GUPIX package (Maxwell 1995). This example shows that weak peaks (Ni) in the vicinity of intense neighbors (Fe) can be reliably determined (<±3% errors) using PIXE with appropriate filter background techniques.

8.3.4 PIXE SYSTEM CALIBRATION

No matter which PIXE code is used for spectrum fitting, one first needs to calibrate the PIXE experimental system carefully. A good approach for the system calibration is to use National Institute of Standards and Technology

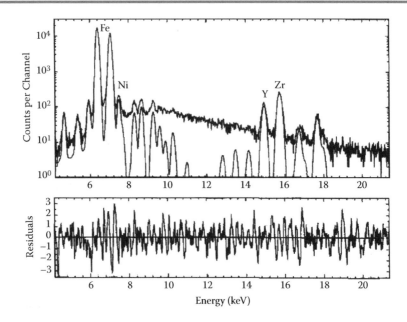

FIGURE 8.11   PIXE spectrum obtained from a garnet containing approximately 120 ppm nickel where the fitting residue is shown in the bottom of the figure using the GUPIX package. (Maxwell, J. A. et al. 1995. *Nuclear Instruments and Methods B* 95:407.)

(NIST) standards that resemble the main matrix of unknown specimens to be measured. Using the manufacturer's values for the detector dimensions and absorber thickness, the H-value for the system can be obtained by substituting the concentration of the first known element in the standard and its Kα (or Lα) x-ray peak area into Equation (8.10); uncertainties in the assumed quantities have little impact at the K (or L) x-ray energy of the interested element. Next, using this H-value with the concentration and peak area of the second known element in the standard, the absorber thickness can be deduced. With this absorber thickness value, the concentration calculation for the first known element can be repeated and a slight final adjustment made to the H-value. If a heavy element (e.g., $E_{k\alpha} > 20$ keV) is present in the standard, this approach can also be used to determine the effective detector thickness based on the high-energy K x-rays of the heavy element. Once a reliable H-value is determined for the system using the standard, one can use Equation (8.10) to quickly figure out elements and their concentrations in unknown samples.

To gauge the success of such a system calibration approach, Cohen and Clayton (1989) measured concentrations ranging from less than 1 µg/g (1 ppm) to over 20% for major components of geological samples. Figure 8.12 shows their comparison of the measured PIXE concentration values and data from standard

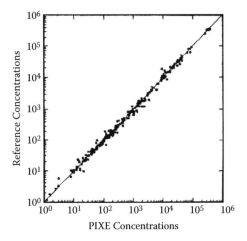

FIGURE 8.12   Comparison of the measured PIXE concentration values and data from standard reference materials. The fitted line has a slope of 1:00 ± 0.20. (Cohen, D. D., and Clayton, E. 1989. In *Ion Beams for Materials Analysis*, eds. J. R. Bird and J. S. Williams, Sydney, Australia: Academic Press.)

reference materials. The fitted line in Figure 8.12 has a slope of 1:00 ± 0.20, indicating that the absolute estimates from Equation (8.10) and measurements relative to both internal and external standards agree remarkably well for such a broad range of concentrations.

8.3.5   ACCURACY OF PIXE ANALYSIS

Many factors can influence the overall accuracy of a PIXE analysis:

- Specimen preparation and the consequent homogeneity and smoothness
- The nature and homogeneity of standards
- Beam-specimen-detector geometry
- Charge measurement
- Detector line shape and efficiency
- Spectrum fitting
- Correction for matrix effects
- Correction for secondary fluorescence
- Background subtraction and overlap peaks resolution
- Accuracy of the database for beam stopping powers, ionization cross sections, fluorescence yields, x-ray branch ratios, and x-ray absorption coefficients

**TABLE 8.1　Analysis of PIXE of Thin Films (BCR 128)
Containing Certified Fly Ash (BCR 38)**

Element	Reference Value[a,b]	PIXE Value[c]	PIXE/Reference[d]
Mg[e]	9.54 ± 0.34[f]	8.3 ± 1.1	0.88 ± 0.12
Al[e]	127.5 ± 4.9[f]	104.0 ± 1.6	0.82 ± 0.01
Si[e]	227 ± 12[f]	176.0 ± 2.1	0.78 ± 0.01
P[e]	—	1.60 ± 0.15	
S[e]	3.89 ± 0.13[f]	4.17 ± 0.03	1.07 ± 0.01
K[e]	34.1 ± 1.6[f]	30.6 ± 0.7	0.90 ± 0.02
Ca[e]	13.81 ± 0.63[f]	13.35 ± 0.16	0.97 ± 0.01
Ti[e]	—	5.07 ± 0.12	
V	334 ± 23	340 ± 2	1.02 ± 0.07
Cr	178 ± 12	161 ± 16	0.90 ± 0.09
Mn	479 ± 16	454 ± 10	0.95 ± 0.02
Fe[e]	33.8 ± 0.7	32.9 ± 0.5	0.97 ± 0.01
Ni	194 ± 26	192 ± 8	0.99 ± 0.04
Cu	176 ± 9	169 ± 8	0.96 ± 0.05
Zn	581 ± 29	572 ± 18	0.98 ± 0.03
Ga	—	56.2 ± 4.5[g]	
Ge	—	16.3 ± 4.7[g]	
As	48.0 ± 2.3	49 ± 11[g]	1.02 ± 0.23
Br	—	29 ± 7[g]	
Rb	—	183 ± 18[g]	
Sr	—	187 ± 18[g]	
Zr	—	169 ± 23[g]	
Pb	262 ± 11	225 ± 26[g]	0.86 ± 0.10

Source:　Maenhaut, W. 1987. *Analytica Chimica Acta* 195:125.

[a]　Concentrations are given in micrograms per gram unless indicated otherwise.

[b]　Reference values and associated errors are for the certified fly ash (BCR 38) unless indicated otherwise.

[c]　PIXE data are averages and standard deviations, based on the analysis of five thin-film samples unless indicated otherwise.

[d]　Values were obtained by dividing both the PIXE result and its associated error by the reference concentration.

[e]　Concentration in milligrams per gram.

[f]　Concentration and standard deviation obtained by averaging round-robin values for the fly ash.

[g]　Result derived from the sum spectrum, obtained by summing the PIXE spectra of the five films analyzed; the associated error is the error from counting statistics.

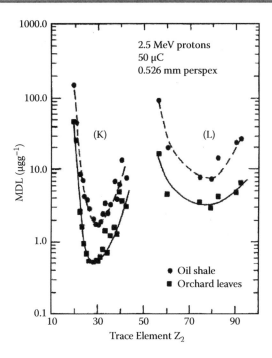

FIGURE 8.13   Minimum detection limits calculated for routine PIXE analysis of a biological (orchard leaves) sample and a geological (oil shale) sample, where 2.5 MeV protons are used, the total accumulated charge is 50 µC, and 0.526 mm Perspex serves as an absorber. (From Cohen, D. D., and Clayton, E. 1989. In *Ion Beams for Materials Analysis,* eds. Bird, J. R. and Williams, J. S. Sydney, Australia: Academic Press.)

Table 8.1 summarizes PIXE analysis results of a certified fly ash in comparison with the reference values (Maenhaut 1987). Ratios of PIXE and reference values indicate that the accuracy of PIXE analysis for most elements is within 5%.

8.3.6   PIXE DETECTION LIMITS

Generally, an x-ray peak is considered detectable (limit of detection [LOD]) if its intensity exceeds a three-standard-deviation fluctuation of the underlying background, which may comprise contributions from bremsstrahlung, nuclear reaction gamma rays, overlapping x-rays, etc. Detection limits for PIXE are much lower than for electron-induced x-rays (EDS, approximately 0.1%) due to the much lower background of the PIXE, but comparable to x-ray-induced fluorescence (XRF). PIXE is more advantageous than XRF when a small amount of sample is available for analysis. Figure 8.13 shows minimum detection limits calculated for routine PIXE analysis of a biological (orchard leaves) sample and a geological (oil shale) sample (Cohen and Clayton 1989), where 2.5 MeV protons are used, the total accumulated charge is 50 μC, and 0.526 mm Perspex is used as an absorber. Figure 8.13 indicates that there are two groups of elements in which PIXE shows good detection limits: sub-parts per million sensitivity for transition metals (Sc to Zn) when K x-rays are detected and tens of parts per million sensitivity for refractive metals (Hf to Pt) when L x-rays are detected.

Other, heavier ion beams than protons have also been used in PIXE analysis. Generally, heavier ions produce less gamma background through nuclear reactions, and thus detection limits could be even lower, provided that the heavier ion energies are carefully selected. If the energy is too high, then the gamma background from nuclear reactions is inevitable, especially for biological specimens where light elements of C, N, or O are their major constituents. If the energy is too low, the x-ray production cross section is reduced, and beam specimen heating could become an issue, especially for thick targets. Figure 8.14 shows measured minimum detection limits for 100 μC each of protons (1.8 and 3 MeV) and helium ions (5 MeV) and for 5 μC of carbon ions (10 MeV) for thin organic specimens (Johansson 1992). Since 5 MeV helium ions provide roughly the same ionization cross sections as do 2 MeV protons, the fact that the former produces lower detection limits indicates that the helium ions produce lower radiation background (gamma ray Compton scattering and bremsstrahlung).

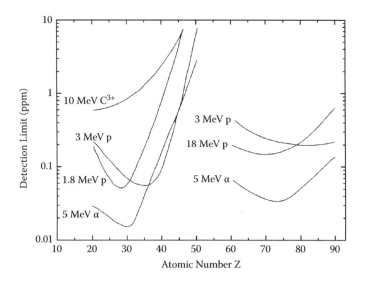

FIGURE 8.14   Measured minimum detection limits for 100 μC each of protons (1.8 and 3 MeV) and helium ions (5 MeV) and for 5 μC of carbon ions (10 MeV) for thin organic specimens. (Johansson, S. A. E. 1992. *International Journal PIXE* 2:33.)

8.4   INSTRUMENTATION FOR PIXE ANALYSIS

8.4.1   ION BEAM CHARACTERISTICS

Most PIXE work is done with small single-end or tandem electrostatic accelerators that are able to produce 2–4 MeV protons. A modest but growing amount of use is made of He ions in PIXE, sometimes in conjunction with RBS. Depending on the sample uniformity, the beam uniformity may become an issue in PIXE analysis. For example, fluid drops dried upon a substrate are a type of thin sample that may not have spatially homogeneous element distribution, so uniform beam intensity over the cross-sectional area is required. This is accomplished by use of a very thin diffuser foil or by defocusing the beam with a pair of quadrupole magnets; in each case a subsequent aperture provides a well-defined beam area. For thick, homogeneous samples, the issue of beam uniformity is less important, but accurate charge collection on the sample requires electrical isolation, along with effective suppression of secondary electrons, and that nonconductive samples may require carbon coating.

Analyzing small-size specimens or individual micrometer-sized particles or domains in the field of environmental sciences, earth sciences, biological/medical sciences, and microelectronics requires focused microbeams as projectiles. Micron-sized proton beams can be made using commercially available quadrupole magnet arrangements such as the Oxford microbeam triplet lenses. An optical microscope equipped with a CCD camera is a necessary adjunct for observing the beam spot on a fluorescent target. Nuclear microprobes usually contain magnetic or electrostatic rastering arrangements so that the PIXE analysis can be spatially resolved in one or two dimensions, and concentration maps are produced.

Certain objects that are too large or vacuum noncompatible or too valuable to be placed in a vacuum chamber require an external beam for PIXE analysis: The beam or microbeam is extracted outside the vacuum chamber through a very thin window. The external PIXE has found very important applications in art and archaeology where precious art work has been studied in ambient conditions. While the polymer Kapton was a popular early choice for the exit window, external PIXE practitioners now often use commercially available thin silicon nitride windows. For example, use of the 100 nm thick silicon nitride means that the mean energy loss for 3 MeV protons is only 3 keV, comparable to the intrinsic beam energy spread from the accelerator itself. With external beams, direct charge integration is no longer feasible; sometimes a second detector records argon K x-rays from the intervening air as a proxy for beam charge; otherwise, a rotating vane coated with a metal film can be used in conjunction with a second detector.

8.4.2   X-RAY DETECTION

While both energy dispersive x-ray spectrometers (EDSs) and wavelength dispersive x-ray spectrometers (WDSs) are used in EPMA, EDS is almost solely used to measure x-rays in PIXE analysis, mainly due to its higher efficiency in multielemental analysis. Higher resolution (~5 eV for Mn K x-rays) is obtainable with WDS techniques—albeit at the expense of efficiency—in which scattering of x-ray photons of wavelength λ from a crystal of spacing d results in constructive interference at an angle θ according to Bragg's law: $2d \sin\theta = n\lambda$, where n is an integer, and a flow proportional gas counter running in a pulse mode located at the specific angle θ detects individual x-rays. The EDS detector used in PIXE is usually Si(Li), but Ge and silicon drift (SDD) detectors are increasingly used.

When an x-ray photon is absorbed by a Si atom in the Si(Li) detector, an energetic electron (photoelectron) is created that scatters inelastically and eventually dissipates its energy through the formation of electron–hole pairs. The number of the created electron–hole pairs is proportional to the x-ray photon energy. A bias across the detector separates the electrons and holes. An external circuit with a field effect transistor amplifier measures the charge and converts it into a voltage pulse. The small voltage pulse is further amplified and shaped by a main amplifier before converting into a digital pulse through an analog-to-digital convertor (ADC). The digital signal is finally registered into a computer to generate an x-ray spectrum of counts versus energy.

There are two main contributions to the full width half maximum (FWHM) of Si(Li) or Ge type photon detectors:

(8.11)
$$(\text{FWHM})^2 = \Delta_e^2 + 2.355^2 (F\varepsilon E)$$

where

Δ_e, the width associated with the electronics, is a function of the main amplifier shaping time constant

the Fano factor, F, is a constant of 0.11–0.13, depending on detector crystal material

ε is the average electron–hole pair formation energy ($\varepsilon_{Si} = 3.81$ eV and $\varepsilon_{Ge} = 2.98$ eV)

E is the x-ray photon energy

Equation (8.11) suggests that the detector resolution increases with photon energy and can change by a factor of two or more over the useful photon energy range of most semiconductor detectors.

Figure 8.15(a) shows a typical K x-ray spectrum from Mn measured with an energy dispersive Si(Li) solid-state detector. The Mn K_α and K_β lines have energies of 5.89 and 6.49 keV, respectively, and the energy resolution of the detector (full width at half maximum) is 148 eV. X-Ray energy resolution is set by the statistical variations associated with the electron–hole pair creation process as described earlier. Figure 8.15(b) shows transmission efficiency curves of a typical Si(Li) detector for various types and thicknesses of windows where the polymer window curve does not show the effect of the support grid on the overall efficiency. The detector efficiency is the product of the geometric efficiency (solid angle fraction) and the intrinsic efficiency of the crystal shown

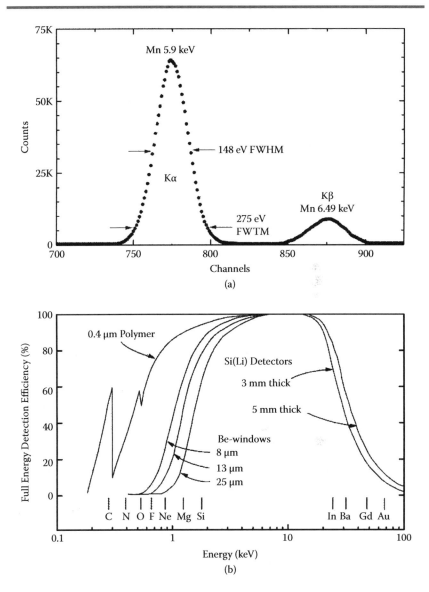

FIGURE 8.15   (a) Typical K x-ray spectrum from Mn measured with an energy dispersive Si(Li) solid-state detector. The Mn K_a and K_b lines have energies of 5.89 and 6.49 keV, respectively, and the energy resolution of the detector (full width at half maximum) is 148 eV. (b) Intrinsic transmission efficiency curves of a typical Si(Li) detector for various types and thicknesses of windows where the polymer window curve does not show the effect of the support grid on the overall efficiency.

in Figure 8.15(b). The intrinsic efficiency is affected by the detector dead layer that includes the Be-window, a possible layer of ice, the gold electrode on the Si surface, a surface layer of incomplete charge collection (ICC), and the loss of events via Si K x-ray (1.74 keV) escape (~5%). The thinnest beryllium window (8 μm) allows detection down to sodium K x-rays, but the polymer window of 0.4 μm gives a useful increase in the efficiency at even lower x-ray energies. Thus, the lower energy detection limit is dictated by the window material and its thickness, whereas the higher energy detection limit is controlled by the thickness of the Si crystal. For a Si(Li) detector, its useful x-ray detection range is roughly from 1 to 25 keV. To detect higher energy x-rays, a Ge x-ray detector is recommended.

8.4.3   X-RAY ABSORBING FILTERS

Accuracy in PIXE analysis demands the best possible counting statistics, so the limited counting rate capabilities of the Si(Li) system should not be wasted on special regions carrying information of no interest. Therefore, an x-ray absorber (filter) between the sample and detector is often used in PIXE analysis. This absorber suppresses the low-energy portion of the secondary electron bremsstrahlung, which would otherwise dominate the signal processor throughput and thereby raise the contribution of higher energy characteristic x-rays in the detected spectrum. It also suppresses the intense x-rays of light matrix elements, allowing the spectrum to become dominated by the much less intense x-rays of higher Z trace elements. Common filter materials used include Kapton, Be, and Al foils of tens to hundreds of micrometers due to their good uniformity, known thickness, and easy accessibility. The idea of a "funny filter" was sometimes used to enable simultaneous recording in one detector of both the matrix and the trace elements. A filter thick enough to suppress the former is chosen and a very small aperture is drilled in it to permit the passage of a small fraction of the intense major element x-rays. When a "funny filter" is used, it is recommended that a magnetic or electrostatic beam deflector is interposed between the target and detector to eliminate or reduce the direct registry of the scattered protons into the detector.

8.5   SUMMARY

While PIXE has been commonly used to analyze the major or minor elements in specimens, the real advantages of PIXE are trace element analyses in medium to heavy mass and good mass resolution for heavy elemental identification. Thus, PIXE is often conducted together with RBS to help identify similar mass elements in the specimens. As an ion beam analysis technique, PIXE shares with RBS the advantageous simplicity of having the elements of the sample appear in the spectra in a monotonic and predictable way, while differing from particle-induced gamma emission (PIGE) where the employed nuclear reactions can differ drastically from one element to the next. Compared to RBS and nuclear reaction analysis (NRA), PIXE offers only very modest possibilities for depth profiling of elements within a target. More than half of current PIXE work is done with microbeams, where the ability to produce quantitative trace element distributions in one or two dimensions has been valuable in diverse fields of application, including applications in materials science (corrosion rate measurements), biological and medical research (human blood, serum fluids, hair, fingernail, and teeth analysis, etc.), environmental sciences (tree rings, atmospheric aerosols for pollution monitoring, etc.), the Earth and planetary sciences (lunar rocks, fluid inclusions, sulfide ore deposits, etc.), and art and archaeology (ancient glasses, pigments and paintings, gem stones, etc.). Some of these applications are described in more detail in the Applications part of this book.

PIXE also belongs to the family of x-ray emission techniques and shares with these the property of being an elemental analysis method, with no isotopic sensitivity and, except in special cases, very little sensitivity to chemical bonding. These techniques include electron probe microanalysis (EPMA) and x-ray fluorescence analysis (XRF). As a highly developed technique that for several decades has been a workhorse in many application areas, EPMA characterizes a sample region approximately 1 μm below the surface and its detection limits are a typically a little better than one part per thousand; it is thus a method for major and minor elements. PIXE interrogates a depth that is element dependent through the attenuation coefficient of the x-rays of the element of concern, and this depth can reach a few tens of micrometers. Its detection limits can reach down to the parts per million level, and thus it often complements EPMA with trace element analysis, especially in geochemical applications. XRF is popular due to its obvious advantage of small footprint and portability, with an increasing number of handheld XRF devices becoming available. It also has the advantage of a modest degree of "tunability" of the exciting source to the specimen through choice of the x-ray tube's anode element. In its synchrotron variant (SXRF), the beam can be focused to spot sizes that compete with micro-PIXE. In addition, the exciting x-ray energy can be adjusted to lie just above the absorption edge of an element of specific interest, thereby optimizing its detection limit; this tunability confers a significant advantage over micro-PIXE. However, XRF or SXRF requires bulk-like specimens to achieve high sensitivities since x-ray-induced fluorescence yield is significantly smaller than charged-particle-induced x-ray emission yield. In summary, one should consider pros and cons of these x-ray emission techniques and use them in a complementary way to solve specific challenges in quantitative elements analysis.

PROBLEMS

8.1. Compare x-ray emission analysis techniques in terms of mass resolution, detection sensitivity, spatial resolution, and sample requirement when a monoenergetic proton, electron, or x-ray beam is used as an excitation source.

8.2. Estimate the cross section for Ni K-shell ionization production by 3 MeV protons and 5 MeV alphas. Compare proton and alpha velocities with those of the Bohr velocity of Ni K-shell electrons.

8.3. Compare the Rutherford scattering cross section with the K_α x-ray production cross section for 5 MeV alphas incident on Ag. Compare two techniques on the basis of mass resolution and depth resolution.

8.4. We want to determine Ag to Zn ratio in $Ag_xZn_{1-x}Mo$ compound but suspect that the sample may contain 1 at.% of Fe as an impurity. We mount a 100 nm thick film in an IBA chamber and use a 4 MeV proton beam for PIXE analysis. The Si(Li) x-ray detector has an energy resolution of 145 eV.
   a. For a given composition $x = 0.25$ of $Ag_xZn_{1-x}Mo$, estimate the ratio of Ag to Zn and Ag to Mo K x-ray intensities from K x-ray production cross sections.
   b. Would you expect interference among K, L_α, or M x-ray lines from Ag, Zn, or Mo?
   c. If the Fe/Ag K x-ray intensity ratio is measured to be 5%, what is the atomic ratio of Fe in $Ag_xZn_{1-x}Mo$ film when $x = 0.25$?

REFERENCES

Bambynek, W., Craseman, B., Fink, R. W., Freund, H. U., Mark, H., Swift, C. D., Price, R. E., and Rao, P. V. 1972. X-ray fluorescence yields, auger, and Coster-Kronig transition probabilities. *Reviews of Modern Physics* 44:716.

Bird, J. R., and Williams, J. S. 1989. *Ion beams for materials analysis.* Sydney, Australia: Academic Press.

Campbell, J. L. 2009. Particle induced X-ray emission: PIXE. In *Handbook of modern ion beam materials analysis,* 2nd ed., Chapter 11. eds. Wang, Y. and Nastasi, M., Warrendale, PA: Materials Research Society Publisher.

Chadwick, J. 1912. The γ rays excited by the β rays of radium. The Philosophical Magazine 24:594.

Chen, M. H. and Crasemann, B. 1985. Relativistic cross sections for atomic K- and L-shell ionization by protons, calculated from a Dirac-Hartree-Slater model. *Atomic Data and Nuclear Data Tables* 33:217.

Chen, M.H. and Crasemann, B. 1989. Atomic K-, L-, and M-shell cross sections for ionization by protons: A relativistic hartree-slater calculation. *Atomic Data and Nuclear Data Tables* 41:257.

Cohen, D. D., and Clayton, E. 1989. Ion induced X-ray emission. In *Ion beams for materials analysis,* eds. Bird, J. R. and Williams, J. S., chapter 5. Sydney, Australia: Academic Press.

Garcia, J. D. 1970. X-ray production cross sections. *Physics Reviews* A 1:1402.

Ishii, K. 1987. Scaling law for a continuum of X-rays produced by light-ion-atom collisions. *Nuclear Instruments and Methods B* 22:68.

Johansson, S. A. E. 1992. *International Journal PIXE* 2:33.

Johansson, S. A. E., and Campbell, J. L. 1988. *PIXE: A novel technique for elemental analysis.* Chichester, UK: Wiley.

Johansson, S. A. E., Campbell, J. L., and Malmqvist, K. G. 1995. *Particle-induced x-ray-emission spectrometry (PIXE).* Chichester, UK: Wiley.

Johansson, S. A. E., and Johansson, T. B. 1976. Analytical application of particle induced X-ray emission. *Nuclear Instruments and Methods* 137:473.

Liu, Z., and Cipolla, S. J. 1996. ISICS: A program for calculating K-, L- and M-shell cross sections from ECPSSR theory using a personal computer. *Computer Physics Communication* 97:315.

Maenhaut, W. 1987. Particle-induced x-ray emission spectrometry: An accurate technique in the analysis of biological environmental and geological samples. *Analytica Chimica Acta* 195:125.

Maxwell, J.A., W.J. Teesdale, J.L. Campbell, 1995. The Guelph PIXE software package II. *Nuclear Instruments and Methods B* 95:407.

Maxwell, J. A., and Campbell, J. L. 2002. *X-Ray Spectrum* 34:320.

Maxwell, J. A., Campbell, J. L., and Teesdale, W. J. 1988. The Guelph PIXE software package. *Nuclear Instruments and Methods B* 43:218.

Moseley, H. G. J. 1913. *Philosophical Magazine* 26:1024.

Ryan, C. G., Clayton, E., Griffin, W. L., Sie, S. H., and Cousens, D. R. 1988. SNIP, a statistics-sensitive background treatment for the quantitative analysis of PIXE spectra in geoscience applications. *Nuclear Instruments and Methods B* 34:396.

Ryan, C. G., Cousens, D. R., Sie, S. H., Griffin, W. L., and Suter, G. F. 1990. Quantitative pixe microanalysis of geological matemal using the CSIRO proton microprobe. *Nuclear Instruments and Methods B* 47:55.

Ryan, C. G., van Achterberg, E., Griffin, W. L., Pearson, N. J., O'Reilly, S. Y., and Kivi, K. 2001. Nuclear microprobe analysis of melt inclusions in minerals: Windows on metasomatic processes in the earth's mantle. *Nuclear Instruments and Methods B* 181:578.

Thomas, J. P., and Cachard A. 1978. *Materials characterization using ion beams.* New York: Plenum Press.

Wang, Y. Q., and Nastasi, M. 2009. *Handbook of modern ion beam materials analysis,* 2nd ed., Warrendale, PA: Materials Research Society Publisher.

Ziegler, J. F., Biersack, J. P., and Ziegler, M. D. 2008. *The stopping and range of ions in matter.* Morrisville, NC: LuLu Press.

SUGGESTED READING

Campbell, J. L. 2009. Particle Induced X-ray emission: PIXE, in *Handbook of modern ion beam materials analysis,* 2nd ed. Chapter 11 eds. Y. Wang, and M. Nastasi, Warrendale, PA: Materials Research Society Publisher.

Cohen, D. D., and Clayton, E. 1989. Ion induced X-ray Emission, Chapter 5 In *Ion beams for materials analysis,* eds. J. R. Bird and J. S. Williams. Sydney, Australia: Academic Press.

Johansson, S. A. E., and Campbell, J. L. 1988. PIXE: *A novel technique for elemental analysis*. Chichester, UK: Wiley.

Johansson, S. A. E., Campbell, J. L., and Malmqvist, K. G. 1995. *Particle-induced x-ray emission spectrometry (PIXE)*. Chichester, UK: Wiley.

Ion Channeling

9

9.1 INTRODUCTION

In the earlier chapter on backscattering spectrometry, backscattering analysis was treated on the basis that the targets were either amorphous or composed of randomly oriented polycrystallites. That approach ignores one of the important effects available in ion beam analysis: the perception of structural and crystalline order by use of the channeling effect. Some solids exist in a single crystalline form, which allows ion beam analysis in channeling mode to reveal their crystal structures. Figure 9.1 shows a model of lattice atoms and suggests the pronounced effects that would occur when the crystal orientation is shifted with respect to the ion beam. There can be a hundred-fold decrease in the number of backscattered particles when the crystal is rotated so that the beam is no longer incident along a random orientation of the crystal (Figure 9.1a) and becomes incident along a planar direction (Figure 9.1b) or along an axial direction (Figure 9.1.c). For both cases, the particles that are steered along the "channels" in the crystal do not approach the lattice atoms in the axial planes and rows close enough to undergo the large angle scattering process that results in a backscattering event. It should be pointed out that while most of the channeling applications involve the Rutherford backscattering spectrometry (RBS) technique, other ion beam analysis (IBA) techniques such as nuclear reaction analysis (NRA), particle-induced x-ray emission (PIXE), and elastic recoil detection (ERD) also are applicable in channeling mode.

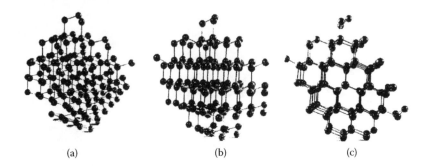

FIGURE 9.1 Model of lattice atoms showing the atomic arrangement in the diamond-type lattice viewed along (a) random, (b) planar, and (c) axial directions.

A channeling experiment requires three basic components: a source of collimated ions, a detector to sense the backscattered particles, and an accurate crystal manipulator, a goniometer. The goniometer (Figure 9.2) provides rotation so that the crystal axes can be aligned with the collimated beam of particles. Channeling effects are observed when the crystal axes are aligned to within ~1° of the beam. The energy spectrum of particles scattered back from

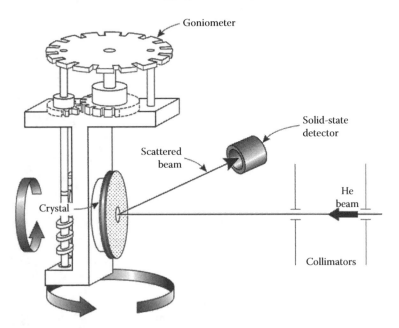

FIGURE 9.2 Schematic drawing showing the experimental setup for channeling experiments. The ion impinges on the crystal held in a goniometer, an alignment apparatus. The solid-state detector, set at a specific scattering angle, is sensitive to the energy of the backscattered ions.

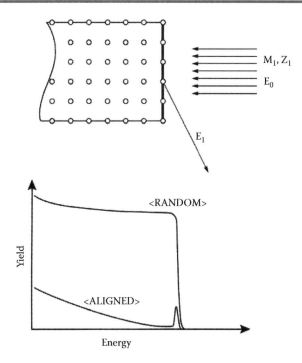

FIGURE 9.3   Schematic of ion channeling in a single crystal and the corresponding backscattered energy spectra for a beam aligned with a symmetry direction (see Figures 9.1b and 9.1c) and in a random direction (Figure 9.1a).

an aligned crystal is dramatically different from that of a nonaligned (random) or noncrystalline solid, as shown in Figure 9.3. In the aligned spectrum the scattering yield from the bulk of the solid is reduced by almost two orders of magnitude, and a peak occurs at a position corresponding to scattering from surface atoms. Both the surface peak and the reduction of yield are due to shadowing—the ability of the outermost atoms to shadow the underlying atoms and hence shield these atoms from direct interaction with the beam.

In a perfect crystal, as the ions penetrate deeper, they only make small-angle collisions with the atomic rows at distances greater than 0.01 nm from the atomic cores. The initial motion is oscillatory, with the particles bouncing from row to row, with a wavelength between bounces of some tens of nanometers (Figure 9.4). The channel particles cannot get close enough to the atoms of the solid to undergo close encounters such as large-angle Rutherford scattering. Hence, the yield of backscattered particles decreases by almost two orders of magnitude. However, lattice imperfections, such as impurities on the surface, substitutional atoms in the lattice, or interstitial atoms residing in the channel, can interact with the channeling particles, resulting in backscattering events. The interaction will be much greater for displaced host atoms or interstitial impurities compared with substitutional atoms.

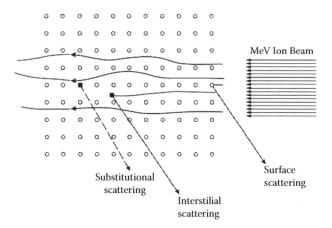

MeV Ion Beam

Substitutional
scattering

Interstilial
scattering

Surface
scattering

FIGURE 9.4 Schematic showing ions channeling into a single crystal and inter-acting with defects and surface atoms.

9.2 CHANNELING IN SINGLE CRYSTALS

Channeling of energetic ions occurs when the beam is carefully aligned with a major symmetry direction of a single crystal (Lindhard 1965). By major symmetry direction we mean one of the open directions in a single crystal as viewed down a plane or row of atoms (Figure 9.1). Figure 9.4 shows a side view of this process in which most of the ion beam is steered (channeled) through the channels formed by the string of atoms. A more detailed picture of this process is presented in Figure 9.5, which shows a single particle's trajectory through a series of sequential binary collisions with the atoms in an atomic row. The angular deflection associated with each binary collision is governed by a screened Coulomb scattering potential, with the screening arising from the electrons surrounding the nucleus. The net effect on the particle incident at a low angle with respect to the row is to steer the particle gently so that it leaves the row with the same angle as the incident angle. As seen in the inset, the channeled particle undergoes a series of correlated collisions such that each individual collision gives rise to a change in angle $\Delta\psi$ that is small compared to the total angle through which the particle may be steered ($\Delta\psi \ll \psi$). In the absence of additional scattering processes, this angle would be maintained for each encounter with an atomic row.

Since the steering of the channeled particle involves successive collisions with many atoms, it is possible to represent the process with a continuum model that describes the nuclear charges of the atoms in a row (or plane) as being uniformly averaged along the row (or plane). The interaction of channeled particle with an atomic row can now be described in terms of a single continuum potential $U_a(r)$, where r is the particle's perpendicular distance from the row. The potential $U_a(r)$

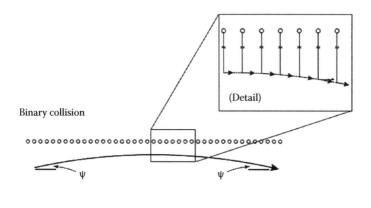

Binary collision

(Detail)

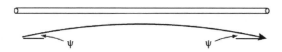

ψ ψ

Continuum

ψ ψ

FIGURE 9.5 Schematic showing the trajectory of a channeled particle, which is defined by a correlated sequence of individual collisions. This can be described by a continuum model.

is the value of the atomic potential averaged along the atomic row with atomic spacing d. For the axial channeling case, the potential is expressed as

(9.1)
$$U_a(r)=\frac{1}{d}\int_{-\infty}^{\infty}V(\tilde{r})dz=\frac{1}{d}\int_{-\infty}^{\infty}V\left(\sqrt{z^2+r^2}\right)dz$$

where
$V(\tilde{r})$ is the screened Coulomb potential
$\tilde{r}$ is the spherical radial coordinate, $\tilde{r}^2=z^2+r^2$
where
r is the distance of the ion from the string of atoms
z is the distance traveled along the string
d is the spacing between atoms in the string

To facilitate computations, we will use a more mathematically convenient form of the screened Coulomb potential that is extensively used in channeling theory. This *standard potential* is given by

(9.2)
$$V(\tilde{r})=Z_1Z_2e^2\left(\frac{1}{\tilde{r}}-\frac{1}{\sqrt{\tilde{r}^2+C^2a^2}}\right)$$

where C^2 is usually taken as equal to three and a is the Thomas–Fermi screening distance (Equation 4.3 in Chapter 4). Then we obtain the following for the axial continuum potential:

(9.3)
$$U_a(r) = \frac{Z_1 Z_2 e^2}{d} \ln\left[\left(\frac{Ca}{r}\right)^2 + 1\right]$$

For r = 0.01 nm and He along the <110> rows of Si (d = 0.384 nm), the value of $U_a(r)$ is 223 eV.

9.3 TRANSVERSE ENERGY AND CRITICAL ANGLE FOR CHANNELING

The continuum potential and conservation of energy allow us to find the critical angle for channeling. The conservation of the incident and exit angles, ψ, as shown for a row of atoms in Figure 9.5, is an important consequence of what is referred to as the conservation of transverse energy. The total energy of a particle inside the crystal is the sum of the potential and kinetic terms:

(9.4)
$$E = \frac{p_\parallel^2}{2M} + \frac{p_\perp^2}{2M} + U_a(r)$$

where $p_\parallel$ and $p_\perp$ are, respectively, the parallel and perpendicular components of the momentum with respect to the string direction (Figure 9.6a) and M is the ion mass. Then, from Figure 9.6(a) we have $p_\parallel = p\cos\psi$, $p_\perp = p\sin\psi$, and since the channeling angle with respect to the rows or planes is sufficiently small, $\sin\psi \cong \psi$ and $\cos\psi \cong 1$, we have

(9.5)
$$E = \frac{p^2}{2M} + \frac{p^2\psi^2}{2M} + U_a(r)$$

This total energy is essentially constant over the dimension involved for the steering of a channeled particle. The last two terms in Equation (9.5) equate as a *transverse energy*, $E_\perp$, which will also be constant and is given by

(9.6)
$$E_\perp = E\psi^2 + U_a(r)$$

where $E = p^2/2M$.

As the particle penetrates close to an atomic row it senses the roughness of the string potential because of the discrete nature of the atoms in the row and the thermal vibrations (Figure 9.6b). This suggests a minimum distance of approach to the string, r_{min}, for which the continuum model and the channeling concept are valid. There is a corresponding maximum incident angle, the

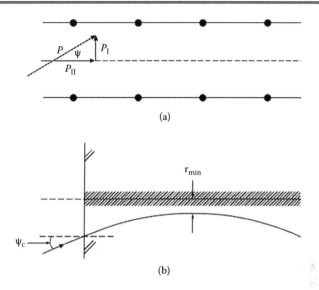

FIGURE 9.6 (a) Components of the initial momentum vector for a particle incident on a string of atoms at an angle y. (b) Schematic of a channeled trajectory incident at a crtical angle y_c and having a distance of closest approach, r_{min}.

critical angle, ψ_c, for which the incident particle can be steered by the atomic row (Figure 9.6b). This critical angle is defined by equating the transverse energy at the turning point $U(r_{min})$ (sole potential energy) to the transverse energy at the midpoint (sole kinetic energy), so that

$$(9.7) \qquad E\psi_c^2 = U_a(r_{min})$$

or

$$(9.8) \qquad \psi_c = \sqrt{\frac{U_a(r_{min})}{E}}$$

Since the continuum potential is on the order of 100 eV and typical channeling experiments use MeV particles, ψ_c is on the order of 0.01 rad or ~ 1°.

In reality, thermal smearing of the atoms' positions on the lattice sets a lower limit for which a row can provide the necessary correlated sequence of scattering required for the channeling condition. The most useful first approximation to the critical angle is obtained by substituting $r_{min} = \rho$ in Equations (9.3) and (9.8), where ρ^2 is two-thirds of the mean square thermal vibration amplitude, which gives

$$(9.9) \qquad \psi_c(\rho) = \frac{\psi_1}{\sqrt{2}} \left| \ln\left[\left(\frac{Ca}{\rho}\right)^2 + 1 \right] \right|^{1/2}$$

where

$$(9.10) \qquad \psi_1 = \sqrt{\frac{2Z_1 Z_2 e^2}{Ed}}$$

where d is the atomic spacing along the string and ψ_1 is the *characteristic angle for axial channeling.*

The calculated values of $\psi_c(\rho)$ are within 20% of the experimental measurements and follow measured temperature dependence. For 1 MeV He incident on <110> Si at room temperature, $\psi_c(\rho) = 0.65°$ while the experimentally measured value is 0.55°.

The continuum description can be applied to planar channeling as well as to axial channeling. For planes, two-dimensional averaging of the atomic potential results in a sheet of charge, and a corresponding planar continuum potential, $U_p(y)$ (Feldman, Mayer, and Picraux 1982). An analysis shows that, similarly to axial channeling, a critical angle for planar channeling can be defined as

$$(9.11) \qquad \psi_p = \sqrt{\frac{U_p(y_{min})}{E}}$$

where y is the distance from the plane and $y_{min} \cong \rho/\sqrt{2}$ is the one-dimensional vibration amplitude. In a manner similar to the axial case (Equation 9.10), we define the *characteristic angle for planar channeling:*

$$(9.12) \qquad \psi_1 = \sqrt{\frac{2\pi Z_1 Z_2 e^2 a N d_p}{E}}$$

where d_p is the spacing between planes, N is the number of atoms per unit volume, and Nd_p is the average number of atoms per unit area in the plane. In general, planar critical angles are a factor of two to four times smaller than characteristic critical angles for axial channeling.

Experimentally, the critical angle for channeling is determined by measuring the backscattered yield from just beneath the surface as a function of angle between the beam and symmetry direction of the crystal, as shown in Figure 9.7. The half angle, $\psi_{1/2}$, is taken as a measure of the critical angle.

9.4   MINIMUM YIELD

The concept of distance of closest approach, r_{min}, allows a simple geometric derivation of the fraction of particles that will be channeled for incidence ions aligned parallel to the crystal axis, $\psi = 0°$. Figure 9.8 presents a head-on view of

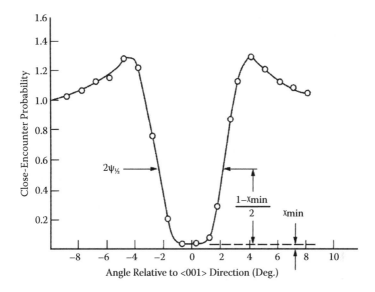

FIGURE 9.7 Schematic showing backscattered data from an angular scan between the beam and the symmetry direction of the crystal. This schematic shows the experimental definition of the angular width $\Psi_{1/2}$ and the minimum yield χ_{min}.

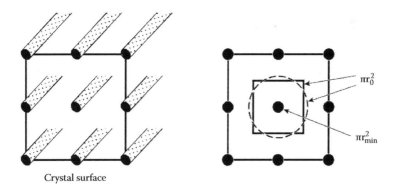

FIGURE 9.8 View of an ideal crystal showing continuous rows of atoms and defining the area/row, πr_{min}^2.

an ideal crystal showing continuous rows (strings) of atoms. Around each string of atoms is an area πr_{min}^2 in which particles cannot channel. For particles incident at $r > r_{min}$, channeling is permitted. If we define r_0 as the radius associated with each string, we can now define the fraction of particles not channeled as

$$(9.13) \qquad \text{fraction not channeled} = \frac{\pi r_{min}^2}{\pi r_0^2} \equiv \chi_{min}$$

where r_0 is the radius associated with each string and χ_{min} is called the *minimum yield* (see Figure 9.7 for experimental definition). In backscattering experiments χ_{min} is the yield of close-encounter events as a probability of the channeling process. For $r_{min} = \rho \sim 0.01$ nm, the minimum yield is on the order of 1% or the fraction of particles channeled is ~99%.

The variable r_0 can be defined from the atomic areal density as

$$(9.14) \qquad \pi r_0^2 = \frac{1}{Nd}$$

where d is the atom spacing along the row and N is the atomic density. Combining Equations (9.13) and (9.14) allows the minimum yield to be expressed as

$$(9.15) \qquad \chi_{min}(\rho) = Nd\pi\rho^2$$

The interesting feature of Equation (9.15) is that it is independent of scattering parameters Z_1, Z_2, and E and depends solely on the properties of the host crystal.

A geometric picture of planar channeling indicates that the minimum yield, χ_{min}(planar), is approximately given by

$$(9.16) \qquad \chi_p = \frac{2y_{min}}{d_p},$$

where $y_{min} \cong \rho/\sqrt{2}$ is the one-dimensional vibration amplitude and d_p is the atom spacing between planes. The value of the minimum yield for good planar channeling directions is typically on the order of 10%–25%, which is substantially larger than the corresponding value for axial channeling.

9.5 ENERGY LOSS

The special trajectories of channeled ions not only reduce their probability of interactions with the atomic nuclei of the solid but also change their interactions with the electrons of the solid. Since MeV ions lose energy primarily through electronic interactions, the rate of energy loss, dE/dx, of channeled ions is a measure of the channeling effect.

At high energies the electronic stopping of energetic particles in amorphous solids is well described by Equation (4.18b) in Chapter 4:

$$(9.17) \qquad \frac{dE}{dx} = \frac{2\pi Z_1^2 e^4}{E} N Z_2 \left(\frac{M_1}{m_e}\right) \ln \frac{2 m_e v^2}{I}$$

The complete formulation of this equation showed that energy loss process can be divided into two roughly equal types of particle–electron collisions: distant resonant collisions with small momentum transfer and close collisions with large momentum transfer. Since the close collisions are directly proportional to the local electron density, Lindhard (1965) modified Equation (9.17) to

$$(9.18) \qquad \frac{dE}{dx}(\tilde{r}) = \frac{2\pi Z_1^2 e^4}{E} \left(\frac{M_1}{m_e}\right) \ln \frac{2 m_e v^2}{I} [(1-\alpha) N Z_2 + \alpha \varepsilon(\tilde{r})]$$

where $\varepsilon(\tilde{r})$ is the density of electrons at point $\tilde{r}$ through which the particle moves and $\alpha \cong 1/2$. Clearly, $\varepsilon(\tilde{r})$ is a minimum in the center of the channel. These basic considerations suggest that the best channeled particles will have a stopping power on the order of one-half the stopping power in the corresponding solid.

Experimental data of the ratio channeling to random stopping loss of ^{4}He ions along <100>, <111>, and <110> Si axes are presented in Figure 9.9 (Shao et al. 2006). These data show that the channeling stopping power increases with decreasing incident energy and reaches a maximum in the energy region of 400–800 keV. In theory, this is due to the decreased contribution of the Si L-shell electrons to the random stopping power at lower ion energies, while the L-shell electrons are inaccessible to channeled ions. When the energy reaches around 500 keV, both the random stopping power and channeling stopping power are dominated by interactions with valence electrons.

Figure 9.9 also shows that, in the energy region of 400–800 keV, the stopping ratios for the <100> and <111> axes approach 1, but the value for the <110> axis is significantly less than 1. This can be understood by the variation of the valence-electron distribution for <110>, the widest channel of Si crystal. For <100> and <111> axes, the valence electrons are almost homogeneously distributed across the channels. For <110>, the density of valence electrons is low in the middle of the channel, which results in a low channeling stopping loss of He ions.

The y axis of Figure 9.9 (the factor R) is defined by the following equation:

$$(9.19) \qquad \left.\frac{dE}{dx}\right|_{channeling} = R \left.\frac{dE}{dx}\right|_{random}$$

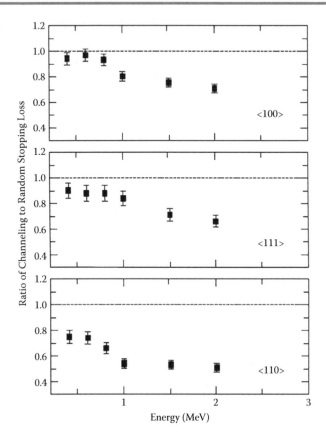

FIGURE 9.9   Ratio of the channeling to random stopping loss of He ions in Si along different crystallographic axes as a function of incident energy.

Using the scattering geometry presented in Figure 5.2 in Chapter 5, the conversion of the energy axis of the channeling data to depth is thus obtained from

$$(9.20) \qquad x(E_1) = \left(\cos\theta_1 \int_{E_0}^{E_1} \frac{dE}{R(E)\, dE/dx} + \frac{\cos\theta_2 \int_{kE_1}^{E_{det}} \frac{dE}{dE/dx}}{2} \right)$$

where $x(E_1)$ is the depth of backscattering event after the incoming ion has lost an energy $E_0 - E_1$ and has an energy after the scattering event of kE_1, where k is the kinematic scattering factor. The first term on the right side of Equation (9.20) represents the distance traveled by the incoming channeled ion, and the second term represents the distance traveled by the outcoming nonchanneled ion. Equation (9.20) assumes that the ion's outward path is not in a channeling direction.

9.6   DECHANNELING BY DEFECTS

In the previous sections of this chapter we have assumed a perfect crystal and that particles once channeled remained in channeling trajectories. However, in practice, most materials contain a variety of defects such as displaced atoms, point defects, dislocations, stacking faults, twins, defect clusters, small precipitates, and amorphous regions. Defects that introduce distortions to the lattice can lead to the scattering of channeled particles into nonchanneling trajectories. This scattering process is referred to as dechanneling. Each type of defect has a particular influence on the trajectory of a passing particle and can be associated with a corresponding dechanneling factor, σ_D.

The presence of defects can greatly enhance dechanneling over the perfect crystal. For example, displaced atoms in the center of the channel provide much stronger scattering to the projectile than the electrons. Electrons can only cause small deflections of the channeled particle relative to the critical angle for channeling. However, displaced atoms can cause greater deflections and, for sufficiently close collisions, can directly deflect particles beyond the critical angle. Extended defects, such as dislocations, can cause distortion or curvature of the channeled wall, which gives greater dechanneling than that due to wall roughening caused by thermal vibrations. Thermal vibrational displacements are ~0.01 nm; distortions induced by defects may be comparable over a single atomic spacing, but the cumulative effect of the distortion is larger when viewed over the many tens of atomic spacing required for the atomic row or plane to deflect a channeled particle. Dechanneling will occur when the defect-induced distortions of the channel become significant relative to the channeling critical angle for the channeled particle.

The probability of dechanneling per unit depth, dP_D/dz, is given by the defect dechanneling factor and the defect density n_D:

$$(9.21) \qquad \frac{dP_D}{dz} = \sigma_D n_D(z)$$

The units of σ_D and n_D depend on the class of defect (point, areal, and volumetric defects) so that the probability of dechanneling per unit depth $\sigma_D n_D$ has units of cm^{-1}. For point-scattering centers, such as interstitial atoms, σ_D can be thought of as a cross section for dechanneling and n_D is the density of interstitial atoms per unit volume at a given depth. The units for the dechanneling factors and defect densities for different classes of defects are presented in Table 9.1.

In the following sections some examples of the different defects will be discussed. A complete discussion can be found in Feldman et al. (1992).

TABLE 9.1   Definitions of Defect Dechanneling Factors σ_D and Densities n_D

Defect	Class	σ_D Units	n_D Units
Interstitial	Point	Area (cm²)	No./unit volume (cm⁻³)
Dislocation	Line	Area/defect length (cm)	Projected length/unit volume (cm⁻²)
Stacking fault	Area	Area/defect area (no unit)	Projected area/unit volume (cm⁻¹)
Twin	Volume	Area/defect volume (cm⁻¹)	Defect volume/unit volume (no unit)

9.6.1   POINT DEFECTS

Figure 9.10 schematically shows the scattering of a beam of channeled particles by interstitial atoms. Here we are interested in scattering angles greater than the critical angle ($\psi_{1/2} \sim 1°$), so the impact parameter is relatively small ($\sim 10^{-3}$ nm). Under these conditions we can use the unscreened Coulomb potential to estimate the dechanneling factor for an isolated atom in the channel. In this calculation, the dechanneling is a result of binary scattering by isolated displaced atoms in an otherwise perfect crystal. For these conditions the defect dechanneling factor is the cross section for a close impact collision, scattering a particle through an angle θ greater than $\psi_{1/2}$. In center of mass coordinates, the Rutherford scattering cross section with $M_1 \ll M_2$ is given approximately by (see Equation 3.22 in Chapter 3):

(9.22)
$$\frac{d\sigma}{d\Omega} = \left[\frac{Z_1 Z_2 e^2}{4E \sin^2 \dfrac{\theta_{cm}}{2}} \right]^2$$

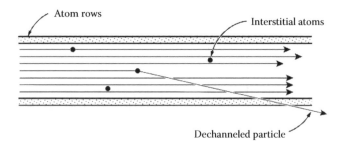

Atom rows

Interstitial atoms

Dechanneled particle

FIGURE 9.10   Schematic showing scattering of channeled ions by interstitial atoms.

An integration of Equation (9.22) over all angles greater than $\psi_{1/2}$, assuming the axial symmetrical case of axial channeling, leads to

$$(9.23) \qquad \sigma_D(\psi_{1/2}) = \int_{\psi_{1/2}}^{\pi} \frac{d\sigma}{d\Omega} d\Omega \approx \frac{\pi Z_1^2 Z_2^2 e^4}{E^2 \psi_{1/2}^2}$$

where $d\Omega = 2\pi \sin\theta \, d\theta$, and the small angle expansion for $\sin\theta$ has been used. If the critical angle $\psi_{1/2}$ is approximated by ψ_1 (Equation 9.10), the following is obtained

$$(9.24) \qquad \sigma_D \approx \frac{\pi Z_1 Z_2 e^2 d}{2E} = 2.3 \times 10^{-22} \frac{Z_1 Z_2 d}{E} \, cm^2$$

where the value of the atomic spacing, d, along the axial direction is given in nanometers and the incident energy E is in MeV. For 2 MeV He ions incident along the <100> axis of Si, d = 0.543 nm, and $\sigma_D = 1.7 \times 10^{-19} \, cm^2$. This cross section value is sufficiently small compared to the area of crystal channels ($10^{-15} \, cm^2$) and indicates that one isolated atom cannot close off channeling. Thus, the number of channeled particles remains essentially unchanged even for a collection of several isolated atoms in a channel.

9.6.2   DISLOCATION LINES

Line imperfections, or dislocations, in a crystalline solid are defects that produce lattice distortions centered about a line. Such distortions in the lattice enhance dechanneling. An extrinsic edge dislocation introduces an extra half-plane, ABCD, into the crystal, as shown in Figure 9.11. The Burgers vector, b, is normal to the line CD, the dislocation axis. A screw dislocation involves a lateral shifting of the crystal planes as shown in Figure 9.12. In this case the Burgers vector is parallel to the dislocation axis. The introduction of the dislocation results in a distorted crystal where the center of the distortion is along the dislocation axis. Atomic distortions around the dislocation decrease with decreasing distance from the dislocation axis.

To understand the effect of dislocations on dechanneling requires determining the influence of lattice distortions on the trajectories of channeled particles. Examining Figure 9.11(b) and imagining a beam of particles incident from the top, parallel to the center extra half-plane, it is easy to see how the distortions near the dislocation core can steer a particle beyond the critical angle. An estimate of the amount of dechanneling due to a dislocation can be obtained by determining the maximum amount of distortion a channeled particle can traverse and still remained channeled. This sets a boundary around the dislocation within which the particle can be dechanneled. The distance from the center of the dislocation to the boundary depends on the magnitude of the Burgers

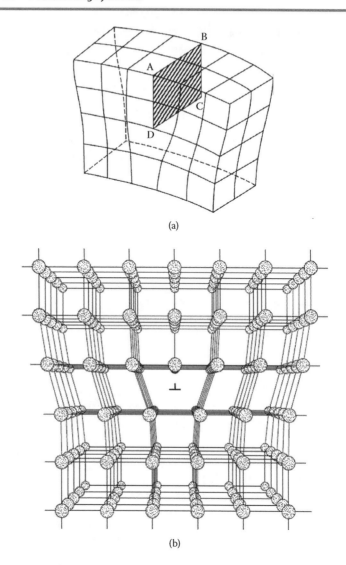

(a)

(b)

FIGURE 9.11   Schematics showing an edge dislocation in a crystalline solid.

vector, b, the elastic properties of the crystal, and the critical angle for channeling, $\psi_{1/2}$. Large burgers vectors and small critical angles will increase the region over which dechanneling occurs.

An analysis of a screw dislocation (Figure 9.12) results in a dechanneling factor (Feldman et al. 1982) given by

$$(9.25) \qquad\qquad \sigma_D = \frac{b}{\pi \psi_{1/2}}$$

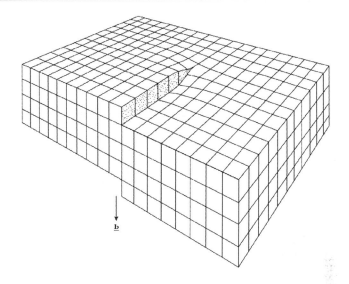

FIGURE 9.12   Schematic showing a screw dislocation in a crystalline solid.

A similar expression can be derived for an edge dislocation. Equation (9.25) shows that the dechanneling factor is inversely proportional to the critical angle, which indicates that dechanneling will be an order of magnitude greater for planar than for axial channeling. Using Equation (9.10) for the analytical description of the critical angle yields

(9.26) $$\sigma_D \propto \psi_{1/2}^{-1} \propto (E/Z_1)^{1/2}$$

Equation (9.26) states that the dechanneling factor σ_D increases with increasing ion energy, which is observed experimentally and theoretically for dislocation loops. At sufficiently high energies, there is no further increase and the dechanneling factor saturates.

The dechanneling probability per unit length, dP_D/dz, is obtained from Equation (9.25) by multiplying σ_D by the projected length n_D per unit volume of the dislocation. For a typical value of the critical angle (~1°), the dechanneling factor is 10 nm for both the screw and edge dislocations. In real crystals, dislocation densities range from zero in dislocation-free silicon to values of 10^{10} to $10^{11}/cm^2$ in cold worked metals. For $\sigma_D = 10$ nm and $n_D = 10^{10}/cm^2$, approximately 10% of the incident beam will be dechanneled by dislocations after penetrating 100 nm into the crystal. This amount of dechanneling provides a clearly measurable signal.

The orientation of the dislocation relative to the channeling direction of the incident beam will determine the amount of lattice distortion experienced by the channeled particles. For example, for the edge dislocation shown in Figure 9.11(a), particles incident along the dislocation axis (along line DC) will have minimal dechanneling because there are effectively no planar distortions in the channeling direction. In contrast, particles incident perpendicular to the dislocation axis (along line AD) will experience greater planar distortions.

9.7 DEFECT SCATTERING

As shown in Section 9.6, defects can contribute to dechanneling. Defects can also contribute to the backscattering spectra by direct backscattering. The problem can be considered in the following way: At some depth in the crystal, the beam consists of a channeled component and a nonchanneled component. The channeled beam cannot scatter from atoms on normal lattice sites, but may backscatter from defect atoms that lie within the channel. In addition, the interaction of the channeled beam with defects can result in small-angle collisions that deflect particles out of the channel and gradually increase the nonchanneled component of the beam.

Figure 9.13 shows how these two processes can contribute to a backscattering spectrum. In this figure χ_R corresponds to the random fraction of the

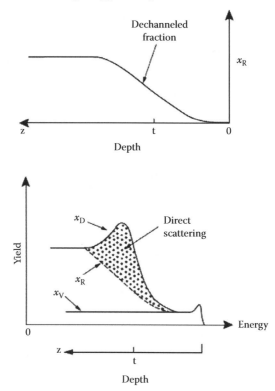

FIGURE 9.13 Schematic of the change in backscattering yield with depth through a region of disorder that gives rise to both dechanneling and direct backscattering.

backscattered beam, χ_V is the fraction backscattered from a perfect virgin crystal, and χ_D is the fraction of the beam backscattered by defects. The depth-dependent relationship between χ_D and χ_R is given as

(9.27)
$$\chi_D(z) = \chi_R(z) + (1 - \chi_R(z)) \frac{f\, n_D(z)}{n}$$

where $\frac{f\, n_D(z)}{n}$ is the effective concentration of atoms in defects normalized by the total atomic concentration n, and f is the defect scattering factor. The units of n_D, the defect density, depend on the defect geometry (Table 9.1). However, $f\, n_D(z)$ has units of cm^{-3} in all cases.

Equation (9.27) simply states that the normalized yield in a crystal containing defects is the sum of the contributions from the scattering of the random component, $\chi_R(z)$, with all atoms, and the contribution of the scattering of the channeled component, $(1 - \chi_R(z))$, with the defect atoms within the channel.

The first term in Equation (9.27), the random fraction, $\chi_R(z)$, is given by (Feldman et al. 1982)

(9.28)
$$\chi_R(z) = \chi_V(z) + [1 - \chi_V(z)] \left[1 - \exp\left(-\int_0^z \sigma_D n_D(z')dz'\right) \right]$$

where $\chi_V(z)$ is the aligned yield at depth z for a virgin (perfect) crystal, and σ_D is the dechanneling cross section for defects (see Table 9.1)

For small defect concentrations we can rewrite Equation (9.28) as

(9.29)
$$\chi_R(z) \approx \chi_V(z) + [1 - \chi_V(z)]\, \sigma_D N_D$$

where the defect density per unit area, N_D, is defined as

$$N_D = \int_0^z n_D(z')dz'$$

The random fraction, Equation (9.29), can be thought of as a sum of the random fraction for the perfect crystal, χ_V, plus the dechanneling fraction resulting from the total number of defects traversed. In the limit of $\chi_V \ll 1$, the contribution to the dechanneling fraction at depth z is linearly related to the total number of defects between the surface and depth z.

9.8   DEFECT SCATTERING FACTOR

The defect scattering factor, f, was introduced in Equation (9.27) because the direct backscattering contributions of defects depends on their geometry. Not all defects contribute equally to direct backscattering and different defects can have different effective numbers of scattering centers. The factor f accounts for any difference between the number of defects and the effective number of scattering centers per defect (Figure 9.14). For example, for defects that consist of

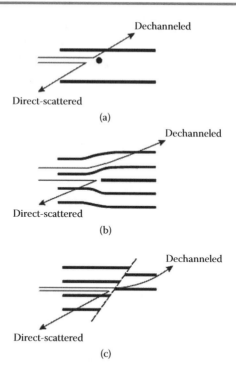

FIGURE 9.14   Direct scattering and dechanneling of particles due to three types of defects: (a) an interstitial atom; (b) a dislocation; and (c) a stacking fault.

single, randomly displaced atoms (Figure 9.14a), $f = 1$. Here the defects contribute to measured backscattered signal both through dechanneling and direct scattering. On the other hand, when the defect consists only of small distortions from the ideal lattice, such as for a dislocation (Figure 9.14b), the value of f can be near zero. In this case the dechanneling can occur in the distorted volume and the presence of the defect is detected through the dechanneling process rather than by direct scattering by the defect itself. In contrast to isolated interstitials, the direct scattering contribution for dislocations is much less significant. Dislocations can be viewed as a heavily distorted core surrounded by a large volume of strained crystal. The value of f is given by the number of direct scattering centers per unit length of dislocation. Examining Figure 9.14(b) we see that the atoms that terminate the extra plane of the edge dislocation are the ones that contribute to the direct backscattering. This indicates that the direct backscattering contribution to the defect scattering factor comes from the displaced atoms in the core region of the dislocation. This contribution is small compared to the dechanneling fraction σ_D that arises from the much larger strained volume around the core. Consequently, the sensitivity of channeling to dislocations comes primarily from the dechanneling contribution, with $f \sim 0$.

Stacking faults (Figure 9.14c) represent an intermediate case between interstitials and dislocation in terms of the contribution of the direct scattering to the observed backscattering signal. From Figure 9.14(c), a fault can be viewed as a new internal surface along directions other than the stacking directions with rows displaced into the channel. These first atoms in each row at the fault boundary give rise to direct scattering and thus are similar in this contribution to isolated interstitials. However, the contribution to dechanneling is much larger for the exposed row in the stacking fault case than for a point scattering center. The scattering factor for stacking faults has units of cm^{-2}.

The dechanneling contribution to the observed signal χ_D is given by the first term χ_R in Equation (9.27). As seen in Equation (9.29), the random fraction χ_R is determined by dechanneling fraction times the defect density $\sigma_D N_D$. For MeV channeling, one monolayer of stacking fault area has a dechanneling contribution, $\sigma_D N_D \cong \chi_{min} = 10^{-2}$, whereas one monolayer of isolated interstitials has a much smaller contribution, $\sigma_D N_D \cong 10^{-4}$. Thus, the dechanneling contributions for stacking faults are significantly greater than for isolated interstitials.

9.9   DECHANNELING AND DIRECT SCATTERING

The simplest case for channeling analysis occurs when only defects contribute to dechanneling, and there are no contributions from particles directly scattered from the defects. Under these conditions the defect scattering factor $f \approx 0$. For this case, the dechanneling fraction in Equation (9.27) determines the channeling yield (see top of Figure 9.13) and

$$(9.30) \qquad \chi_D(z) = \chi_R(z)$$

For small defect concentrations (Equation 9.29), the total disorder $N_D(t)$ between the surface and depth $z = t$ is

$$(9.31) \qquad N_D(t) = \frac{1}{\sigma_D} \left[\frac{\chi_D(t) - \chi_V(t)}{1 - \chi_V(t)} \right]$$

For small χ_V,

$$\chi_{min}(t) \approx \chi_D(t) \approx N_D(t)\sigma_D(t)$$

The more general case for channeling analysis of disordered profiles is for $f \neq 0$. Both dechanneling and direct scattering contribute to the measured aligned yield, as shown at the bottom of Figure 9.13. The objective is to determine the dechanneling fraction, the dashed line χ_R in the bottom of Figure 9.13, and subtract it from the total aligned yield to obtain the direct scattering contribution. Both terms in Equation (9.27) must be used, and in general there is no simple closed-form solution. An iterative approach developed by Shao and Nastasi (2005) is shown as a channeling example in Chapter 11.

PROBLEMS

9.1.   What is the Thomas–Fermi screening distance for He ions incident on Si? Compare it with the Bohr radius.

9.2.   What is the atomic potential at r = 0.02 nm for He ions along the <110> rows of Si? (d = 0.384 nm).

9.3.   (a) What is the characteristic angle for axial channeling angle for 1.0 MeV He incident on Si <110>, (d = 0.384 nm)? (b) What is the critical angle?

9.4.   Cu is an fcc metal with lattice constant of 0.3615 nm. Calculate the minimum yields and characteristic angles of the axial and planar channeling for 2 MeV He ions incident along the 100 axial and planar directions with a thermal vibration amplitude $\rho = 0.012$ nm.

9.5.   What is the wavelength for 1 MeV electrons and 1 MeV He ions?

9.6.   What is σ_D for point defects in Si along the <100> direction for 2 MeV H, He, and Li ions? How does this compare to the area of the crystal channels?

REFERENCES

Feldman, L. C., Mayer, J. W., and Picraux, S. T. 1982. *Materials analysis by ion channeling.* New York: Academic Press.

Lindhard, J. 1965. Comprehensive theory of channeling. *Matematisk–Fysiske Meddelelser Konglige Danske Videnskabernes Selskab* 34(14).

Shao, L., and Nastasi, M. 2005. Methods for the accurate analysis of channeling Rutherford backscattering spectrometry. *Applied Physics Letters* 87:64103

Shao, L., Wang, Y. Q., Nastasi, M., and Mayer, J. W. 2006. Measurements of the stopping powers of He ions incident along the different channel axes and channel planes of Si. *Nuclear Instruments and Methods in Physics Research B* 249:51.

SUGGESTED READING

Alford, T. L., Feldman, L. C., and Mayer, J. W. 2007. *Fundamentals of nanoscale film analyses.* New York: Springer.

Chu, W.-K., Mayer, J. W., and Nicolet, M.-A. 1978. *Backscattering spectrometry.* New York: Academic Press.

Feldman, L. C., Mayer, J. W., and Picraux, S. T. 1982. *Materials analysis by ion channeling.* New York: Academic Press.

Wang, Y., and Nastasi, M. eds. 2009. In *Handbook of modern ion beam materials analysis,* 2nd ed., chap. 12. Warrendale, PA: Materials Research Society.

Applications

Thin Film Depth Profiling

10

Chris Jeynes and Richard L. Thompson

10.1 INTRODUCTION

This chapter will explore some uses of IBA to determine depth profiles in *real materials*, mainly using RBS but also involving EBS (including resonant EBS) and PIXE in some examples.[*] In this chapter the examples are all of analyses that can be carried out using a 1.7 MV tandem accelerator (for example), so that there is no discussion of MEIS or LEIS examples. We have also deliberately not included any examples of ERD since heavy ion ERD uses (typically) much larger accelerators and completely different instrumentation, and the light ion (helium) ERD frequently carried out on these small accelerators for hydrogen profiling is interpreted in a way very similar to RBS (or EBS). We also do not discuss analyses carried out on sample cross sections (as in, for example, Jenneson et al. 1998 or Riggs et al. 1999) since this is the trivial case of using a sequence of (essentially) bulk analyses to build up a depth profile.

We aim to help the reader to an appreciation of the great power of IBA methods, especially when they are

---

[*] Glossary: ***IBA*** = "ion beam analysis"; ***BS*** = "backscattering spectrometry"; ***RBS*** = "Rutherford BS"; ***EBS*** = "elastic (non-Rutherford) BS"; ***ERD*** = "elastic recoil detection"; ***NRP*** = "nuclear reaction profiling"; ***PIXE*** = "particle-induced X-ray emisssion"; ***MEIS (or LEIS)*** = "medium (or low energy ion scattering"; ***differential PIXE*** = "spectra taken with different geometries or beam energies or species to make depth information explicit"; ***XRF*** = "X-ray fluorescence"; ***IBM*** or ***Cornell*** geometries = detector in (respecttively) horizontal or vertical planes (neither IBM in Yorktown Heights nor Cornell now have IBA laboratories.

combined (as in the later examples). The attentive readers will find examples representing a very wide range of problems in materials science, including simple elemental diffusion, reaction kinetics, intermixing of multilayers, corrosion, etc. An interdisciplinary example of materials science and geology is presented in the end to show the real power of the so-called "Total-IBA" (Jeynes et al. 2012) with three-dimensional (3D) mapping capability. Readers will find examples where extraordinarily detailed information can be extracted from quite complicated samples; they will also occasionally find examples where the solution given by IBA is known to be wrong, pointing to shortcomings in the models (or the databases) used by the IBA codes. We hope that analysts will increasingly publish model analyses in the technical journals and thereby encourage our community to improve its ability to handle correctly and in increased detail sample types, which are more complex in general.

10.2   EXAMPLES OF OBTAINING ELEMENTAL AND MOLECULAR DIFFUSION PROFILES

10.2.1   FICKIAN DIFFUSION

While there are many methods available to determine the local motion of molecules in solid materials, ion beam analysis is perhaps uniquely well suited to measure the center of mass diffusion over longer ranges. Characterizing long-range diffusion is important since it enables us to understand processes such as drug release and delivery (Smith, Massingham, & Clough 2002), barrier properties (Clough et al. 2006), and absorption (Gall, Lasky, & Kramer 1990).

The best-known example of diffusion is Fickian diffusion, in which the flux material across an interface is proportional to the concentration gradient. By considering the flux of material in and out of an infinitesimal slice, it is clear that the rate of change of composition within that slice is given by the second derivative of the concentration with respect to distance. The constant of proportionality, D, is a diffusion coefficient defined by Fick's second law:

(10.1)
$$\frac{\partial c}{\partial t} = D \frac{\partial^2 c}{\partial x^2}$$

where c is concentration, t is time, and x is depth. Depending on the boundary conditions and the response of the material to the diffusing species, various solutions to Equation (10.1) are possible. When the penetrating species comes from a relatively mobile bulk phase, such as a liquid or vapor in contact with a solid host material, diffusion into the solid has no effect on the diffusant concentration at the surface. If the diffusion coefficient in the host material is independent of the concentration, then D is given by

(10.2)
$$c(x,t) = c(0,0)\,\text{erfc}\left(\frac{x}{2\sqrt{Dt}}\right)$$

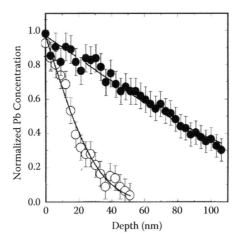

FIGURE 10.1   RBS concentration profiles of Pb in-diffusing into a synthetic mon-azite (closed circles) and natural monazite (open circles), with fitted erfc function curves (lines). (Reproduced from Figure 1a of Cherniak et al. 2004.)

This method was used by Cherniak et al. (2004) to characterize the diffusion coefficient of Pb into $CePO_4$ monazites (see Figure 10.1). Good agreement between the experimental profiles and Fickian diffusion was confirmed by inverting the data through the error function; that is, a plot of $erf^{-1}(\frac{c(0,0)-c(x,t)}{c(0,0)})$ versus x does in this case successfully yield a straight line plot of slope $\frac{1}{2\sqrt{Dt}}$.

10.2.2   COMPLEXITIES: LABELING

In reality, the diffusion process is often more complex. The diffusing species may have a profound effect on the properties of the material into which it is diffusing and, consequently, the process cannot be described by a single simple diffusion coefficient. The ability of ion beam analysis to capture an entire concentration profile in a single measurement makes these techniques ideal for studying one-dimensional diffusion problems. When diffusion occurs over ranges that exceed the conveniently accessible range of ion beams (a few micrometers) it is equally possible to quantify diffusion by scanning a microfocused ion beam across the fracture surface (Riggs et al. 1999). Provided that it can be frozen or vitrified for the duration of an experiment, any material can be studied in this way. Gall et al. (1990) pioneered the use of ion beam analysis to study diffusion of liquids into glassy polymers using liquid nitrogen to trap the diffusing species for the duration of the measurements. In this kind of experiment, the chemistry of the diffusants is usually well defined—for example a series of iodoalkanes, so this information can be applied to infer molecular diffusion via the backscattering from just the constituent iodine atoms. These were key experiments in testing the case II diffusion theory put forward by Thomas and Windle (1982) some years earlier. Prior to ion beam analysis experiments this model was based on observations of mass change with time, but it was not possible to observe the initial concentration profile of the penetrant.

When diffusion occurs from a thin surface layer into the host material, the total volume of the diffusing species must be independent of time, and the concentration profile is given by

$$(10.3) \qquad c(x,t) = \frac{c(0,0)}{2}\left[\mathrm{erf}\left(\frac{h-x}{2\sqrt{Dt}} \right) + \mathrm{erf}\left(\frac{h+x}{2\sqrt{Dt}} \right) \right]$$

where h is the thickness of the surface layer. This functional form has been widely used to study rates and mechanisms of diffusion in polymers. There is in fact frequently sufficient detail within the ion beam analysis data to reveal that this functional form is not completely appropriate and therefore IBA has successfully been employed to challenge and improve our understanding of diffusion processes (Shearmur et al. 1998).

Rates of diffusion are acutely sensitive to the thermodynamics of mixing but very hard to predict *a priori* without quantitative experimental data. The term "thermodynamic slowing" has been used to describe the dramatic decrease in diffusion coefficient that may result from small positive enthalpy of mixing between two species (Geoghegan et al. 1999; Composto et al. 1986). This problem applies particularly to interdiffusion, where the concentration of the diffusant remains significant even after diffusion has commenced.

If, on the other hand, the surface layer of the diffusing species is initially small, then it effectively becomes diluted as soon as diffusion begins, and under these circumstances the tracer diffusion in this component is independent of the thermodynamics of mixing. Deuterium labeling has commonly been applied to determine the diffusion of one organic material into another (Riggs et al. 1999; Geoghegan et al. 1999; Composto et al. 1986; Thompson et al. 2005). Here, some or all of the hydrogen atoms of one component in the mixture are substituted for deuterium, which has minimal impact on the chemical and physical properties, but enables direct determination by elastic recoil detection and nuclear reaction analysis. However, even this most innocuous of labeling systems can subtly change the thermodynamics of mixing in a manner that can be detected by ion beam analysis studies of diffusion coefficients. One rather ingenious solution to this problem was to label the diffusing component selectively with bromine only after diffusion was complete (Composto & Kramer 1991).

What is perhaps most remarkable about the ion beam analysis study of diffusion is that it applies not only to simple systems but also equally to arbitrarily complex materials. Our ability to label just a single component within any material makes this technique perfect for studying anything from nanocomposites to biomaterials.

10.2.3   NUCLEAR WASTE

Moncoffre et al. (1998) from the Lyons group have made an interesting review of Fickian diffusion in various types of materials (including calculational details, which is why we mention it), including (1) obtaining (by 1.5 MeV He-RBS) the diffusion coefficients of La diffusing in hydroxy-apatite, (2) obtaining (by NRP using the $^{15}N(p, ag)12^C$ reaction near the 429 keV resonance) the coefficients of nitrogen diffusion in aluminum, and (3) obtaining (by 2.5 MeV He-ERD with an E **X** B filter instead of an absorber foil for the primary scattered beam) the coefficients of H diffusion in SiC:H under D-implantation. The first example is relevant to the immobilization of radioactive waste, the second to the growth properties of an important refractory material aluminum nitride (AlN), and the third to investigating fusion vessel (tokomak) wall materials.

Alonso et al. (2009) are also interested in the problem of nuclear waste repositories—this time from the geological point of view. Measurement of diffusion coefficients by "classical" methods usually involves waiting for the diffusion to give measurable concentrations over length scales of centimeters. But this would involve unfeasibly long experiments for the geological timescales appropriate to these applications. Therefore, they turned to nuclear methods (RBS) to measure diffusion in clays over submicrometer length scales. Their paper is very explicit (helpfully to us!) and we will consider it in some detail. Figure 10.2(a) shows the RBS spectrum from the clay in Alonso and colleagues' study, and Table 10.1 shows the analysis they made of this spectrum. This table

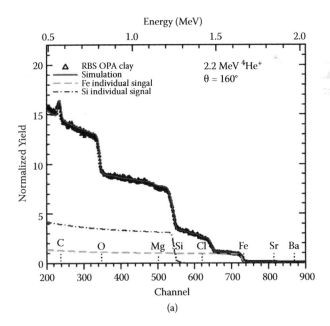

FIGURE 10.2(a) RBS spectrum from a clay mineral with composition given in Table 10.1. (Reproduced from Figure 2 of Alonso et al. 2009.)

TABLE 10.1 Composition of a Clay Mineral: Chemical Analysis Compared to RBS[a]

Element	Composition in at. mass%[b]	RBS
C	4.6	4.4
O	67	65
Na	5.2×10^{-1}	6.1×10^{-1}
Mg	1.1	2.1
Al	10	7.9
Si	10	16
S	1.6×10^{-1}	1.0×10^{-1}
Cl	1.4×10^{-2}	1.4×10^{-2}
K	1.8	2.2
Ca	1.2	1.7
Ti	3.1×10^{-1}	1.2×10^{-1}
Mn	2.9×10^{-1}	1.2×10^{-2}
Fe	2.6	1.4
Sr	6.0×10^{-3}	1.6×10^{-2}
Ba	4.0×10^{-3}	1.8×10^{-2}

[a] Alonso, U. et al. 2009. Table 4. *Applied Clay Science* 43:477–484.
[b] Fernandez. 2007.

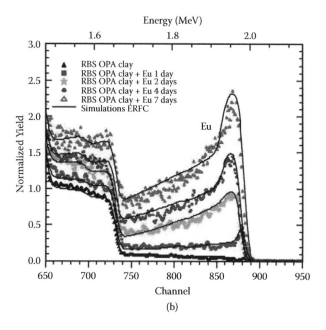

FIGURE 10.2(b) RBS spectra from a clay mineral including diffusion profiles. (Reproduced from Figure 4b of Alonso et al. 2009.)

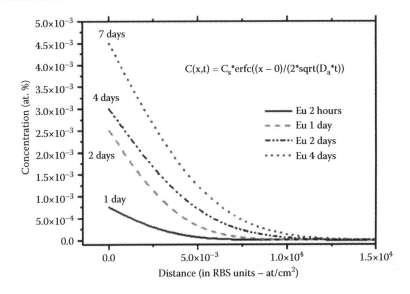

$$C(x,t) = C_s * erfc((x - 0)/(2*sqrt(D_a*t)))$$

legend:
——— Eu 2 hours
-- -- -- Eu 1 day
—·--·-· Eu 2 days
· · · · · Eu 4 days

Distance (in RBS units – at/cm²)

FIGURE 10.3 Diffusion constants extracted from RBS spectra. (Reproduced from Figure 5 of Alonso et al. 2009.)

is an example of overinterpretation of data (see Problem 10.1). Figure 10.2(b) shows one of the diffusion profiles studied by these authors, and Figure 10.3 shows the error function profiles fitted to the RBS data. The diffusion equation constants are obtained directly from these fits (see Table 5 of their paper) since the code used to interpret these data supports fitting with **erfc** (the complementary error function). These data are considered more closely in Problem 10.1.

10.2.4   DEFLUORINATION: AMBIGUITY

Fickian diffusion processes are also quantified by Ross et al. (2001) in an RBS study of defluorination mechanisms in poly(vinylidene) fluoride when treated with alkalis in the presence of phase transfer catalysts (see Figure 10.4). This work depends on the interpretation of the spectra in terms of *molecular* depth profiles (although this is not explicitly stated). The chemical state of the elements is known because the defluorination process has been determined by other work (including XPS [X-ray photoelectron spectroscopy]), and this chemical knowledge is used to relieve the ambiguity of these spectra. For a further, and thorough, discussion of ambiguity, see Jeynes et al. (2003).

10.2.5   REAL TIME MEASUREMENTS

Real-time processes can be followed by IBA since the analysis time can be very short. An early example of this is an *in situ* study of the photodissolution of silver in an amorphous germanium selenide by Rennie, Elliott, & Jeynes (1986) (see

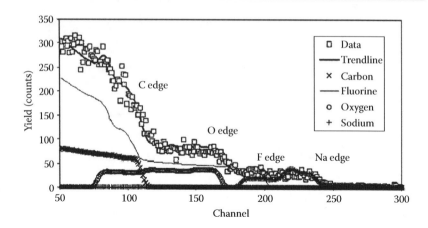

FIGURE 10.4 RBS of poly(vinylidene fluoride) exposed to alkali treatment in the presence of catalysts. Defluorination followed by oxidation reactions occur. (From Figure 4 of Ross et al. 2001.)

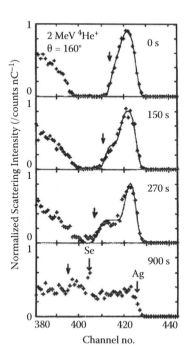

FIGURE 10.5 RBS of a 38 nm Ag film on a-GeSe$_2$ exposed *in situ* to 550 nm illumination at 4 mW/cm^2 for various times. (From Figure 1 of Rennie et al. 1986.)

Figure 10.5). This work shows both the kinetics of the process and the way the mechanism switches when the thin continuous surface Ag film breaks up into islands before dissolving completely. The analysis is tricky because the RBS itself causes dissolution of the Ag, and the total beam fluence must be restricted to 250 nC/mm², with a fresh analysis spot used after every illumination. The spectra are then simply fitted with rectangular profiles, the low counting statistics in any case precluding any more detailed treatment. In Figure 10.5 the position of the surface signals of Ag and Se are shown, as is the position of the depth of the reaction interface at each time step. For short illumination times, the figure clearly shows the thickness of the surface Ag film reducing as the silver dissolves into the selenide. For long illumination times, the reaction is complete. What is not shown are the intermediate times where the surface Ag signal keeps the same width but reduces its height, indicating an islanding reaction.

Reaction kinetics are best followed in true real time, as has been done in a series of recent reports. We will discuss the latest of these by Demeulemeester et al. (2010) (see Figures 10.6–10.8), who were investigating the kinetics of the formation of nickel silicides. What is of interest in such studies is the detection of the onset of the formation of phases, in this case the mono- and disilicides. Conventionally, a series of anneals on different samples would be made, and then off-line analysis (RBS and other techniques) carried out to determine what had happened. The difficulty is that many samples are needed to determine annealing kinetics and many more to determine precisely the position of the phase boundaries. This group has solved this first problem by collecting RBS spectra in real time (every 30 seconds) during the ramp annealing. The question then is what to do with the voluminous data collected. Figure 10.6 shows a composite of all

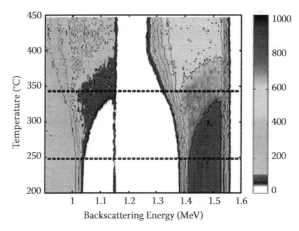

FIGURE 10.6   Composite of 250 RBS spectra taken during a Ni silicidation reaction of an 80 nm film on silicon, capped with 7 nm of Si and annealed at 2°C/min. The gray scale on the right shows spectral intensity in arbitrary units. The dashed lines indicate the initiation of the mono- and disilicides. (From Figure 2 of Demeulemeester et al. 2010.)

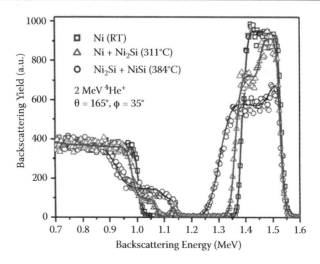

FIGURE 10.7   Three spectra from Figure 10.6 with solution by ANN. (From Figure 4 of Demeulemeester et al. 2010.)

these data, which already looks suggestive. But of course, a quantitative analysis is desired. So these researchers *trained* an artificial neural network (ANN) to recognize each spectrum! We should point out that this is considerably easier said than done, since a training set of 18,000 spectra was required. This training set covered all possible variations of the spectra in this materials system, including reasonable

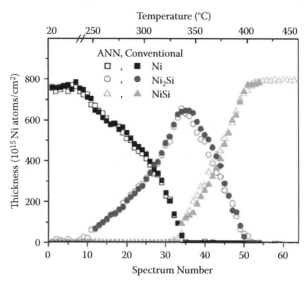

FIGURE 10.8   Solutions of spectra from Figure 10.6 over the whole annealing range, showing proportions of different phases, obtained both conventionally and by ANNs. (From Figure 3 of Demeulemeester et al. 2010.)

roughness of the layers. (We will discuss roughness in more detail later.) The quality of the solutions obtained by the ANN can be seen in Figure 10.7, and the solution of the whole data set is shown in Figure 10.8 in terms of the proportion of each phase present as a function (effectively) of anneal temperature.

10.2.6   ARTIFICIAL NEURAL NETWORKS

We pause here for a digression on ANNs. If an artificial neural network can be trained to analyze RBS (and indeed, IBA data) with such precision, why do we need analysts at all? Surely we can now do "IBA without humans" (Barradas, Vieira, & Patricio 2002a)? ANNs give the (correct) answer, completely automatically, and in effectively zero time. But an ANN is the ultimate black box. We do not (and probably *cannot*) know what exactly an ANN is doing. Therefore there are the dual problems of validation and of traceability (not to mention the problem of training the ANN in the first place). We have to be able to check the correctness of the answer, and we have ultimately to be able to say what confidence we have in it.

Fortunately, these are both very easy to supply: All we need to do is to give the ANN's solution back to a regular IBA code for refinement, and then we are back on solid ground since we know exactly how the regular codes work. This is what these workers did. Figure 10.8 shows the raw ANN result (for the whole data set) directly compared with the result refined by the regular IBA code; the two are very similar. Thus, the ANN results can be treated as a "first-and-rather-good guess" to help the regular code with the solution. As a matter of fact, in this example the ANN could not precisely locate the phase boundaries, but what it could easily do is to determine the reaction kinetics (the activation energies and preexponential factors). And, of course, these data are available in real time, while the annealing is in progress.

Artificial neural networks are actually rather difficult to set up, and even when they are trained they are tricky to use. One always has to check that they are not giving nonsense. In principle, one can train an (independent) ANN to recognize whether the input to an ANN is valid—that is, whether it is covered by the training set (Barradas, Vieira, & Patricio 2002b). It is important to understand that an ANN can only *interpolate* within its training set; it *cannot* extrapolate. But this is a significant example because it shows that handling industrial quantities of data is entirely feasible and, moreover, that it can be done at state-of-the-art accuracy if necessary. If we wanted to, we could extend this sort of treatment to wider and wider classes, dramatically reducing the tedium of the routine analyses required from every IBA lab.

10.3   EXAMPLE OF AN OPTICAL MULTILAYER

The next example highlights something previously only touched on: the usefulness of chemical information we may have about a sample. We also elaborate on our knowledge of the reliability of our results. Note that roughness is also important in the following examples.

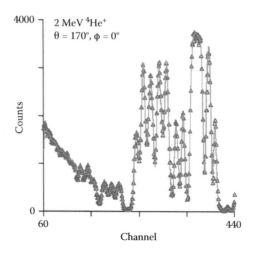

FIGURE 10.9   Antireflection coating with alternate zirconia and silica layers on float glass. Normal incidence beam: The fit is shown as a line. (From Figure 2 of Jeynes et al. 2000.)

Figure 10.9 shows a complex spectrum from a complex sample (see Jeynes et al. 2000). It is an antireflection coating on glass—that is, alternating zirconia and silica (ZrO_2 and SiO_2) layers on float glass: These oxides have strongly differing refractive indices giving the optical effect. The glass has many elements, and the zirconia naturally has a small amount of hafnium. This analysis cannot be made unless the problem is simplified to three *logical elements,* as opposed to the dozen or so real elements present. Again, we have the problem of *ambiguity* (discussed at length in Jeynes et al., 2003, and touched on in our discussion of Figure 10.4); that is, the information we would like to access is *not* unequivocally present in the data. One spectrum cannot usually determine the sample structure. Figure 10.9 is actually one spectrum from a pair: It was taken at normal incidence and there was another taken at 45° incidence (not shown here). The two spectra look entirely different, and the two together determine the sample very much better than just one spectrum. But analysts must beware a gaping pitfall at their feet: The fact that the fit to the data is perfect is not a reason to think that the sample is perfectly known. The spectrum is *ambiguous.* A perfect fit means that the proposed profile is *valid*—not that it is *true.*

Even with two spectra, the structure of the sample is completely ambiguous—until, that is, chemical information is introduced. Butler (1990) pointed out almost 25 years ago that our knowledge of the chemistry of the sample can always supplement our objective interpretation of IBA spectra obtained from the sample. In this case, of course, we are certain that the Zr and Si are fully oxidized because the optical properties tell us that unequivocally. And we are also certain of the glass composition; centuries of glass technology assure us of that. Therefore, we confidently use molecules as the *logical elements* of the analysis and note that in our discussion of Figures 10.6–10.8 we already silently introduced molecules (the two phases of the silicide) to effect the analysis.

The data in Figure 10.9 are actually analyzed by the authors in an interesting way. They use *simulated annealing,* a mathematical global minimization algorithm described elegantly in a *Science* article (Kirkpatrick, Gelatt, & Vecchi 1983). Simulated annealing was introduced into physics in 1953 by Metropolis et al. and into IBA analysis in 1997 by Barradas, Jeynes, and Webb; also see the careful discussion in Jeynes et al. (2003). Simulated annealing is powerful in IBA since the user simply gives the data to the computer, *without* any initial guess of the depth profile, which is extracted completely automatically. The method uses *Markov chain Monte Carlo* (MCMC) mathematics; this is described in the cited literature and can be summarized for this application as *a systematic way of searching the space of possible solutions.* And it is because the search is systematic that we can do statistics on it, using a Bayesian inference method (Barradas et al. 1999). This can be seen in Table 10.2, which shows extraordinarily precise layer thicknesses extracted from the RBS data down to depths of over a micrometer (and see Problem 10.2 for further meditations on the table).

10.4   EXAMPLE OF INTERMIXING ANALYSIS

Figure 10.10 shows an interesting example of Fe/Si multilayer samples with and without ion-beam-induced intermixing, and Problem 10.3 explores this further. The idea is to try to find some feasible way to make amorphous iron disilicide, which is a semiconductor that could become an important thin film solar photovoltaic material if technological problems can be overcome (Leong et al. 1997). Much ion beam analysis is aimed at elucidating the behavior of materials in a variety of processes to enable some approaches to be ruled out and others to be developed effectively. This example is included as "one that didn't work," since materials scientists usually need to understand these cases at least as well as the successful ones!

We intended to lay down pure Fe and Si layers (by magnetron sputtering) of the correct thicknesses and then intermix these by ion bombardment at elevated temperatures. The nuclear displacements during ion implantation ("ballistic mixing") will stir up (mix) the sample to an extent that can be calculated by the standard Monte Carlo codes (stopping and range of ions in matter [SRIM]; see www.srim.org). But *extra* radiation-enhanced diffusion is also expected—that is, diffusion in *excess* of the equilibrium diffusion expected at the implantation temperature. Ion implantation is a strongly nonequilibrium process, and radiation-enhanced diffusion is a consequence of this. Indeed, strong intermixing is observed. However, the analysis showed that the presence of Ar (from the sputter deposition) poisons the development of the silicide.

The coating includes layers of only a few nanometers thick, which is challenging for standard RBS. Fe and Si are sequentially sputter-deposited, but although the closely packed metal does not incorporate significant Ar, the more open Si layers

TABLE 10.2 Solution of Antireflection Coating with Uncertainties from a Bayesian Inference Analysis[a]

Layer no.	Material Air	ZrO$_2$ Content %	Last Figure Error	Measured Equivalent Thickness for Simulated Annealing 10^{15} Atoms cm^{-2}	nm	Measured Equivalent Thickness for MCMC 10^{15} Atoms cm^{-2}	nm	Error (nm)
0	Au	35.3	5	24.3	4.1	35.2	5.96	0.02
1	ZrO$_2$	91.5	13	154.0	18.8	141.2	17.2	0.6
2	SiO$_2$	2.7	2	436.2	72.7	411.9	68.6	0.8
3	ZrO$_2$	94.0	11	357.7	43.6	384.6	46.8	1.0
4	SiO$_2$	4.0	5	401.3	66.9	394.4	65.7	1.4
5	ZrO$_2$	95.0	6	1045.2	127.3	1044.4	127.2	1.7
6	SiO$_2$	1.3	4	706.6	117.8	689.8	115.0	2.4
7	ZrO$_2$	73.9	43	169.5	20.6	182.8	22.3	1.8
8	SiO$_2$	3.5	7	533.6	88.9	533.6	88.9	2.5
9	ZrO$_2$	53.1	25	169.2	20.6	168.1	20.5	1.2
10	SiO$_2$	3.0	5	656.0	109.3	688.2	114.7	1.7
11	ZrO$_2$	87.7	23	321.8	39.2	330.0	40.2	2.0
12	SiO$_2$	18.8	12	370.8	61.8	329.1	54.9	2.7
13	ZrO$_2$	80.1	10	319.5	38.9	327.2	39.8	1.5
14	SiO$_2$	7.5	13	428.7	71.4	442.6	73.8	2.0
15	ZrO$_2$	76.0	18	269.9	32.9	299.2	36.4	1.8
16	SiO$_2$	13.0	12	408.6	68.1	389.2	64.9	2.5
17	ZrO$_2$	74.7	13	275.7	33.6	246.5	30.0	1.3
18	SiO$_2$	17.0	8	525.6	87.6	531.8	88.6	1.8
19	ZrO$_2$	26.0	6	106.0	12.9	118.5	14.4	0.5
	Float glass substrate			Total:	1137		1136	

Source: Jeynes, C. et al. 2000. Table 2. *Surface Interface Analysis* 30:237–242.

Notes: Density assumed: ZrO$_2$ = 5.6 g/cm^3 (8.21 × 10^{22} atoms/cm^3). The simulated annealing calculation was the starting point for the MCMC calculation.

[a] See Figure 10.9 and text.

incorporate quite a lot. During the development of the silicide the Ar is swept out to the remaining unreacted Si, preventing the completion of the process.

In this analysis the RBS, of course, objectively sees only the Fe and Si (and Ar) elemental profiles, but it is valuable to be able to interpret these elemental profiles as profiles of the phases present, as we have already done for the data in Figures 10.4 and 10.6–10.9. Note that the only minor element present, with the exception of a little surface oxide, is Ar, which is quite well determined in

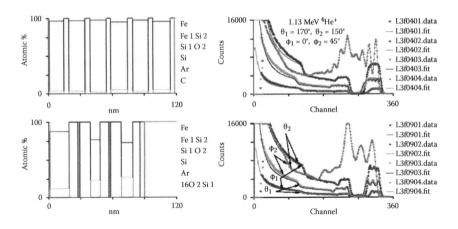

FIGURE 10.10   Above: 12-layer Fe-Si multilayer on quartz as deposited. Below: six-layer Fe-Si multilayer, implanted with 200 keV 5.15^{15} Fe/cm². Left: depth profiles derived from RBS data. Right: collected data for two detectors and two beam incidence angles. For tilted incidence, the exit angle for detector 1 (Cornell geometry) is about 45° but for detector 2 (IBM geometry) it is 75°.

this analysis. But the high quality of the fits to the data seen in Figure 10.10 is only obtained when the straggling is correctly calculated, firstly, by the detailed physics incorporated in Szilágyi's DEPTH code (Szilágyi, Pászti, & Amsel 1995). (This code or its equivalent is incorporated in both the "new generation" codes, SIMNRA and DataFurnace, see Barradas et al. 2007) and, secondly, by taking account of the extra straggling generated by "roughness"—in this case layer thickness inhomogeneity (which the "new generation" codes also incorporate). Note that all four spectra (two separate detectors, two different beam incidence angles) are equally well fitted, showing that these profiles are well determined.

10.5   EXAMPLE OF CORROSION ANALYSIS

Pascual-Izarra et al. (2007) have given a first and powerful demonstration of the simultaneous and self-consistent use of particle scattering and photon emission. They analyzed (for conservation purposes) the earliest photograph in the world, an important cultural artefact from the Louvre Museum, one of the earliest photographs in the world (see Figure 10.11). It was corroding, and precise details were required. The external scanning microbeam of the AGLAE accelerator sited at the Louvre was used to depth profile selected areas of the photograph nondestructively, and the results are shown in Figure 10.12. This figure shows RBS, EBS and PIXE data, and we emphasize that the sample cannot be elucidated by any one of these spectra handled separately; the depth profile also shown in the figure is obtained only when all the data are analyzed self-consistently.

Niépce's first Heliography:
Paysage à Saint-Loup de Varennes (1827)

FIGURE 10.11 Specimen from the collection of the Louvre Museum (Paris).

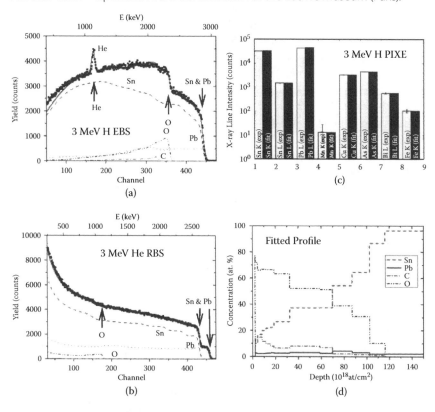

FIGURE 10.12 RBS, EBS, and PIXE data from a corroded region of Niépce's helio-graph (see Figure 10.11) together with the fitted profile extracted from the data using DataFurnace (NDF). PIXE data are preprocessed by GUPIX to obtain charac-teristic line areas. (From Pascual-Izzara et al. 2007.)

First (not shown), an uncorroded region was analyzed to obtain the bulk composition (homogenous in depth). The authors demonstrated that the PIXE and RBS/EBS results agreed. In this case, of course, either the particle spectra or the PIXE spectrum can separately yield the bulk composition. In this case the PIXE detector must be properly calibrated, the RBS gives the Pb/Sn composition at the surface, and the EBS demonstrates for the analysis depth (over 15 μm) that the composition is uniform. Differential PIXE (whose information depth exceeds 30 μm) can be used to show the composition is uniform to greater depth. XRF can go even deeper. Note that, in this case, although the Pb K lines could not be excited, both the Sn K and L lines were used, which itself is a form of differential PIXE.

For the corroded sample, however, the modified surface layer thickness is not great enough to use differential PIXE easily; in any case, GUPIX (the widely used Guelph PIXE program; Campbell & Maxwell 2010) does not currently support this sort of depth profiling. But in this case the EBS immediately shows that the sample is oxidized, with an oxidation front penetrating the sample irregularly to a depth of around 10^{20} atoms/cm^2 (what linear thickness this represents depends on the oxidation states present). The trouble is that the EBS cannot tell which component is oxidizing, the RBS can only distinguish Pb and Sn at the surface, and the PIXE has no depth resolution. The spectra, separately, are entirely ambiguous, but when all the data are put together, the profile is unambiguous. The clean sample determines the substrate (underlying) composition (a boundary condition), the RBS determines the Pb:Sn ratio at the surface (the other boundary condition), the EBS determines the O profile (strongly non-Rutherford, see Gurbich 2010), and the PIXE determines the relative Pb:Sn in the profile calculated self-consistently. Of course, the PIXE also determines the total quantity of the minor and trace elements, which can also be profiled in the same way. In this way some oxidation models can be demonstrated to be consistent with the data and other models to be inconsistent.

We point out here some interesting features of these data. An external beam was used; that is, the proton beam was brought out to air to analyze these valuable objects, which, of course, cannot be cut to fit inside a vacuum chamber and, in any case, are usually considered too delicate to be subjected to a vacuum system. Megaelectron volt charged particle beams are very penetrating and have useful working distances of centimeters. To increase working distances and reduce effects both of X-ray absorption (giving more sensitivity for the lower X-ray energy lines) and of backscattered particle energy loss, a He flush around the beam paths is used. This He can be seen in the EBS spectrum. Note that the He signal is fitted very well, showing that the particle detector is aligned and collimated correctly. If He is not used, then Ar (present at about 1% in air) will be seen in the PIXE spectrum and O + N in the BS spectra.

Note that, at these energy proton beams, all elements to at least Fe have non-Rutherford scattering cross sections, and note also that evaluated cross sections (that is, known very well for any geometry) are now available for most of the important elements (Gurbich 2010). Clearly, to use EBS the scattering cross sections must be known, but at present the RBS–EBS boundary is not well known, even in principle. The MRS handbooks (Tesmer & Nastasi 1995; Wang & Nastasi 2010) have chapters on the Coulomb barrier that are potentially extremely misleading since the actual "Coulomb barrier" is not a well defined idea (Gurbich 2004). But at this energy, the backscattering spectrometry (BS) signals for the heavy elements Sn and Pb are Rutherford. Note that, with a 3038 keV ^{4}He beam, there is a strong EBS resonance for surface O, so that if a slightly increased beam energy had been used, the surface O would have been visible even for the "RBS" spectrum.

In Figure 10.12, clearly the 3 MeV RBS/EBS and the 3 MeV PIXE spectra were collected simultaneously using the proton beam. For 3 MeV He the PIXE spectrum (available as well as the BS spectrum) is not shown for clarity, but it is worth pointing out that this is another way to do differential PIXE since the information depth for 3 MeV He-PIXE is much less than for 3 MeV H-PIXE.

10.6   EXAMPLE OF 3D ANALYSIS FOR GEOLOGY

Our last example is of the so-called "Darwin glasses" (see Bailey et al. 2009; Howard et al. 2013), which are impact glasses resulting from a meteor strike 800,000 years ago near Mt. Darwin in Tasmania. The geologist studying these subsequently used one of his Darwin glass samples as an amorphous standard for setting up his X-ray diffraction (XRD) kit and was astonished to see the diffraction spots of quartz. These crystals—unexpected in a glass—turned out to be inside *inclusions* in the glass. But the nature of these inclusions was entirely unknown.

In Figure 10.13, μPIXE shows that the samples are very heterogeneous, and EBS analysis unequivocally demonstrated them to be carbonaceous from the ^{12}C(p,p)^{12}C resonance at 1734 keV (Figure 10.14). This result initially baffled the geologists, for whom such a sample was unprecedented. Figure 10.15 shows that the IBA data could be completely and quantitatively analyzed without any presupposed model despite the heterogeneity.

Microbeam PIXE/BS demonstrated the great heterogeneity of the samples both laterally and in depth; this sort of mapping microbeam data effectively gives two 3D data cubes, with (say) 128 × 128 pixels and a PIXE and BS spectrum pair for each pixel. The data cube pair, intractable as they stand, can be analyzed into principal components by using some multivariate image analysis program; then each pair of BS/PIXE spectra can be analyzed to give an explicit depth profile, from which the 3D representation of the sample could be reconstructed.

What is also interesting about these data is that the heterogeneities are radical; that is, there are precipitates of one material (quartz) in another (the carbonaceous matrix). On the face of it, one does not expect to be able, by IBA, to distinguish a material with precipitates from a material with a uniform (average) composition. But Stoquert & Szörenyi (2002) demonstrated that, in

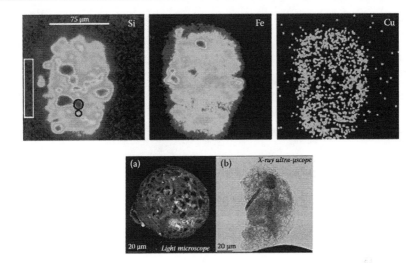

FIGURE 10.13   Above: PIXE maps for Si, Fe, and Cu of Darwin glass set in resin, showing radical inhomogeneity. (Figure 1 of Bailey et al. 2009.) Below: light microscope and X-ray radiographs of typical Darwin glass inclusions (See Howard et al. 2013). (Courtesy of KT Howard.)

fact, the density variation in a material will measurably affect the straggling of the probe ion beam as it penetrates the sample. And we can use sharp, non-Rutherford resonances in the elastic scattering cross-section as markers for the straggle as a function of depth. This was used by Tosaki (2006) to distinguish different forms of carbon. It is the behavior of the EBS resonance for $^{12}C(p,p)^{12}C$ at 1734 keV that allows us to prove the presence not only of carbon but also of SiO_2 precipitates in these inclusions.

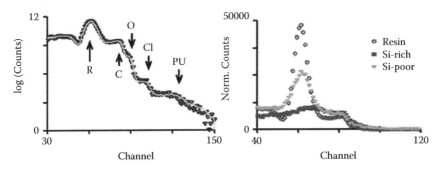

FIGURE 10.14   Left: EBS for resin—see rectangle in Figure 10.13 (Si PIXE). The surface signals for C, O, and Cl are marked, as are the pileup signal and the signal due to the $^{12}C(p,p)^{12}C$ resonance at 1734 keV (note $\log_e$ scale). (Figure 4 of Bailey, M. J. et al. 2009. *Nuclear Instruments and Methods B* 267:2219–2224.) Right: charge-normalized EBS for all marked areas in Figure 10.13. (From Figure 3 of Bailey et al. 2009.)

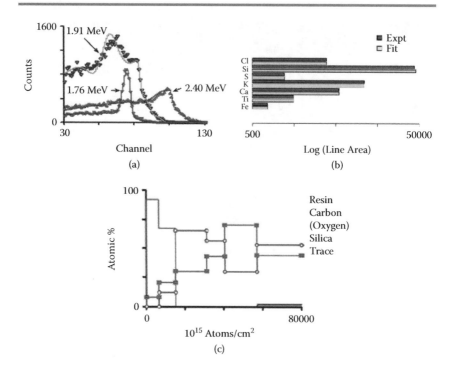

FIGURE 10.15   Analysis of a Si-rich area of a Darwin glass (see region marked in Figure 10.13 and spectrum in Figure 10.14. (From Figure 6 of Bailey et al. 2009.) (a) EBS spectra at various energies; (b) 2.4 MeV PIXE; (c) fitted depth profile consistent with the data.

It is worth elaborating on this example, since the details are rather arcane. Figure 10.14 shows, first, the EBS spectrum from the resin in which the geological sample is embedded in the standard mineralogical procedure. This spectrum is readily interpreted, and the most prominent feature in it is clearly the signal of the 1734 keV resonance in the EBS cross-section for carbon. The spectrum is shown on a log scale, which makes visible the signal of Cl, a minor element. Note also that the pileup signal is not negligible, even for a microbeam analysis, which usually uses a very restricted beam current. The C resonance gives both a greatly enhanced and (in the tail of the resonance) a greatly reduced EBS signal, following the cross-section function. But when the spectra from the inclusion itself are analyzed, the resonance shape seems smeared out. In the Si-rich region we might expect that the reason for this is that there is just not that much C present, but this is clearly not true in the Si-poor region, which still has a very pronounced C resonance signal. But it is the wrong shape!

This behavior is emphasized when the Si-rich region is analyzed in more detail (Figure 10.15). Clearly, C is present as a major element judging from the 1.76 MeV EBS spectrum, where the EBS cross section for surface C is enhanced (although this is best interpreted as some resin present on the surface). But at some depth the C concentration is markedly reduced, as the 1.91 MeV EBS spectrum shows (this is the same spectrum shown in Figure 10.14). But the spectral shape is also very different from what we expect. The correct interpretation of this spectrum is that extra straggling is introduced by the fact that the large amount of Si present is in the form of silica precipitates, giving a strongly inhomogeneous density. The algorithm implemented in the IBA code approximates this extra straggling quite well, even for this extreme case, and, when the information in the PIXE data on the Si content is included, an unequivocal depth profile is obtained. Note that a self-consistent analysis is made of the multienergy, multitechnique data.

This last example points toward tomography. X-ray tomography (XR-T) is already established (see an example in Figure 10.14), and STIM-T is an almost equivalent (and solved) problem (Satoh, Oikawa, & Kamiya 2009). Great strides have also been made toward a PIXE-T (Ryan 2011), which is qualitatively much more complex than either STIM-T or XR-T. We have shown that, in principle, IBA-T (that is, using the BS signals as well as the PIXE signals) is already achievable and should be significantly more efficient (and therefore much faster) than pure PIXE-T since a single slice already has (nearly) complete 3D information. This is important since tomography is rather slow, and its importance is increased since it seems that beam damage limits the use of a pure PIXE-T for important classes of samples.

10.7   CONCLUSIONS

Backscattering analysis (Rutherford or not) is a naturally powerful technique for thin film depth profiling. There are modern IBA codes powerful enough to untangle the most complicated spectra, and these methods have been used effectively throughout thin film materials science. But the power of IBA increases dramatically when PIXE is added to the analyst's toolbox. Note that, for all IBA, PIXE and BS data are always present together, so the data collection itself does not take longer. PIXE data give the analyst an integral view of the sample, with high elemental specificity and sensitivity, almost perfectly complementing the depth-profiling BS, which mixes up mass and depth information. We have shown two examples of an integrated PIXE/BS analysis: the first of a rather simple case, but the second of a highly complex case involving not only elemental inhomogeneity but also partition into discrete phases inside the sample. We showed how even this case could be handled successfully by IBA.

Looking to the future, we have also shown that an IBA tomography that is likely to be more than an order of magnitude faster than a pure PIXE tomography is feasible. Whether or not this is ever realized, it is clear that microbeam IBA is capable of a 3D analysis that is today largely untapped and that shows immense promise.

We conclude that, where backscattering and PIXE individually are powerful and versatile techniques, "Total-IBA" (that is, an integrated treatment of BS/PIXE data: see Jeynes et al. 2012) has a synergy that effectively completely generalizes depth profiling IBA. Twenty-first century thin film depth profiling needs total IBA.

PROBLEMS

10.1.   a.   Considering Figure 10.1, estimate the sensitivity to S.
      b.   Is the nice match between RBS and the chemical analysis for S in Table 10.1 any more than wishful thinking?
      c.   Comment on the rest of Table 10.1. What contribution could PIXE make in this analysis?
      d.   Considering Figure 10.2, (1) explain the high concentration of Eu at the surface. (2) Explain the background signal above channel 750—that is, in the (black) spectrum of the unexposed material. (3) Explain the rise of the clay signal (below channel 730) as the Eu exposure increases. (4) How could the precision of these data be increased?

10.2.   a.   In Table 10.2, the precision of the capping Au layer thickness measurement is about half a percent. What is the absolute accuracy of this measurement? Comment on the 50% discrepancy of the simulated annealing (columns 5, 6) and the MCMC (columns 7, 8, 9) results for this layer.
      b.   In Table 10.2, the precision of the last zirconia layer thickness measurement is about 3%. What is its accuracy?
      c.   In Table 10.2, is it likely that silica contaminates the zirconia and vice versa (see column 3)? Comment.

10.3.   a.   In the example of Figure 10.10, why use a tilted incidence beam? Why use 1.13 MeV?
      b.   In Figure 10.10, why use two detectors? Where does (or could) information about the Ar content come from? Comment on the Ar distribution claimed.
      c.   In Figure 10.10, the depth profiles are expressed (1) in nanometers, and (2) in terms of layers with well-defined interfaces. Is this representation of the result *required* by the IBA data? What assumptions (if any) are implied? Comment.

REFERENCES

Alonso, U., Missana, T., García-Gutiérrez, M., Patelli, A., Siitari-Kauppi, M., and Rigato, V. 2009. Diffusion coefficient measurements in consolidated clay by RBS micro-scale profiling. *Applied Clay Science* 43:477–484.
Bailey, M. J., Howard, K. T., Kirkby, K. J., and Jeynes, C. 2009. Characterization of inhomogeneous inclusions in Darwin glass using ion beam analysis. *Nuclear Instruments and Methods B* 267:2219–2224.

Barradas, N. P., Jeynes, C., Jenkin, M., and Marriott, P. K. 1999. Bayesian error analysis of Rutherford backscattering spectra. *Thin Solid Films* 343–344:31–34.

Barradas, N. P., Jeynes, C., and Webb, R. P. 1997. Simulated annealing analysis of RBS data. *Applied Physics Letters* 71:291.

Barradas, N. P., Vieira, A., and Patricio, R. 2002a. RBS without humans. *Nuclear Instruments and Methods B* 190:231–236.

———. 2002b. Artificial neural networks for automation of Rutherford backscattering spectroscopy experiments and data analysis. *Physics Review E* 65 (6): 066703.

Barradas, N. P., Arstila, K., Battistig, G., Bianconi, M., Dytlewski, N., Jeynes, C., Kótai, E., Lulli, G., Mayer, M., Rauhala, E., Szilágyi, E., Thompson, M. 2007. IAEA intercomparison of IBA software, *Nucl. Instr. Methods B*, 262: 281–303.

Butler, J. W. 1990. Criteria for validity of Rutherford scatter analysis. *Nuclear Instruments and Methods B* 45:160–165.

Campbell, J. L., Boyd, N. I., Grassi, N., Bonnick, P., Maxwell. J.A. 2010. The Guelph PIXE software package IV, Nucl. Instrum. Methods B, 268: 3356–3363.

Cherniak, D. J., Watson, E. B., Grove, M., Harrison, T. M. 2004. Pb diffusion in monazite: A combined RBS/SIMS study. *Geochimica et Cosmochimica Acta* 68:829.

Clough, A. S., Collins, S. A., Gauntlett, F. E., Hodgson, M. R., Jeynes, C., Rihawy, M. S., Todd, A. M., and Thompson, R. L. 2006. *In situ* water permeation measurement using an external $^3He^{2+}$ ion beam. *Journal of Membrane Science* 285:137–143.

Composto, R. J., and Kramer, E. J. 1991. Mutual diffusion studies of polystyrene and poly(xylenyl ether) using Rutherford backscattering spectrometry. *Journal of Materials Science* 26:2815–2822.

Composto, R. J., Mayer, J. W., Kramer, E. J., and White, D. M. 1986. Fast mutual diffusion in polymer blends. *Physics Review Letters* 57:1312–1315.

Demeulemeester, J., Smeets, D., Barradas, N. P., Vieira, A., Comrie, C. M., Temst, K., and Vantomme, A. 2010. Artificial neural networks for instantaneous analysis of real-time RBS spectra. *Nuclear Instruments and Methods B* 268:1676–1681.

Fernandez, A.M., 2007. Physical, chemical and mineralogical characteristics of the Opalinus clay, Callovo-Oxfordian and boom clay minerals. CIEMAT Technical Report P.I.D. 3.2.1B Madrid, Spain.

Gall, T. P., Lasky, R. C., and Kramer, E. J. 1990. Case II diffusion: Effect of solvent molecule size. *Polymer* 31:1491–1499.

Geoghegan, M., Jones, R. A. L., van der Grinten, M. G. D., Clough, A. S. 1999. Interdiffusion in blends of deuterated polystyrene and poly(a-methylstyrene). *Polymer* 40:2323–2329.

Gurbich, A. F. 2004. On the concept of an actual Coulomb barrier. *Nuclear Instruments and Methods B* 217:183.

———. 2010. Evaluated differential cross-sections for IBA. *Nuclear Instruments and Methods B* 268:1703–1710.

Jenneson, P. M., Clough, A. S, Hollands, R., Mulheron, M. J., and Jeynes, C. 1998. Profiling chlorine diffusion into ordinary Portland cement and pulverized fuel ash pastes using scanning MeV proton micro-PIXE. *Journal of Material Science Letters* 17 (14): 1173–1175.

Jeynes, C., Barradas, N. P., Marriott, P. K., Boudreault, G., Jenkin, M., Wendler, E., and Webb, R. P. 2003. Elemental thin film depth profiles by IBA using simulated annealing—A new tool. *Journal of Physics D Applied Physics* 36:R97–R126.

Jeynes, C., Barradas, N. P., Rafla-Yuan, H., Hichwa, B. P., and Close, R. 2000. Accurate depth profiling of complex optical coatings. *Surface Interface Analysis* 30:237–242.

Jeynes, C., Bailey, M. J., Bright, N. J., Christopher, M. E., Grime, G. W., Jones, B. N., Palitsin, V. V., Webb. R. P. 2012. Total IBA – Where Are We?. *Nucl. Instr. and Methods B,* 271: 107-118.

Kieren T. Howard, Melanie J. Bailey, Deborah Berhanu, Phil A. Bland, Gordon Cressey, Lauren E. Howard, Chris Jeynes, Richard Mathewman, Zita Martins, Mark Sephton, Vlad Stolojan, Sasha Verchovsky. 2013. Biomass preservation in distal impact melt ejecta. *Nature Geoscience* 6: 1018–1022.

Kirkpatrick, S., Gelatt, C. D., Jr., and Vecchi, M. P. 1983. Optimization by simulated annealing. *Science* 220:671–680.

Leong, D., Harry, M., Reeson, K. J., and Homewood, K. P. 1997. A silicon/iron-disilicide light-emitting diode operating at a wavelength of 1.5 μm. *Nature* 387:686–688.

Metropolis, N., Rosenbluth, A. W., Rosenbluth, M. N., Teller, A. H., and Teller, E. 1953. Equation of state calculations by fast computing machines. *Journal of Chemical Physics* 21:1087–1092.

Moncoffre, N., Barbier, G., Leblond, E., Martin, Ph., and Jaffrezic, H. 1998. Diffusion studies using IBA. *Nuclear Instruments and Methods B* 140:402–408.

Pascual-Izarra, C., Barradas, N. P., Reis, M. A., Jeynes, C., Menu, M., Lavedrine, B., Ezrati, J. J., and Röhrs, S. 2007. Towards truly simultaneous PIXE and RBS analysis of layered objects in cultural heritage. *Nuclear Instruments and Methods B* 261:426–429.

Rennie, J., Elliott, S. R., and Jeynes, C. 1986. Rutherford backscattering study of the photodissolution of Ag in amorphous GeSe$_2$. *Applied Physics Letters* 48:1430–1432.

Riggs, P. D., Clough, A. S., Jenneson, P. M., Drew, D. W., Braden, M., and Patel, M. P. 1999. ^{3}He ion-beam analysis of water uptake and drug delivery. *Journal of Controlled Release* 61:165.

Ross, G. J., Barradas, N. P., Hil, M. Pl., Jeynes, C., Morrissey, P., and Watts, J. F. 2001. RBS and computer simulation for the in-depth analysis of chemically modified poly(vinylidene fluoride). *Journal of Materials Science* 36:4731–4738.

Ryan, C. G. 2011. PIXE and the nuclear microprobe: Tools for quantitative imaging of complex natural materials. *Nuclear Instruments and Methods B* 269:2151–2162.

Satoh, T., Oikawa, M., and Kamiya, T. 2009. Three-dimensional measurement of elemental distribution in minute samples by combination of in-air micro-PIXE and STIM. *Nuclear Instruments and Methods B* 267:2125–2127.

Shearmur, T. E., Clough, A. S., Drew, D. W., van der Grinten, M. G. D., and Jones, R. A. L. 1998. Interdiffusion of deuterated and protonated poly(methyl methacrylate). *Polymer* 39:2155–2159.

Smith, R. W., Massingham, G., and Clough, A. S. 2002. In *10th Symposium on Radiation Measurements and Applications* Elsevier Science Bv: Ann Arbor, MI, 582.

Stoquert, J. P., and Szörenyi, T. 2002. Determination of the number and size of inhomogeneities in thin films by ion beam analysis. *Physics Review B* 66:144108.

Szilágyi, E., Pászti, F., and Amsel, G. 1995. Theoretical approximations for depth resolution calculations in IBA methods. *Nuclear Instruments and Methods B* 100:103–121.

Tesmer, J. R., and Nastasi, M., eds. 1995. *Handbook of modern ion beam analysis.* Pittsburgh: Materials Research Society.

Thomas, N. L., and Windle, A. H. 1982. A theory of case II diffusion. *Polymer* 23:529–542.

Thompson, R. L., McDonald, M. T., Lenthall, J. T., and Hutchings, L. R. 2005. Solvent accelerated polymer diffusion in thin films. *Macromolecules* 38:4339–4344.

Tosaki, M. 2006. Energy-loss straggling caused by the inhomogeneity of target material. *Journal of Applied Physics* 99 (3): 034905.

Wang, Y. Q., and Nastasi, M., eds. 2010. *Handbook of modern ion beam analysis,* 2nd ed. Pittsburgh: Materials Research Society.

Defects Measurements of a Crystalline Solid

Lin Shao

11.1 INTRODUCTION

11.1.1 ION IMPLANTATION

In the 1950s, Bell Lab researchers invented the technique of ion implantation for device doping (Shockley 1949, 1954; Ohl 1950; Moyer 1954). The technique was widely accepted by the semiconductor industry in the 1970s for various electronic device fabrications (Wolf and Tauber 1986; Gibbons 1987; Rimini 1995; Nastasi and Mayer 2007). Prior to that, doping was realized by thermal drive-in diffusion of dopants from a surface source, for which both doping level and doping depth were controlled by thermal annealing parameters. In ion implantation, isotope pure dopants are precisely introduced into a designed depth, often with a peak concentration beyond its solid solubility. The technique, combined with photolithography and later, rapid thermal annealing, make it possible to fabricate complicated microelectronics with great device quality control and precise site-selective doping.

A doped semiconductor is very sensitive to defects initially introduced during ion implantation. Nuclei–nuclei collisions during ion bombardments lead to displacement of semiconductor lattice atoms and formation of high-density interstitials and vacancies along ion tracks (Nastasi and Mayer 2007). The majority of these point defects are removed by self-annealing in a short period of time (e.g., 1 ps after displacement creation) through interstitial-vacancy recombination. A small fraction of these point defects, however, survive the recombination process and a high-temperature annealing step is necessary to remove them completely through defect annihilation on the surface. An incomplete defect removal will lead to either electron/hole creation or trapping by the defects and cause device failure. This does not mean, however, that a high thermal budget annealing is always welcome. Complexity arises due to defect–dopant interactions (Stolk et al. 1997; Cowern et al. 1990, 1991, 1992, 1999). Although the majority of dopants already take substitutional sites during interstitial-vacancy recombination soon after damage creation, their subsequent interactions with defects during thermal annealing make them fast diffusers. The anomalous profile spreading has been observed for all dopants in Si, the most important semiconductor substrate. Thus, thermal annealing needs to be optimized to minimize such irradiation-induced dopant diffusion.

Studies on both defect removal and dopant–defect interactions require a tool being able to characterize defects quantitatively within a submicrometer depth. Atomic scale characterization using transmission electron microscopy is powerful to reveal the nature of large defect clusters, but direct observation of precursors to large defect clusters is not possible. Also, a quantitative analysis of defect populations is both manpower and machine time costly for such a technique. Information on defect evolution is important for a semiconductor engineer to obtain kinetics of defect removal to optimize thermal budgets and to develop modeling capability to predict both dopant spreading and defect removal.

Fortunately, during almost the same period of the birth and development of ion implantation techniques, channeling ion beam analysis was developed. It immediately opened a door for materials scientists to understand both radiation responses and defect developments in a wide range of crystalline materials. The simultaneous developments in both ion implantation and ion beam analysis are not accidental. Both were consequences of research in the pioneering days to understand fundamentals of ion–solid interactions, which can be traced back to the early work done by Rutherford's group in their famous backscattering experiments. The interests in studying radiation damage in crystals were also motivated by the research needs of nuclear engineers to understand irradiation-induced structural failures. That part of the research expands the channeling analysis from semiconductor materials to metals and ceramics.

11.1.2   ION CHANNELING

A crystal lattice can influence ion trajectories. This phenomenon, called channeling, was first predicted by Stark in 1912. During the latter 1950s and early 1960s, the channeling was accidently "rediscovered" by several groups. The first experimental evidence was reported by Bredov and Okuneva in studying distributions of 4 keV radioactive Cs implants in germanium. The abnormally long tail distribution observed, however, was then believed to be normal since it agreed with a theoretical prediction, which, unfortunately, was calculated wrong. In 1963, Robinson and Oen (1963) reported an anomalously long range of ion penetration along the open channels of single crystals through Monte Carlo simulations. The modeling results were quickly confirmed experimentally within a few months by several groups (Piercy et al. 1963; Lutz and Sizmann 1963; Nelson and Thompson 1963; Kornelsen et al. 1964). In 1964 and 1965, channeling studies over a wide range of ion energies through combinations with nuclear reaction analysis (Bøgh, Davies, and Nielsen 1964; Thompson 1964), Rutherford backscattering spectrometry (RBS; Bøgh and Uggerhøj 1965), proton-induced x-ray emission analysis (Brandt et al. 1965), and blocking experiments (Domeij 1965) were reported. One milestone in 1965 was Lindhard's ion channeling theory, which predicted well all distinctive features of channeling phenomena. Since then, channeling analysis has rapidly become one of the most important materials characterization tools. The development of ion beam analysis techniques was also boosted by the technology invented in the 1950s: multichannel analyzers and semiconductor detectors including surface barrier detectors and diffused junction detectors (McKay 1951; Mayer and Gossick 1956; Mayer 1959).

Over the history of channeling, fundamental studies in physics and technological applications were driven, more or less, by the needs of the semiconductor industry. The knowledge of channeling, indeed, taught industry how to avoid channeling for a better control on dopant profiles. Channeling analysis, especially by combining Rutherford backscattering spectrometry, provides key information to characterize ion implantation-induced radiation damage quantitatively, which concerns optimization and quality control of ion beam doping procedures. Channeling ion beam analysis has been widely used during the past half-century in materials science as a nondestructive tool to characterize crystalline samples (Davies, Amsel, and Mayer 1992). To name a few, channeling Rutherford backscattering spectrometry can be used for determination of crystal orientation, impurity location, interfacial configuration, lattice imperfections, and radiation damage. These examples are further discussed in this chapter. Due to increasing interest in strain engineering to fabricate transistors and sensors with better performance, an example of using the channeling technique to characterize strains in a heterogeneous structure is also given.

11.2   FINDING CHANNEL AXES

The first step in a channeling analysis is to determine the axial channel or planar channel directions as precisely as possible. Finding axial channel direction needs a goniometer driven by multiple-step motors, which may include as many as five movement modes like rotation, horizontal shift, vertical shift, horizontal tilt, and vertical tilt. The shift usually has a step of 0.01 mm and the tilting movement has a step of 0.005°. The error in reproducibility is less than one step and could be minimized by tilting or translation from one "normalized" direction. For example, for a tilting angle previously recorded during a scan from the left to the right direction, it is better to scan back and rescan from the left and stop at that angle, rather than directly scanning in the backward direction and stopping.

There are two methods of orienting a crystal solid to find channel direction. One is the rotation mode and another is the tilting mode. In the rotation mode, a sample is often tilted a few degrees (e.g., 5°) off the sample normal, and backscattering yields are recorded during sample precession along its normal for 360°. In the tilting model, yields are recorded by fixing horizontal tilt at a few degrees (e.g., 5° and −5°, respectively) off the normal and continuously scanning along the vertical direction (e.g., from −5° to 5° with a stepwise of 0.2°). The procedure is then repeated by fixing the vertical tilt at a few degrees off the normal and continuously scanning along the horizontal direction.

Both modes are essentially the same. The key is to find the positions corresponding to the planar channels by graphic approximation. Figure 11.1 shows a typical yield plotted as a function of rotation angle in a polar coordinate. The rotation is around the <110> axis of a Si sample. The primary yield dips corresponding to the same planar channel are connected by straight lines. Ideally, all the lines should have a common crossing point in the plot, which corresponds to the coordinates of the axial channel.

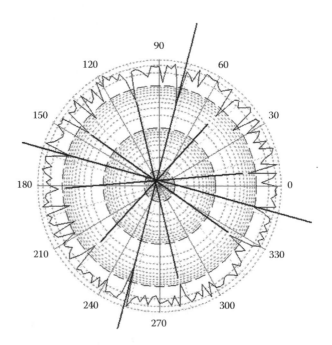

FIGURE 11.1 The plot of RBS yields in a polar coordinate, obtained from <110> Si, as a function of sample rotation with an off normal tilting at 5°.

The graphic approximation, however, only has limited accuracy. A second or even more adjustments may be needed. The position refinement is a stepwise approach featured by fixing one coordinate and changing the other for the yield minimization, in the region closed to cross point. The procedure is repeated by switching the turns of fixing and adjustment between two coordinates. One tip for a fast alignment is always to align one sample cleavage direction with one direction of the goniometer tilting. A crystal often cleaves smoothly along one of its main planar directions. Aligning one cleavage direction (e.g., along the horizontal direction) during sample loading, even guided by eyeballs, can greatly facilitate position refinement. This is due to the fact that changing one coordinate during adjustment does not change the channel position in the other coordinate, which reduces the number of refinement steps required. If a crystal sample has an irregular shape, the sample can be rotated according to the polar plot to let one planar direction parallel one goniometer tilting direction, before starting the second adjustment procedure.

If a sample is not aligned accurately, the yield scanning close to the minimum position will show either a large angular dip or multiple minimums. This is often caused by a scan that misses the minimum and crosses over several planar lines. As for the yield recorded during the scan, it depends on the position and width of the channel window selected for yield integration. If the backscattering yields are integrated narrowly from the near surface region, the yield ratio between the dip bottom to the shoulder should be around 3%–6% for a perfect crystal. This ratio, however, can be much higher if the yield integration window is wide and signals from the deep depth are also accounted for. Such a wide integration window sometimes is necessary to collect signals with reduced statistics fluctuation for a fast alignment when beam damage needs to be minimized on the same spot. Depending on window selections, some dips may or may not appear in the plot.

11.3 CHANNEL-ENERGY CONVERSION

Channel numbers in a backscattering spectrum are related to the energies of backscattered particles, but their conversion relationship is determined by a number of instrument setups including detector bias, preamplifier set, main amplifier, analog-to-digital convertor, and beam-sample-detector configuration. Spectrum simulation requires channel-energy conversion, which can be easily achieved by using a calibration sample, which is often a thin film sample with multiple elements presented on the surface.

In reading channel numbers for each element presented on the surface, it is necessary to consider the limited detection resolution for a thin layer. If the layer is ultrathin, the channel number corresponds to the peak of a delta-like backscattering spectrum. If the layer is thick and the spectra begin to show a box-like shape, the channel number should be read at the half-height of the spectrum front. The energy of the backscattered ions from each element can be easily calculated, given that the element is presented at the sample surface. Therefore, the energy, usually in the unit of kiloelectron volts, equals the product of the beam energy and backscattering kinematic factor, which has been defined in Chapter 2. The kinematic factor is often provided as tabulated data in an ion beam analysis handbook. If the detector angle is not listed in the table, a linear interpolation between the two closest angles serves as a good approximation.

A linear fitting of the plot of channel numbers versus energy gives the following relationship:

$$(11.1) \qquad\qquad E = E_0 + \frac{\Delta E}{\Delta n}n$$

where n is channel number and E is backscattering energy. The fitting will extract both E_0 and $\Delta E/\Delta n$ values. The linear relationship is based on the working mechanism of the detector. Whether it is a p–n junction type semiconductor detector or a Shockley type detector, the energetic particle loses its energy by creating electron–hole pairs in silicon and the carrier number is proportional to the particle energy.

As one example, Figure 11.2 shows a typical RBS spectrum obtained from a Si coated with a 2 nm thick metal layer, which is a mixture of Au, Ag, and Ni elements. The layer is thin enough to assume that all four elements are presented on the surface. Arrows in the figure refer to the position to read the corresponding channel numbers for each element. The numbers are 539.4, 736,

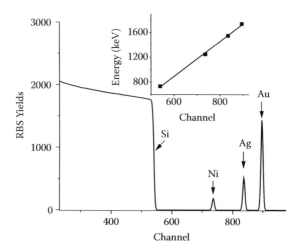

FIGURE 11.2 A typical RBS spectrum from a Si coated with an ultrathin layer of Au, Ag, and Ni. The inset is the corresponding plot of energy versus channel for calibration.

835, and 895 for Si, Ni, Ag, and Au, respectively. Kinematic factors for He as a projectile at a backscattering angle of 165° are 0.3672, 0.6240, 0.7744, and 0.8695 for Si, Ni, Ag, and Au, respectively. For a 2 MeV He beam, these correspond to 734, 1240, 1549, and 1739 keV. The inset in Figure 11.2 shows a linear fitting of energy versus A channel, which suggests that $E_0 = -794$ keV and $\Delta E/\Delta n = 2.81$ keV per channel.

11.4   ENERGY-DEPTH CONVERSION

With the preceding channel-energy conversion, the energy-depth conversion is possible by the following relationship, based on the assumption that the layer of interest is so thin that energy stopping power remains the same within the layer:

$$(11.2) \qquad E(t) = k\left[E_0 - \frac{t}{\cos\phi_1}\left|\left(\frac{dE}{dx}\right)_{E_0}\right|\right] - \frac{t}{\cos\phi_2}\left|\left(\frac{dE}{dx}\right)_E\right|$$

Equivalently, we have

$$(11.3) \qquad t = \frac{kE_0 - E^{(t)}}{\frac{k}{\cos\phi_1}\left|\left(\frac{dE}{dx}\right)_{E_0}\right| + \frac{1}{\cos\phi_2}\left|\left(\frac{dE}{dx}\right)_E\right|}$$

where
   E(t) is the energy recorded by a detector for particles backscattered at the depth t
   ϕ_1 is the beam incident angle with respect to sample normal
   ϕ_2 is the particle exit angle with respect to the sample normal
   $(dE/dx)_{E_0}$ is the energy loss for a particle in its inward path
   $(dE/dx)_E$ is the energy loss for a particle in its outward path

Stopping power of light ions in various elementary solids can be obtained using popular SRIM (stopping and range of ions in matter) code, assuming that the solids are amorphous. These data turn out to be quite reliable for most commonly used beam–target combinations. The data for a compound solid involving multiple elements are less accurate. A good approximation, given that mass differences between different elements are not significant, is to calculate by using a weighted mass and weighted nuclear charge. Much less is known about stopping channeled particles along a channel direction. The channeling stopping power is only a fraction of that of a nonchanneled beam, with the difference depending on ion species, energy, target composition, and specific channel directions. A high accuracy depth conversion in ion channeling spectra analyses must consider such stopping power changes.

For a thick layer requiring consideration on stopping power changes along the path, depth conversion needs an iteration approach. First, the substrate is divided into many discrete thin layers. Energy of ions passing through the layer and the energy of ions backscattered in the layer are calculated. For the layers at deep depths, energy loss accumulated through passing the layers at shallower depths is calculated. The stopping power correction for the inward beam under channeling condition can be easily included in the program. The stopping power for the outward beam can be assumed to take the value under random mode.

11.5   SEPARATION OF DECHANNELING BACKGROUND

A backscattering spectrum includes yield contribution from two kinds of scattering sources: One is the channeled beam backscattered from the displacements within the channel and another is the dechanneled beam backscattered from target lattice atoms. In order to extract information on displacements, the contribution from the dechanneled beam must be separated from the spectrum. Once dechanneling background is determined, the displacement depth profile can be extracted by using

(11.4) $$\chi_D(z) = \chi_R(z) + [1 - \chi_R(z)]n_D(z)/n$$

where n_D is the density of displaced atoms
χ_D is the measured backscattering spectrum
χ_R is the dechanneled background
z is the depth
n is the total atomic density

11.5.1   LINE APPROXIMATION

The line approximation assumes that the dechanneling background is a straight line. This is the easiest way but with lowest accuracy. First, two channel numbers immediately before and after the damage peak are identified. Yields between two points are assumed to increase linearly. This can be easily done by using interpolation function available in any commercial plot software having basic data analysis capability, such as Origin. Note that the point density for interpolation needs to equal the channel difference between two points. Usually, two ending points of the line approximation are selected on the same spectrum. Sometimes, if the damage peak is significantly influenced by the surface peak, the line approximation may start from one point selected from channeling a

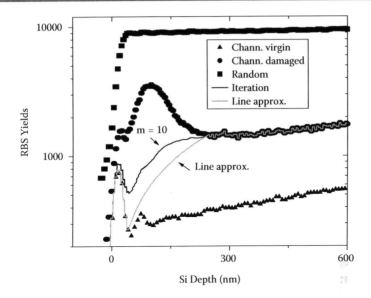

FIGURE 11.3 RBS spectra obtained from virgin Si and 80 keV self-ion irradiated Si. The solid line refers to the dechanneling background obtained from the iterative procedure and the gray line from line approximation.

virgin spectrum and end with another point in the spectrum from the damaged sample. One such example is given in Figure 11.3.

11.5.2 ITERATION PROCEDURE

In this standard approach, the substrate is divided into many thin layers. The iteration starts with the assumption that, in the first layer, the dechanneling background is zero. Thus, yield in the first layer is converted to displacements. In the second layer, the dechanneling caused by the displacements in the first layer is subtracted, and the yield after subtraction is converted to the displacements. The procedure is repeated for the subsequent layers and, for each layer, yield contribution due to accumulated dechanneling by displacements in all shallower layers is subtracted.

This procedure requires knowledge on the dechanneling cross section, which can be approximated by using the equation from classical channeling theory (Chapter 9), obtained by integration of cross sections for the scattering causing beam deflection larger than the critical channeling angle. However, the classical formula does not consider the structural relaxation around a point defect, which can significantly underestimate dechanneling probability. For example, in H-implanted Si, the small H-defect complex formation causes a large strain, and dechanneling cross section per displacement could be one order of magnitude higher in comparison with classical theory (Shao and Nastasi 2005).

11.5.3   DOUBLE ITERATION PROCEDURE

A double iteration procedure can be used to extract both dechanneling probability and the dechanneling background (Shao and Nastasi 2005). This method requires no knowledge on dechanneling cross sections. The iteration starts by selecting a depth (denoted by z_1) beyond a disordered region. We have

$$(11.5) \qquad \chi_D(z_1) = \chi_V(z_1) + [1 - \chi_V(z_1)] \left\{ 1 - \exp\left[-\int_0^{z_1} \sigma_D \frac{\chi_D(z') - \chi_R(z')}{1 - \chi_R(z')} n dz' \right] \right\}$$

where
> χ_D is the backscattering spectrum experimentally measured from a defective solid
>
> χ_R is the dechanneled fraction of incident ions, which is not directly measurable
>
> z is corresponding depth
>
> n is crystal atom density
>
> σ_D is dechanneling cross section per displacement
>
> χ_V is the backscattering spectrum experimentally measured from a virgin crystal

At a depth shallower than z_1, we have

$$(11.6) \qquad \chi_R(z) = \chi_V(z) + [1 - \chi_V(z)] \left\{ 1 - \exp\left[-\int_0^z \sigma_D \frac{\chi_D(z') - \chi_R(z')}{1 - \chi_R(z')} n dz' \right] \right\}$$

If an estimated $\chi_R(z)$ is used in Equation (11.5), the dechanneling cross section σ_D can be estimated since all other values in the equation are experimentally measurable. Once σ_D is obtained, it can be used in Equation (11.6) for calculation of $\chi_R(z)$ at each depth point. The recalculated $\chi_R(z)$ is substituted into Equation (11.1) in order to get a new value for a continuous calculation. This procedure will continue until finally both $\chi_R(z)$ and σ_D converge. This procedure is not sensitive to the initially assumed $\chi_R(z)$. For example, it can be simply estimated to be constant zero at the beginning.

The preceding procedure applies to a crystal with low defect concentration for which the dechanneling is caused by single scattering; this means that dechanneling the cross section per displacement does not increase with increasing defect numbers. In a highly defective crystal, the dechanneling cross section will increase nonlinearly with displacement numbers, due to the accumulated

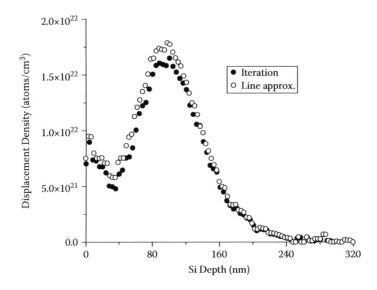

FIGURE 11.4 Extracted displacement profiles from Figure 11.3, with dechanneling background calculated through the iterative procedure and line approximation, respectively.

effect from small-angle deflection caused by multiple scattering. In that case, both Equations (11.5) and (11.6) need to be changed (Shao 2008). But the principles of the previous iteration are still valid.

As an example (Shao 2008), Figure 11.3 shows RBS spectra obtained from virgin Si and 80 keV Si self-ion-irradiated Si. The solid line refers to the calculated dechanneling backgrounds using the preceding iterative procedure, obtained after 10 iteration loops. Further iteration with higher loop numbers causes negligible difference. The iteration starts by selecting depth z_1 at 350 nm. In a comparison, the dotted line in the figure shows the dechanneling background from line approximation. Figure 11.4 compares the displacement profile extracted from Figure 11.3 by using the dechanneling background obtained through a double iterative procedure and line approximation, suggesting that a correct dechanneling background subtraction is important to obtain a reliable displacement profile.

11.6   DETERMINATION OF IMPURITY DISPLACEMENTS

Yield angular scan from impurities is one important tool to determine their lattice positions, although a precise determination often requires scans along different axes since some interstitial sites might be hidden in the atomic row under certain channel directions, and be viewed as substitutional sites by an analysis beam. A conclusive analysis often requires a comparison between experimental data and modeling results.

If the projected location of impurities is in the center of the channel, yields will show a peak in the middle of the angular scan. If displacements are only a fraction of one angstrom away from the atomic rows, the angular scan will show a narrower dip. The half-width ($\psi_{1/2}$) of the channeling angular scan is approximately the critical angle for channeling and is related to lattice vibration amplitudes and atomic displacement. Analysis of $\psi_{1/2}$ changes can be used to estimate impurity displacements.

In one example (Zhu et al. 2007), Figure 11.5 compares the angular scans around <110> from Si and from Sb in a Sb-doped Si, after different thermal treatments. The Sb atoms are doped by using low-temperature molecular beam epitaxial growth. With increasing postgrowth annealing temperature, the Sb χ_{min} increases and Sb $\psi_{1/2}$ decreases, which suggests instability and displacements of Sb atoms. The $\psi_{1/2}$ values for Si are measured to be around 0.45°. The $\psi_{1/2}$ values for Sb are measured to be 0.34°, 0.30° and 0.27° for samples after growth, after postgrowth annealing at 700°C and 900°C, respectively.

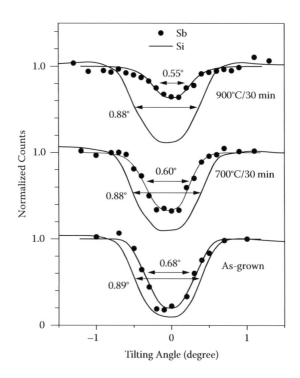

FIGURE 11.5   RBS yields of angular scans around <110> axis of Sb-doped Si sample, before annealing and after annealing at 700°C and 900°C, respectively. (Zhu, L. et al. 2007. *Journal of Applied Physics* 101:086110.)

Based on Linhard's continuum model, $\psi_{1/2}$ is expressed as

(11.7)
$$\psi_{1/2} = \left[\frac{1}{2}\ln\left(\frac{3a^2}{r_{min}^2}+1\right)\right]^{1/2}\sqrt{\frac{2Z_1Z_2e^2}{Ed}}$$

with

(11.8)
$$r_{min} = (\rho^2\ln 2 + u^2)^{1/2}, \quad a = 0.8853a_0\left(Z_1^{1/2}+Z_2^{1/2}\right)^{-2/3}$$

where
   Z_1 and Z_2 are atomic numbers of projectile (helium ion with $Z_1 = 2$) and tar-
       get (Si with $Z_2 = 14$) atoms, respectively
   d is the atomic spacing along the atomic string (d = 0.384 nm for <110>)
   E (E = 2 MeV) is the energy of the analysis beam
   ρ is the thermal vibration amplitude
   u is the projected displacement of atoms in the view along the channel-
       ing direction
   a_0 (= 0.0528 nm) is the Böhr radius

For lattice Si atoms, u = 0 and ρ = 0.08 Å at room temperature. Equation (11.7) predicts that $\psi_{1/2}$ = 0.70°, while experimentally $\psi_{1/2}$ = 0.45°. The theoretical value needs to be adjusted by a factor of 0.64. The same factor is applied to adjust theoretical prediction for Sb. That is, the right side of Equation (11.7) is multiplied by 0.64. Substituting the experimentally obtained $\psi_{1/2}$ into the left side, we have determined that Sb displacements are 0.25 Å for the as-grown sample, 0.29 Å for the sample annealed at 700°C, and 0.33 Å for 900°C. The calculation of Equation (11.7) uses $Z_1 = 2$ for helium ions, $Z_2 = 51$ for Sb atoms, and ρ = 0.08 Å, based on the assumption that Sb thermal vibration amplitude is the same as that of Si.

11.7 DETERMINATION OF STRAIN IN A HETEROGENEOUS STRUCTURE

When an epitaxial layer is deposited on a substrate with different but almost the same lattice parameters, a heterogeneous structure is fabricated. Under a small lattice parameter mismatch, the deposited layer follows the lattice parameter of the substrate in the horizontal direction with vertical elastic strain introduced as accommodation. Such a perfect epitaxy exists for an ultrathin layer. Channeling analysis is powerful to characterize the strain by studying shifts of angular scan dip around an off-normal axis.

As an example, for a heterogeneous (100) Si/Si$_{1-x}$Ge$_x$/Si structure with a thin layer of strained Si$_{1-x}$Ge$_x$ buried in Si, the relatively larger lattice parameter in the vertical direction of Si$_{1-x}$Ge$_x$ causes channel bending along the <110> axis. For the top Si layer, the yield minimum corresponding to the <110> channel occurs when the beam is incident at an angle of 45° off the normal. For the buried $Si_{1-x}Ge_x$ layer, the corresponding minimum occurs at an off-normal angle slightly less than 45°.

Tetragonal strain, ε_T, can be determined from the angular offset (Fiory 1984) by

$$(11.9) \qquad \varepsilon_T \equiv \varepsilon_\perp - \varepsilon_\parallel = \frac{\theta_s - \theta_u}{\sin\theta_u \cos\theta_u}$$

where

$\varepsilon_\perp$ is horizontal strain

$\varepsilon_\parallel$ is vertical strain

θ_s and θ_u are angular positions of minimum yield in strained ($Si_{1-x}Ge_x$) and unstrained layers (Si), measured by off-normal channeling analysis

RBS angular scans shown in Figure 11.6 determine an angular offset 0.24°. Substituting this into Equation (11.9) leads to a tetragonal strain of 0.8%. On the other hand, the magnitude of strain can be calculated (Fiory et al. 1984) by

$$\varepsilon_T = f\left(\frac{1+\nu}{1-\nu}\right)$$

$$\nu = 0.278 - 0.005x$$

$$(11.10) \qquad f = 0.042x$$

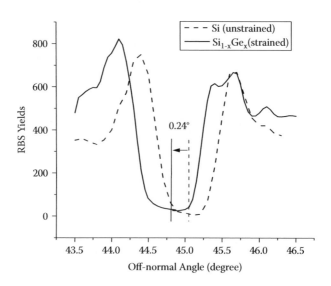

FIGURE 11.6   RBS yields of angular scans around <110> axis of $Si/Si_{1-x}Ge_x/Si$. The solid line refers to Si signals from the top Si layer and the dashed line refers to Ge signals from the strained layer.

where x is Ge atomic concentration ratio in $Si_{1-x}Ge_x$ and v is Poisson's ratio. Thus, a comparison between experimental data and theoretical prediction can help to evaluate strain relaxation. The relaxation is expected if the strain layer has a thickness larger than a critical value, which is usually a few nanometers.

It needs to be pointed out that, if the strained layer has an ultrathin thickness, the beam views the Ge atoms as displacements. Thus, the angular offset will disappear and both Si and Ge signals will have the same minimum, but the Ge angular scan will have asymmetrical yield. Determination of the strain under such conditions requires modeling.

In summary, ion beam channeling analysis has been widely used as one unique surface characterization tool, instead of a complementary method to others, for nondestructive analysis of crystalline solids. Using limited examples, this chapter has explained its basics and the typical applications to obtain information such as defective levels, damage distributions, strains, and lattice locations of impurities and defects. The channeling analysis can provide quantitative information without complicated and costly specimen preparation. The principles of channeled beam interactions with solids have been well known and the knowledge required for data interpretation is generally applicable to arbitrary ion-target systems.

PROBLEMS

11.1. A solid-state detector is located 170° away from the incident direction of a 2 MeV alpha particle. Signals from an Au surface are recorded at channel number 800, and a Ni surface at channel number 700. What is the expected channel number for ions scattered from an Al surface?

11.2. A depth conversion of an RBS spectrum shows a damage peak located at the depth of 500 nm. The conversion uses random stopping power to describe ion stopping, but the spectrum is actually obtained under channeling conditions. Should the true depth of damage peak be deeper or shallower than 500 nm? Explain your answer.

11.3. Estimate the half-width of angular scan dips around the <100>, <111>, and <110> Si axes, respectively, for a 2 MeV alpha beam.

11.4. Using a 2 MeV alpha beam, channeling analysis shows that the half-width of an angular scan dip around the <110> axis of virgin Si is 0.5°. This half-width is reduced to 0.3° in an ion-irradiated Si sample. Estimate the displacement of atoms away from lattice rows.

11.5. A channeling spectra of ion-irradiated Si shows a damage peak at the half-height of the random spectra. Assuming that the dechanneling background is negligible, what is displacement density in the peak? (A crystalline Si has an atomic density of $5 \times 10^{22}/cm^3$.)

11.6. An angular scan around the <110> axis of a <100> oriented $Si/Si_{1-x}Ge_x/$ Si sample shows that the dip offset between Ge signals and Si signals from the top layer is about 0.2°. What is the tetragonal strain of the $Si_{1-x}Ge_x$ layer?

APPENDIX: QUICKBASIC PROGRAM FOR DOUBLE ITERATIVE PROCEDURE TO EXTRACT DEFECT PROFILE

Note: For each input spectrum, yields are arranged in the order of their depths, starting from the surface, and their corresponding depths are not included. Raw RBS data as a yield versus channel profile cannot be used. Channel-to-depth conversion must be done first. A spectrum must be saved as a DAT profile.

```
CLS

OPEN "c:\temtest\virgin.dat" FOR INPUT AS #1 'specify
location for channeling spectrum fron virgin sample
OPEN "c:\temtest\rbs.dat" FOR INPUT AS #2 ' 'specify
location for spectrum from damaged sample
OPEN "c:\temtest\random.dat" FOR INPUT AS #3 'specify
location for random spectrum
OPEN "c:\dechannel.dat" FOR OUTPUT AS #5 'specify location
to save extracted dechanneling background
OPEN "c:\calcudef.dat" FOR OUTPUT AS #6 'specify location to
save extreacted defect profile
OPEN "c:\decross.dat" FOR OUTPUT AS #7 'specify location to
save dechanneling cross section

density = 5E22 'atomic density of substrate
depthmax = 1000 'the maximum depth of interest
depthcut = 614 'selected point beyond damage peak
maxit = 10 'the maximum number of interation loop

DIM virgin(depthmax) 'create space to store virgin spectrum
DIM rbs(depthmax) 'create space to store spectrum from
damaged sample
DIM random1(depthmax) 'create space to store random spectrum
DIM dechanneling(depthmax) 'create space to save
dechanneling background to be calculated
DIM defect(depthmax) 'create space to save defect profile to
be calculated
```

```
FOR i = 0 TO depthmax 'input virgin spectrum
  INPUT #1, virgin(i)
NEXT i

FOR i = 0 TO depthmax 'input spectrum from the damaged sample
  INPUT #2, rbs(i)
NEXT i

FOR ii = 0 TO depthmax 'input random spectrum
  INPUT #3, random1(ii)
NEXT ii

FOR j = 1 TO maxit 'start iteration
  totaldefect = 1 'assume the integrated defect number is
1, as starting condition
  FOR iii = 0 TO depthcut 'start integration of defect
profile to get total defect
  totaldefect = defect(iii) *.0000001 + totaldefect 'the
initial defect profile is empty. The factor 0.0000001
corresponds to ratio of 1nm to 1cm, for integration
  NEXT iii
  decross = -1/totaldefect * LOG(1 - (rbs(depthcut) -
virgin(depthcut))/(random1(depthcut) - virgin(depthcut)))
'finding corresponding dechanneling cross section
  PRINT #7, USING "##.###^^^^^ ##.###^^^^^";-j; decross
'save dechanneling cross section as a function of
interation loop number

FOR iii = 0 TO depthcut
  totaldefect = 0
  FOR i = 0 TO iii 'integrate defects up to a depth which
is shallower than the cutting point
  totaldefect = defect(i) *.0000001 + totaldefect
  NEXT i
  dechanneling(iii) = virgin(iii) + (random1(iii) -
virgin(iii)) * (1 - EXP(-decross * totaldefect)) 'extract
dechanneling background at a specific depth
  NEXT iii

  FOR i = 0 TO depthcut
    defect(i) = (rbs(i) - dechanneling(i))/(random1(i)
- dechanneling(i)) * density 'extract defect numbers after
subtraction of dechanneling background at a specific depth
```

```
    IF defect(i) < 0 THEN defect(i) = 0 'to avoid negative
defects causing catastrophical failure
  NEXT i
  PRINT j, decross 'print extracted dechanneling cross
section as a function of interation number
NEXT j

FOR i = 0 TO 614
  PRINT #5, USING "##.###^^^^^ ##.###^^^^^"; i;
dechanneling(i) 'save dechanneling background
NEXT i

FOR i = 0 TO 614
  PRINT #6, USING "##.###^^^^^ ##.###^^^^^"; i; defect(i)
'save defect profiles
NEXT i
PRINT "calculation finished"
PRINT "total defect integrated = "; totaldefect

END
```

REFERENCES

Bøgh, E., Davies, J. A., and Nielsen, K. O. 1964. Experimental evidence for the extinction of (p,γ) yields in single crystals. *Physics Letters* 12:129.

Bøgh, E., and Uggerhøj, E. 1965. Experimental investigation of orientation dependence of Rutherford scattering yield in single crystals. *Nuclear Instruments and Methods B* 38:216.

Brandt, W., Khan, J. M., Potter, D. L., Worley, R. D., and Smith, H. P. 1965. Effect of channeling of low-energy protons on the characteristic x-ray production in single crystals. *Physics Reviews Letters* 14:42.

Bredov, M.M., and Okuneva, N.M. 1957. On a depth of penetration of medium energy ions in matter. *Doklady Akad. Nauk SSSR* 113(4):796.

Cowern, N. E. B., Janssen, K. T. F., van de Walle, G. F. A., and Gravesteijn, D. J. 1990. Impurity diffusion via an intermediate species: The B-Si system. *Physics Reviews Letters* 65:2434.

Cowern, N. E. B., Mannino, G., Stolk, P. A., Roozeboom, F., Huizing, H. G. A., van Berkum, J. G. M. Cristiano, F. Claverie, A., and Jaraíz, M. 1999. Energetics of self-interstitial clusters in Si. *Physics Reviews Letters* 82:4460.

Cowern, N. E. B., van de Walle, G. F. A., Gravesteijn, D. J., and Vriezema, C. J. 1991. Experiments on atomic-scale mechanisms of diffusion. *Physics Reviews Letters* 67:212.

Cowern, N. E. B., van de Walle, G. F. A., Zalm, P. C., and Oostra, D. J. 1992. Reactions of point defects and dopant atoms in silicon. *Physics Reviews Letters* 69:116.

Davies, J. A., Amsel, G., and Mayer, J. W. 1992. Reflections and reminiscences from the early history of RBS, NRA and channeling. *Nuclear Instruments and Methods in Physics Research* B 64:12.

Domeij, B. 1965. Crystal lattice effects in the emission of charged particles from mono-crystalline sources. *Nuclear Instruments and Methods B* 38:207.

Fiory, A. T., Bean, J. C., Feldman, L. C., and Robinson, I. K. 1984. Commensurate and incommensurate structures in molecular beam epitaxially grown $GexSi_{1-x}$ films on Si(100). *Journal of Applied Physics* 56:1227.

Gibbons, J. F. 1987. Historical perspectives on ion implantation. *Nuclear Instruments and Methods B* 21:83.

Kornelsen, E. V., Brown, F., Davies, J. A., Domeij, B., and Piercy, G. R. 1964. Penetration of heavy ions of keV energies into monocrystalline tungsten. *Physics Reviews* 136:A849.

Lindhard, J. 1965. Comprehensive theory of channeling. *Matematisk–Fysiske Meddelelser Konglige Danske Videnskabernes Selskab* 34 (14).

Lutz, H., and Sizmann, R. 1963. Super ranges of fast ions in copper single crystals. *Physics Letters* 5:113.

Mayer, J. W. 1959. Performance of germanium and silicon surface barrier diodes as alpha-particle spectrometers. *Journal of Applied Physics* 30:1937.

Mayer, J. W., and Gossick, B. R. 1956. Use of Au-Ge broad area barrier as alpha-particle spectrometer. *Review of Scientific Instruments* 27:407.

McKay, K. G. 1951. Electron–hole production in germanium by alpha-particles. *Physics Reviews* 84:829.

Moyer, J.W. 1954. Method of making p-n junction semiconductor unit. US Patent 2, 842, 466, filed June 15, 1954.

Nastasi, M., and Mayer, J. W. eds. 2007. *Ion implantation and synthesis of materials.* Berlin: Springer-Verlag.

Nelson, R. S., and Thompson, M. W. 1963. The penetration of energetic ions through the open channels in a crystal lattice. *Philosophical Magazine* 8:1677.

Ohl, R. S. 1950. Semiconductor translating device. US Patent 2,750,541, filed Jan. 31, 1950.

Piercy, G.R., Brown, F., Davies, J.A., and McCargo, M. 1963. Experimental evidence for the increase of heavy ion ranges by channeling in crystalline structure. *Physics Reviews Letters* 10:399.

Rimini, E. 1995. *Ion implantation: Basics to device fabrication.* Alphen aan den Rijn, Netherlands: Kluwer Academic Publishers.

Robinson, M. T., and Oen, O. S. 1963. Computer studies of the slowing down of energetic atoms in crystals. *Physics Reviews* 132:2385.

Shao, L. 2008. Towards high accuracy in channeling Rutherford backscattering spectrometry analysis. *Nuclear Instruments and Methods in Physics Research B* 266:961.

Shao, L., and Nastasi, M. 2005. Methods for the accurate analysis of channeling Rutherford backscattering spectrometry. *Applied Physics Letters* 87:064103.

Shockley, W. 1949. Semiconductor translating device. US Patent 2,666,814, filed Apr. 27, 1949.

Shockley, W. Forming semiconductive devices by ionic bombardment. US Patent 2,787,564 filed Oct. 28, 1954.

Stark, J. 1912. Bemerkung fiber Zerstreuung und Absorption. *Zeitschrift für Physik* 13:973.

Stolk, P. A., Gossmann, H.-J., Eaglesham, D. J., Jacobson, D. C., Rafferty, C. S., Gilmer, G. H., Jaraiz, M., Poate, J. M., Luftman, H. S., and Haynes, T. E. 1997. Physical mechanisms of transient enhanced dopant diffusion in ion-implanted silicon. *Journal of Applied Physics* 81:6031.

Thompson, M. W. 1964. Effect of proton channeling at 2.8 MeV on the Cu^{65} (p, n) Z_n^{65} reaction rate in a single crystal of Cu. *Physical Review Letters* 25:756.

Wolf, S., and Tauber, R. N. 1986. *Silicon processing for the VLSI era. Volume 1—Process technology.* Sunset Beach, CA: Lattice Press.

Zhu, L., Thompson, P. E., Zhang, X., Hollander, M., and Shao, L. 2007. Displacements of Sb atoms in supersaturated Sb-doped Si layer formed by molecular beam epitaxy growth. *Journal of Applied Physics* 101:086110.

SUGGESTED READING

Chu, W.-K., Mayer, J. W., and Nicolet, M.-A. 1978. *Backscattering spectrometry.* New York: Academic Press.

Feldman, L. C., Mayer, J. W., and Picraux, S. T. 1982. *Materials analysis by ion channeling.* New York: Academic Press.

Nuclear Energy Research Applications

Yongqiang Wang and Amit Misra

12.1 INTRODUCTION

To address energy shortages and environmental issues that the world faces today and in the future, nuclear energy needs to be part of the solution. One of the great challenges in nuclear energy research and development is to develop new materials with radically extended performance limits in irradiation, high temperature, and corrosive environments occurring in nuclear reactors or fusion devices. Thus, the overarching goal in materials discovery for nuclear energy applications is to design and perfect atom- and energy-efficient syntheses of revolutionary new materials that maintain their desired properties while being driven very far from equilibrium. One approach for meeting this challenge is by manipulating atomic structures and properties of interfaces in nanoscale materials using directed synthesis, guided by predictive design principles that have been developed by advances in theoretical understanding of the role of interfaces in the evolution of defects and damage produced by irradiation.

Neutron damage in materials in a reactor environment includes not only atomic displacements through collision cascades, but also the concurrent generation of new elemental species through transmutation or nuclear reactions. In particular, helium (He) and hydrogen (H) species produced by (n,α) and (n,p) reactions in nuclear reactors or fusion devices may significantly degrade mechanical properties of materials. For instance, as much as 1500 appm He may be generated at 10 appm/displacement per atom (dpa) in the plasma facing walls of tokomaks (Zinkle 2005). The subsequent nucleation of He bubbles at GBs promotes void growth and leads to embrittlement (Trinkaus and Ullmaier 1994). At large, He concentrations and elevated temperature bubbles can grow into voids and blisters (Galindo et al. 2004), resulting in the ejection of unwanted impurities back into the plasma that could disrupt or even quench the plasma. Therefore, helium management in plasma facing components is a very important issue. One way to address this problem may be to trap He at nanoscale inclusions (e.g., YTiO clusters in oxide-dispersed-strengthened [ODS] steels) (Yamamoto et al. 2007). Another may be to synthesize materials containing interfaces with increased resistance to He-assisted degradation. He bubble formation and subsequent unstable void growth may be delayed at interfaces that can trap high He concentration without forming bubbles, allowing such materials to remain in service longer than existing structural alloys. Furthermore, if these interfaces can promote He outgassing—as demonstrated in Cu–Nb (Demkowicz et al. 2007)—then He-assisted degradation may be further delayed. The distribution of favorable sites to trap He at face-centered cubic (fcc)–body-centered cubic (bcc) heterophase interfaces depends on the atomic structure of the interface as determined by the interface geometry: lattice misfit and the orientation relationship between the two phases and the habit plane of the interphase boundary. Thus, it may be possible to tailor interface He interactions by adjusting the distribution of interface constitutional vacancies via judicious selection of interface crystallography to control the configuration of misfit dislocation intersections (Demkowicz, Wang, and Hoagland 2008).

In general, accelerator ion beams play an important role in advancing nuclear energy research in three different ways. First, ion irradiation provides displacement damage in materials that bears similarity to neutron damage but with significantly greater damage rates (displacements per atom per second). Higher damage rates by ion irradiation make high radiation damage tests (e.g., 10s to 100s of dpa) feasible on advanced materials or new microstructures at a laboratory timescale. Second, He or H ion implantation allows for incorporating He or H atoms in research targets in a very controlled way in terms of doping concentration and depth distribution. In fact, there are a number of ion beam facilities available in the world that allow for simultaneous heavy ion irradiation (dpa) and He/H implantation (appm) so that He/dpa and/or H/dpa ratios (appm/dpa) in a given neutron environment are better emulated (Serruys et al. 2008). Third, ion beam analysis (IBA) techniques, particularly nuclear reaction analysis (NRA) and elastic recoil detection (ERD) analysis, provide excellent characterization tools to measure He and H depth profiles in materials of interest for nuclear energy applications (Wang and Nastasi 2009).

While there are a handful of analytical techniques for elemental identification and depth profiling, few are available or effective for detecting H and He. For example, most x-ray techniques for compositional analysis are "blind" to hydrogen and insensitive to helium. Sputtering techniques for depth profiling, such as secondary ion mass spectrometry (SIMS), may inevitably alter H or He composition due to outgassing during the sputtering process if they are not bonded to other species in the matrix. Ion beam analysis provides a unique, nondestructive, and reliable approach to measure depth profiles of H and He and their isotopes. In this chapter, the first two examples show the application of ion beam analysis techniques to interface stability of ion-implanted metallic multilayers. We will use a third example to demonstrate in detail how (d, p) nuclear reactions are used to determine accurately $^{13}C/^{12}C$ ratios in debris collection experiments that are designed to help diagnose nuclear implosion processes in inertial confinement fusion research.

12.2   STABILITY OF HELIUM ION-IMPLANTED METALLIC MULTILAYERS

Nanoscale metallic multilayers have been studied extensively in recent years because of their unique properties (Clemens, Kung, and Barnett 1999; Hoagland et al. 2002; J. Wang and Misra 2011; Misra and Thilly 2010; Misra 2006). The strengths of metallic multilayers, composed of alternating layers of soft metals such as Cu and Nb, approach the theoretical limit of material strength when the bilayer periods are of the order of a few nanometers. Furthermore, certain nanolayered structures also exhibit very high morphological stability and strength retention following high-temperature annealing (Misra, Hoagland, and Kung 2004) and severe plastic deformation (Misra and Hoagland 2007). The thermal stability of these composites suggests that they may also be stable under ion irradiation environments (Höchbauer et al. 2005; Misra et al. 2007).

Ion irradiation of materials results in both the introduction of the implantation species into the material and the displacement of the target atoms. These two processes are commonly referred to as ion implantation and ion irradiation damage, respectively. One consequence of the displacements of the target atoms is the ion beam mixing of an otherwise immiscible multicomponent or multilayer structure. This example investigates the morphological stability of nanolayered composite materials, subjected to He ion implantation, and their capability to store the ion-implanted He species (Höchbauer et al. 2005).

To demonstrate the usefulness of ion beam analysis in studying the irradiation response and interfacial stability of immiscible metallic multilayers, we chose Cu/Nb multilayer thin films as a model binary metallic system. This system contains essentially immiscible components with the heat of mixing of approximately 4 kJ/mol. Cu–Nb multilayers with equal Cu and Nb layer thicknesses were synthesized by dc magnetron sputtering at room temperature on silicon or carbon substrates. Two different sets of Cu–Nb multilayer thin films with a total film thickness of 240 nm were produced, differing in the individual layer thickness. One set of films consisted of three bilayers, each with a layer thickness of 40 nm; the other set was composed of 30 bilayers, each with a layer thickness of 4 nm. To isolate the role of the interface structure, we also sputter deposited control samples of 240 nm thick Cu and Nb films, respectively. After the deposition, the samples were implanted at room temperature with 33 keV ^{4}He$^+$ ions to a dose of 1.5×10^{17} at./cm^2. Following the He ion implantation the samples underwent vacuum annealing for 1 hour at 600°C. The ion beam energy used is roughly three orders of magnitude higher than the displacement energies of the target atoms (~20 eV); thus, one would expect that the chemical free energy of the system would have a limited influence on ion beam mixing processes that occur far from thermodynamical equilibrium. However, ion beam mixing experiments have shown that the chemical free energy can play a significant role in the radiation resistance of materials, offering a driving force that may compete with the ballistic mixing of target atoms in an immiscible system.

To determine the stability of the nanolayered structure upon He ion implantation, the Cu–Nb nanolayered samples were characterized by Rutherford backscattering spectrometry (RBS) before and after the He ion implantation, as well as after postimplantation annealing. While a typical RBS measurement uses a 2 MeV ^{4}He$^+$ ion beam, to resolve a thin sublayer structure of a Cu–Nb multilayer system, a higher energy of 4.5 MeV ^{4}He$^+$ ion beam was used in this analysis. The RBS detector was located 167° in reference to the beam direction and the incident beam was kept at the normal direction of samples. The detector resolution was approximately 14 keV. Figure 12.1 plots the RBS spectra of the 40 nm sublayer thickness samples (Höchbauer et al. 2005). The figure shows only the energy interval comprising the signal from the Cu–Nb multilayer film. The spectra show six peaks: three large ones at higher energy (channel 620–680) and three lower ones at lower energy (channel 560–620), corresponding to the individual Nb and Cu layers, respectively. The spectrum of the as-implanted sample overlays perfectly with that of the as-deposited sample, indicating the physical stability of the Cu–Nb multilayer film structure upon the He ion implantation. Simulation of these RBS spectra with the computer program RUMP (Doolittle 1985) revealed that both samples exhibit a discrete layer structure. Intermixing of the Cu and Nb atoms near the interfaces due to the He ion bombardment is beneath the detection level of this analysis method. It is worth mentioning that RBS measures atomic concentrations in atoms per square centimeter. Therefore, homogeneous swelling of the He-implanted film, if it occurs, will not be detected by RBS.

We also simulated the He ion implantation with the computer code SRIM (stopping and range of ions in matter) (Ziegler, Biersack, and Littmark 1985). Typically, SRIM calculations do not account for any redistribution of the recoiled target atoms after they come to rest and therefore only identify the unannealed displacement cascades caused by the ion irradiation process. The calculations reveal that considerable ballistic intermixing of Cu and Nb atoms takes place at the Cu–Nb interfaces, extending 5 nm into the adjacent layers. To illustrate how the RBS spectra would change due to such a disturbance of the layered structure near the interface, we superposed

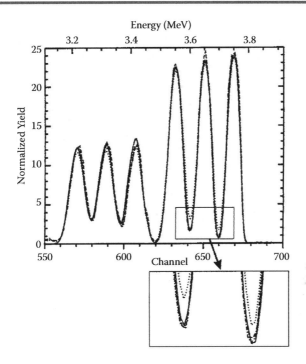

FIGURE 12.1   4.5 MeV ^{4}He^{++} RBS spectra of the Cu/Nb sample with 40 nm layer thickness in the as-deposited state (solid line) and after He ion implantation (dashed line). Also shown for comparison is the simulation of the spectrum assuming ion beam mixing as predicted by SRIM (dotted line). (From Höchbauer, T. et al. 2005. *Journal of Applied Physics* 98: 123516.)

a simulation of the RBS spectrum (dotted line in Figure 12.1), assuming the morphology of the multilayer structure after the He ion implantation as predicted by SRIM. The superposed simulation shows that the mixing due to the ion implantation would have led to a rise of the valleys between the various niobium peaks in the RBS spectra. The fact that this was not observed suggests that dynamic demixing near the interface did take place during the He ion implantation. Cross-sectional transmission electron microscopy (TEM) measurements (not shown here; see Höchbauer et al. 2005) confirmed that even at the film with a layer thickness of 4 nm, the layered structure is still preserved. Furthermore, the morphological waviness of the individual layers, typical of as-deposited films, did not change due to the ion-irradiation process. Furthermore, high-resolution TEM of interfaces and selected area diffraction patterns did not reveal any amorphous phase at the interface due to intermixing of Cu and Nb. (Due to the high concentration of ^{63}Cu and ^{65}Cu atoms in copper, the copper signal consists of a superposition of these two isotopes. This results in high signals between the copper peaks and therefore an increase of the signal between the copper peaks does not occur.)

While the RBS measurement gives important information on Cu and Nb depth profiles and thus the stability of the layered structure, RBS does not detect the implanted ^{4}He atoms in the film. To determine the depth profile of the implanted ^{4}He ions in the film, ERD was used. ERD is advantageous in the detection of light elements, with the relevant information being carried by the target nuclei themselves, not by the backscattering of the bombarding ions, as in RBS. Although in conventional ERD measurements, the ^{4}He$^+$ incident ion beam is normally used to detect hydrogen and its isotopes, for He measurement in this case, a heavier incident beam has to be used. Based on ERD kinematics, we have chosen 12 MeV ^{12}C^{3+} ions as the projectiles to measure implanted ^{4}He atoms in the film. The sample normal is oriented 75° from the incident beam so that the recoil He atoms in the sample can be detected by a silicon surface barrier detector located 30° from the beam direction. To avoid detection of forward scattered ^{12}C particles, a 14.5 μm thick Mylar foil was positioned between the sample and the helium detector.

Figure 12.2 shows the ERD spectra obtained from all He ion implantation samples, before and after annealing (Höchbauer et al. 2005). Superposed with the spectra are the simulations of the spectra assuming the He concentration depth distribution predicted by SRIM (solid line) (Ziegler et al. 1985). Also shown are the He concentration depth profiles for the Cu–Nb multilayer samples, derived from the simulations of the ERD spectra. The data reduction from ERD spectra to He depth profile was carried out with the computer program RUMP (Doolittle 1985). The simulations clearly show the measured as-implanted He concentration depth profiles to be in agreement with the results calculated by SRIM, suggesting that no long-range diffusion leading to the redistribution or escape of the implanted He occurred. In the depth region of the maximum helium content, the He concentration reaches values of 21 at.% in Nb and 13 at.% in Cu. Conversely, ERD analyses on the annealed samples reveal changes in the He concentration depth profiles, which are dependent on the structure of the implanted sample. In the Nb reference sample, no change in the He concentration depth distribution occurred upon annealing. On the other hand, the ERD spectrum of the Cu reference sample reveals that only about one-third of the implanted He is retained in the sample after annealing. For the annealed Cu–Nb multilayer thin films, the ERD spectra show a dependence of the He retention on the individual layer thickness (i.e., the area density of Cu–Nb interfaces in the samples). In the sample with 40 nm layer thickness, almost half of the implanted helium left the sample, while the helium release amounts to only 10% in the sample with a layer thickness of 4 nm.

To complement the ERD results, electron microscopy techniques are used to examine the He bubble formation and growth behavior upon ion implantation and postimplantation annealing. Through-focus imaging in TEM (Höchbauer et al. 2005; Bhattacharyya et al. 2012) (not shown here) revealed the existence of helium bubbles with diameters of about 1–2 nm, extending from the surface to the depth of about 250 nm. He bubbles appear visible throughout the entire He-implanted region, but accumulate with higher density at the Cu–Nb

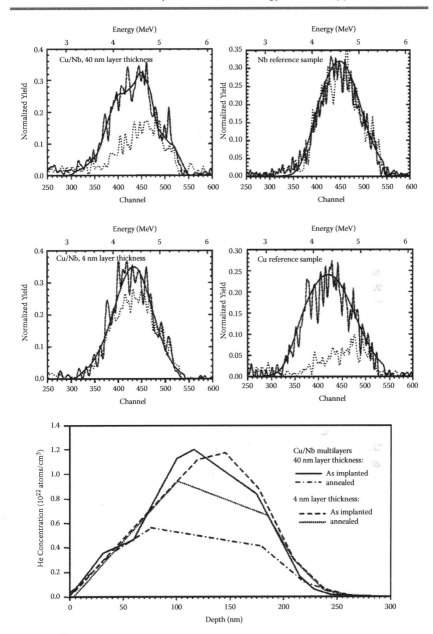

FIGURE 12.2   12 MeV $^{12}C^{3+}$ ERD spectra of all He ion-implanted samples before annealing (dashed lines) and after annealing at 600°C for 1 hour (dotted lines). The simulations of the as-implanted spectra are superposed (solid line). Also shown are the He concentration depth distributions in the Cu/Nb multilayer samples before and after annealing, determined by ERD spectra simulations. (From Höchbauer, T. et al. 2005. *Journal of Applied Physics* 98: 123516.)

interfaces for both Cu–Nb multilayer thin films and at the columnar grain boundaries in the sample with a layer thickness of 40 nm. The TEM images of the samples after the heat treatment show that, upon annealing at 600°C, the He bubbles decrease in density and increase in size. However, the He bubble growth did not destroy the nanolayered structure of the multilayer films. Scanning electron microscopy (SEM) (Höchbauer et al. 2005) did not reveal any surface blisters in the Nb reference sample and the Cu–Nb thin film with a layer thickness of 4 nm, while both the Cu reference sample and the Cu–Nb multilayer thin film with a layer thickness of 40 nm showed a high density of surface blisters with average diameters of 1 and 2.5 μm, respectively.

These results indicate that nanolayered composites may have remarkable resistance to radiation damage at room temperature. Additional studies were performed to evaluate the radiation damage resistance at elevated temperatures. To study the mechanisms of He escape during implantation in Cu–Nb multilayer composites, again we use 12 MeV $^{12}C^{3+}$ ERD to investigate the degree of retention of 33 keV $^4He^+$ implanted to a nominal fluence of 1×10^{17} ions/cm^2 at temperatures ranging from 490°C to 660°C in Cu–Nb multilayer composites with three different layer thicknesses of 4, 40, and 100 nm. Figure 12.3 shows

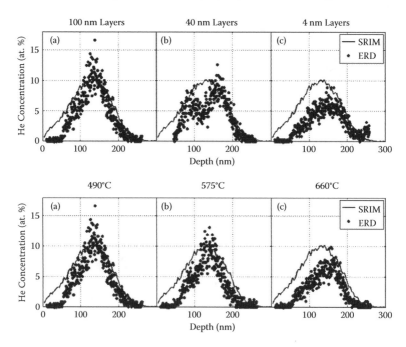

FIGURE 12.3 12 MeV $^{12}C^{3+}$ ERD spectra reduced-depth profiles of retained He concentrations for CuNb samples for two sets of samples. Upper: different layer thickness films under the same implantation temperature of 490°C; lower: different implantation temperatures for the 100 nm thickness samples. (From Demkowicz, M. et al. 2007. *Nuclear Instruments and Methods B* 261:524.)

ERD spectra reduced depth profiles of retained ^{4}He concentrations for Cu–Nb for two sets of samples: (upper) different layer thickness films under the same implantation temperature of 490°C, and (lower) different implantation temperatures for the 100 nm thickness samples (Demkowicz et al. 2007). Several striking features are immediately apparent upon examination of these figures. In all cases, beyond a depth of about 150 nm there is excellent agreement between the He concentration depth profiles predicted by SRIM and those determined by ERD. For depths less than 120 nm, however, the He concentration obtained by ERD is in all cases lower than that predicted by SRIM. This deviation is more pronounced for samples with smaller layer thicknesses and for samples implanted at higher temperatures. Furthermore, the peak He concentrations decrease with decreasing layer thicknesses and with increasing implantation temperatures.

The observations listed here suggest that He implanted under the conditions of interest in this study tends to escape more easily from Cu–Nb targets with smaller layer thicknesses and at higher temperatures. Furthermore, since Figure 12.3 shows that the largest decreases in retained He concentration occur at small depths beneath the surface of the Cu–Nb samples, it can be suspected that the cause of the lowering of retained He concentrations is the escape of implanted He out of the sample's free surface through blistering and bursting. This mechanism is responsible for the He release in the preceding Cu–Nb multilayers implanted by He ions at room temperature and subsequently annealed at 600°C (Höchbauer et al. 2005). Since SEM investigations of the free surface of these Cu–Nb multilayers implanted under high temperatures showed few blisters that did form but exhibited no signs of having burst open to release the He trapped underneath, the likely means of He escape from these Cu–Nb samples is by diffusion out of the samples' free surface. One other prominent feature of the plots shown in Figure 12.3 is that all experimentally determined distributions of retained He concentration are unimodal except for the 40 nm thickness sample, which exhibits two concentration peaks spaced about 80 nm apart, suggesting that He perhaps escapes more easily from layers of one of the two element types.

In view of the fact that atomic diffusion is the likely mechanism of the escape of implanted He from the Cu–Nb multilayer samples, it is not surprising that the amount of retained He should decrease with increasing temperature. The fact that it should decrease with decreasing layer thickness, however, suggests an intriguing deduction: that the interfaces between neighboring Cu and Nb layers—of which there is a greater volume fraction in composites with smaller layer thicknesses—serve as fast diffusion pathways for implanted He leaving the Cu–Nb samples. Atomic-scale simulations of radiation damage cascades indicate that this property stems from the ability of Cu–Nb interfaces to absorb defects created during these cascades (Demkowicz et al. 2007). In addition to being important in its own right as an insight into the behavior of interfaces between Cu and Nb, this finding holds out the possibility of playing a significant technological role in the development of radiation environment-tolerant materials for application in advanced nuclear power systems.

In summary, in conjunction with TEM and SEM, ion beam analysis techniques, in particular RBS and ERD, demonstrate that Cu–Nb multilayer thin films, particularly those with a layer thickness of a few nanometers, have been shown to be extremely resistant against He ion irradiation-induced interfacial mixing and surface blistering. Although no local increase in He concentration was detected by ERD, the as-implanted samples showed higher densities of He bubbles at grain boundaries and Cu–Nb interfaces. For samples with different layer thickness and implantation temperature, ERD demonstrates that the fraction of retained He decreases with increasing implantation temperature and decreasing layer thickness, suggesting that interfaces between neighboring Cu and Nb layers could have served as pathways for fast diffusion of implanted He out of the Cu–Nb multilayer composites. Atomistic simulations indicate that this role played by interfaces in the radiation damage resistance of the Cu–Nb multilayer composites arises from their ability to preferentially trap point defects created during the irradiation process.

12.3   THE EFFECT OF INTERFACE ATOMIC STRUCTURE ON HE BUBBLE FORMATION AT FCC–BCC INTERFACES

Some materials contain constitutional vacancies that are thermodynamically stable at arbitrarily low temperatures—for example, grain boundaries (GBs) in ceramics and semiconductors, compounds with wide phase fields like NiAl, and certain metal hydrides. Since fast neutrons from fusion reactions have much higher probabilities of producing He atoms in reactor materials through (n, α) reactions, interfaces and/or grain boundaries with structures tailored to minimize precipitation and growth of He bubbles may be used to design damage-resistant composites for fusion reactors. In this example, we intend to demonstrate, with the help of nuclear reaction analysis and transmission electron microscopy, that the excess atomic volume at heterointerfaces forming along closely packed fcc–bcc planes has an important effect on He bubble formation characteristics at the interfaces.

All the heterointerfaces in magnetron-sputtered Cu–Nb and Cu–V multilayers form along closely packed fcc and bcc planes and purely in the Kurdjumov–Sachs (KS) orientation relation: A <110> direction in the (111) fcc interface plane is parallel to a <111> direction in the (110) bcc interface plane, as in Figure 12.4(a). Figure 12.4(b) shows a plan view of the Cu–Nb interface in a model that was relaxed using a specially constructed embedded atom method (EAM) potential (Demkowicz and Hoagland 2009). While not strictly periodic, this interface contains a quasi-periodic pattern of sites where a Cu atom and a Nb atom are nearly "on top" of each other. These sites occur at the intersections between two sets of parallel interface misfit dislocations, as elaborated elsewhere (Demkowicz, Wang, et al. 2008; Demkowicz, Hoagland, and Hirth 2008). Figure 12.4(c) demonstrates the effect of removing an atom (creating a vacancy) near a misfit dislocation intersection in the interface Cu plane. The formation

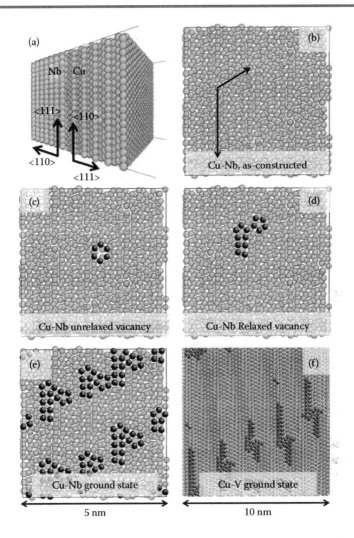

FIGURE 12.4 (a) Cu–Nb bilayer in the experimentally observed crystallography; (b) quasi-periodic pattern formed by interface Cu and Nb planes; (c) unrelaxed interface vacancy; (d) relaxed interface vacancy; (e) ground state Cu–Nb interface with 5 at.% constitutional vacancies; and (f) ground state Cu–V interface with ~0.8 at.% constitutional vacancies. Atoms colored dark have less than six nearest neighbors in the interface Cu plane. (From Demkowicz, M.J. and and Hoagland, R.G. 2009. *International Journal of Applied Mechanics* 1:421.)

energy of the unrelaxed vacancy is 1.5 eV, comparable to fcc Cu (1.3 eV) (Demkowicz et al. 2010; Demkowicz, Misra, and Caro 2012). Upon annealing for 10 ps at 300K followed by energy minimization, the vacancy reconstructs into the configuration in Figure 12.4(d) and its formation energy drops to −0.13 eV. Thus, the structure in Figure 12.4(d) is energetically favorable to the vacancy-free interface in Figure 12.4(b) and the reconstructed vacancy may be viewed as a constitutional vacancy. To find the ground state interface, one can iteratively add vacancies until no negative formation energy sites remain. For Cu–Nb, this structure is shown in Figure 12.4(e). It has ~25 mJ/m² lower energy than that in Figure 12.4(b) and contains 5 at.% constitutional vacancies concentrated near misfit dislocation intersections (i.e., ~2.5 vacancies per intersection). A similar calculation for a Cu–V interface also shows ~2.5 vacancies per intersection and an energy reduction of ~3.4 mJ/m². Due to the differing lattice parameters of Nb and V, however, the areal density of misfit dislocation intersections in Cu–V is smaller, giving an overall ~0.8 at.% constitutional vacancy concentration, shown in Figure 12.4(f).

These theoretical predictions in Figure 12.4 may be indirectly verified by quantifying the He solubility limits at Cu–Nb and Cu–V interfaces. In perfect crystalline metals, formation energies of He interstitials are close to self-interstitial formation energies (Zhang et al. 2008). Consequently, He solubility is below one part per trillion even near the melting temperature (Demkowicz and Hoagland 2009). He therefore preferentially occupies high excess volume sites like vacancies and vacancy clusters. Due to their constitutional vacancies, both Cu–Nb and Cu–V interfaces have high excess volume (Seletskaia et al. 2005) and thus may exhibit elevated He solubility compared to perfect crystalline metals. Furthermore, the solubility limit should be higher in Cu–Nb than in Cu–V interfaces due to a higher constitutional vacancy concentration in Cu–Nb interfaces.

However, experimental measurement of He solubility at interfaces poses serious challenges: Both local He concentration and whether He has precipitated out must be determined. Due to high solubility, the He concentration needed to nucleate bubbles at the interfaces may be considerably increased compared to the bulk, meaning that He could be present in the form of supersaturation without precipitating. Furthermore, TEM might not be able to resolve the smallest, incipient clusters formed in the early stages of precipitation. Thus, rather than attempt a precision measurement of He solubility itself, in this example we propose to measure the critical He concentration at which bubbles are unambiguously detected via through-focus imaging in TEM: a quantity that, like solubility, should scale with interface excess volume.

Here we chose NRA for the He measurement due to its higher sensitivity and better depth resolution than ERD. Since there are no useful nuclear reactions from the major helium isotope, ^{4}He, a deuterium-induced nuclear reaction on ^{3}He isotope was used (Paszti 1992):

(12.1) $d + {}^3He \rightarrow {}^4He + p + 18.352$ MeV

This reaction has one of the highest Q-values (18.352 MeV) among many common nuclear reactions for light element analysis, suggesting that the reaction products to be detected can be easily separated from those of other common interfering light elements and that the elastically scattered deuterium

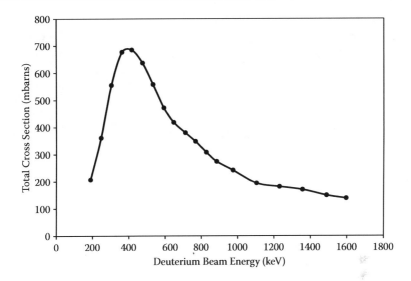

FIGURE 12.5   The total cross section for the ^{3}He (d, p) ^{4}He reaction as a function of deuterium beam energy. (From Bonner, T. W. et al. 1952. *Physics Reviews* 88:473.)

particles can be easily filtered out with a simple range foil put in front of the particle detector. Figure 12.5 shows a total cross section of the ^{3}He (d, p) ^{4}He reaction as a function of the incident deuterium beam energy (Bonner, Conner, and Lillie 1952).

Two Cu–Nb multilayers were magnetron sputtered onto Si substrates, as in previous studies: one 335 nm overall thickness with 5.6 nm thick sublayers and another 340 nm overall thickness with 2.8 nm thick sublayers. These samples were implanted at room temperature with 35 keV ^{3}He ions to a fluence of 1×10^{17} ions/cm^2. To reduce channeling, samples were tilted 7° from the incident ion beam during the implantation. To improve the depth resolution, ^{3}He depth profiles were computed from measured alpha particle energies (instead of proton particle energies) from the ^{3}He (d, α) ^{1}H reaction using the data reduction software SIMNRA (Mayer 1997). The incident deuterium beam energy was chosen at 600 keV and the alpha particle detector was located at 167° from the beam direction. An 8 μm thick Mylar foil was placed in front of the detector to stop the backscattered deuterium particles. The solid angle subtended by the detector was approximately 2.3 msr. The total accumulated charge of the incident deuterium beam for each spectrum was 180 μC with a deuterium beam current of ~50 nA. The stopping power of implanted ^{3}He in the Cu/Nb matrix was neglected in the analysis due to its low concentration. The concentration uncertainty—primarily determined by the reaction cross section near the surface, ~0.2 at./nm^3—was used as a concentration uncertainty estimate throughout the sample depth. The ^{3}He detection limit under our experimental setup was ~0.5 at./nm^3. The depth

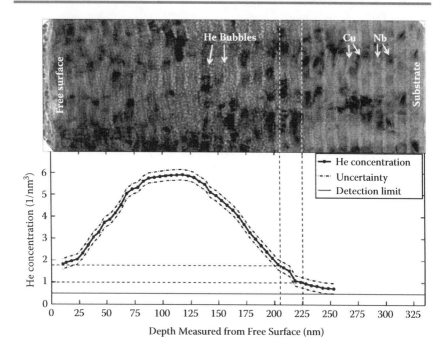

FIGURE 12.6 BFTEM image at ~1 mm under focus of ^{3}He-implanted Cu–Nb multilayers with 5.6 nm thick layers (top) and the corresponding ^{3}He concentration profile measured by NRA (bottom). No bubbles are seen below a critical depth of 215 ± 10 nm (indicated by dashed lines), corresponding to a ^{3}He concentration of 1.4 ± 0.6/nm^3. (From Demkowicz, M. et al. 2010. *Applied Physics Letters* 97:161963.)

resolution near the surface, mainly determined by the beam straggling in the range foil, was ~40 nm in Cu and ~43 nm in Nb.

He bubbles in 100 nm foils taken from the implanted samples were examined using through-focus bright field (BF) imaging in TEM. Figure 12.6 shows an underfocused BFTEM image where bubbles appear as white dots (dark in overfocus, not shown). The diameters of the smallest detectable bubbles were estimated at ~1 nm from images at different defocuses using the approach outlined in Ruhle and Wilkens (1975). In the 5.6 nm sublayer sample numerous bubbles formed at depths containing high ^{3}He concentration, with the number density decreasing in the low concentration tails. Visual inspection shows no bubbles below a critical depth of 215 ± 10 nm, even though the measured ^{3}He concentration there is approximately 1.4 ± 0.6 at./nm^3 (~1.7 at.%), as inferred by comparing the TEM image with the measured ^{3}He concentration profile in Figure 12.6 At greater depths the ^{3}He concentration is too low to form the

~1 nm bubbles resolvable elsewhere in the sample. The critical concentration, $1.4 \pm 0.6/nm^3$, is significantly higher than the (immeasurably small) concentration required to observe bubbles in pure fcc and bcc metals, implying that all the implanted He above the critical depth is trapped at interfaces.

Figure 12.6 also shows that ^{3}He bubbles did not form within ~25 nm of the free surface, even though the ^{3}He concentration there exceeded the critical concentration of 1.4 ± 0.6 at/nm^3. This higher critical concentration may be due to trapping of implanted ^{3}He at radiation-induced vacancies. Such He vacancy complexes are much less mobile than interstitial He and may prevent He from reaching nearby interfaces, especially in multilayers with thick individual layers. The concentration of radiation-induced vacancies decreases with depth faster than that of He, suggesting that the critical concentration measured at the end of the implantation range, albeit possibly still an overestimate, better indicates the amount of He trapped at interfaces than the concentration near the free surface. This interpretation was further supported by the data obtained from the 2.8 nm sublayer sample, whose interface area per unit volume is twice that of the 5.6 nm sublayer sample. If all implanted He above the critical depth is trapped at interfaces, the critical He concentration needed to image bubbles in the 2.8 nm sublayer sample should be twice that of the 5.6 nm sublayer sample (i.e., 2.8 ± 1.2 at./nm^3). Indeed, measurements like those in Figure 12.6 show no He bubbles in the 2.8 nm multilayer above a critical depth of 195 ± 10 nm, corresponding to a ^{3}He concentration of 3.3 ± 0.6 at./nm^3, within the uncertainty range of the predicted value.

For samples of layer thicknesses t, the He concentration per unit interface area C_i can be calculated as $C_i = tC_c$, where C_c is the critical He concentration per unit volume, measured previously. Averaging the results from the 5.6 nm multilayer (i.e., 5.6 nm * 1.4 at./nm^3 = 7.84 at./nm^2) and the 2.8 nm multilayer (i.e., 2.8 nm * 3.3 at./nm^3 = 9.24 at./nm^2), a critical Cu–Nb interface He concentration of 8.5 ± 2.5 at./nm^2 to resolve bubbles is obtained. Similar experiments conducted on magnetron-sputtered Cu–V multilayers gave a critical He concentration of ~1.9 at./nm^2 for Cu–V interfaces (Fu et al. 2010): a factor of ~4.5 times smaller than in Cu–Nb interfaces—in good agreement with the prediction based on constitutional vacancy concentrations obtained from atomistic modeling (i.e., ~5 at.% constitutional vacancies for Cu–Nb versus ~0.8 at.% for Cu–V).

In summary, we have demonstrated experimentally that ^{3}He (d, p) ^{4}He nuclear reaction analysis is a powerful technique to measure ^{3}He depth profiles with excellent detection sensitivity and depth resolution. Combined with He bubble observation with TEM, we were able to verify quantitatively the theoretical predictions that Cu–Nb interfaces can absorb more He atoms than Cu–V interfaces before He atoms conglomerate into bubbles in the multilayer thin films.

12.4   QUANTIFICATION OF $^{13}C/^{12}C$ TO HELP DIAGNOSE THE NUCLEAR IMPLOSION PROCESS IN INERTIAL CONFINEMENT FUSION EXPERIMENTS

Stable isotope ratios such as $^{13}C/^{12}C$ play an important role in many areas, including environmental and energy applications. One of the most convenient techniques for stable isotope ratio measurements is mass spectrometry (MS), including gas chromatography mass spectrometry (Wong, Muir, and Mabury 2003), multicollector induction coupled plasma mass spectrometry (Santamaria-Fernandez, Carter, and Hearn 2008), etc. However, mass spectrometry techniques in general face a unique challenge in reliably measuring $^{13}C/^{12}C$ ratios due to virtually unavoidable hydrogen carbon ($^1H^{12}C$) mass interference with ^{13}C. While some mass spectrometry techniques such as secondary ion mass spectrometry (SIMS) can help resolve this mass interference issue by operating at high-resolution mode, this mode usually means reduced overall detection sensitivities. To account for the instantaneous hydrocarbon contaminations (natural carbon abundance: 98.9% ^{12}C and 1.1% ^{13}C), it is highly desirable that ^{13}C and ^{12}C be measured simultaneously. Although the proton-induced gamma emission (PIGE) technique found a strong and narrow resonance on ^{13}C ($E_p = 1748$ keV, $\Gamma = 0.075$ keV, $\sigma = 340$ mb) (Woodbury, Day, and Tollestrup 1953; Marion and Hagedorn 1956), there are no useful (p, γ) reactions on the ^{12}C isotope near this proton energy. By examining the (p, γ) nuclear reaction database, one will find that likely (p, γ) reactions for simultaneous ^{13}C and ^{12}C measurements could be at a proton beam energy of ~500 keV. At this energy, it is plausible that one may use both $^{13}C(p, \gamma)^{14}N$ and $^{12}C(p, \gamma)^{13}N$ reactions simultaneously due to significant overlaps in their beam energy dependences of cross-section data, since the resonance widths of these reactions are fairly broad near this energy (Bulter 1958): $^{13}C(p, \gamma)^{14}N$ ($E_p = 551$ keV, $\Gamma = 25$ keV, $\sigma = 1.44$ mb) and $^{12}C(p, \gamma)^{13}N$ ($E_p = 457$ keV, $\Gamma = 35$ keV, $\sigma = 0.127$ mb). Unfortunately, the cross sections of these resonances are practically too small in the case for any trace amount of ^{13}C or ^{12}C analysis. Furthermore, compared with charged particle detection efficiency, relatively poor intrinsic detection efficiency for high-energy gamma rays such as 8–9 MeV from $^{13}C(p, \gamma)^{14}N$ also adds to the disadvantage of this technique.

Deuterium beam-induced (particle, particle) nuclear reactions generally provide good sensitivity for light elements analysis. For example, $^{12}C(d, p)^{13}C$, $^{14}N(d, p)^{15}N$, $^{14}N(d, \alpha)^{12}C$, and $^{16}O(d, p)^{17}O$ have been routinely used for trace amounts of carbon, nitrogen, and oxygen measurements (Amsel 1983; Amsel and David 1969; Jiang et al. 2003; Gurbich and Molotdsov 2004; Pellegrino, Beck, and Trouslard 2004). However, $^{13}C(d, p)^{14}C$ has been rarely used for ^{13}C detection mainly due to its relatively small cross section and high vulnerability from other isotope interference. In this example, we demonstrate that this reaction can be used to measure trace amounts of ^{13}C provided that interference from other isotopes can be carefully avoided or minimized. More

TABLE 12.1   Basic Reaction Parameters of $^{12}C(d,p_0)^{13}C$ and $^{13}C(d, p_0)^{14}C$ for 1.4 MeV Incident Deuterium Beam with Particle Detection Angle of 135°

Isotope	^{12}C		^{13}C	
Reaction	$^{12}C(d, p_0)^{13}C$	$^{12}C(d, p_1)^{13}C$	$^{13}C(d, p_0)^{14}C$	$^{13}C(d, \alpha_0)^{11}B$
Q (MeV)	2.72	−0.37	5.95	5.17
Cross section σ (mb/sr) ($E_0 = 1.4$ MeV, $\theta = 135°$)	36.7	7.9	3.6	7.5
Outgoing particle energy ($E_0 = 1.4$ MeV, $\theta = 135°$)	3.31	0.62	6.27	4.00
Outgoing particle energy (after 25 μm Al range foil)	2.74	0	5.94	0

importantly, we will show that when both $^{13}C(d, p)^{14}C$ and $^{12}C(d, p)^{13}C$ are used simultaneously, one can reliably obtain $^{13}C/^{12}C$ ratios conveniently (From Wang, Y. Q. et al. 2010).

Table 12.1 lists the basic parameters of deuterium-induced (particle, particle) reactions provided that a deuterium beam energy of 1.4 MeV and a silicon detector angle of 135° are used (Colaux, Thomé, and Terwagne 2007; Kokkoris et al. 2006). (Note: The beam energy and the detector angle are the optimum conditions for $^{13}C/^{12}C$ ratio measurement in our experimental setup, as discussed later.) Although Table 12.1 shows that $^{13}C(d, \alpha_0)^{11}B$ has a larger cross section than $^{13}C(d, p_0)^{14}C$, it is more difficult to separate these outgoing alpha particles from scattered deuterium particles and reaction proton particles when a standard range-foil and silicon surface barrier detector system is used. For the remainder of the discussion, we will choose the (d, p_0) reaction for both ^{12}C and ^{13}C measurements.

To measure $^{13}C/^{12}C$ ratio sensitively and reliably, we would need to optimize the beam energy and the detector angle such that the measurement sensitivity for both ^{13}C content and $^{13}C/^{12}C$ ratio are maximized when the deuterium beam-produced neutron radiation background is still acceptable in an ordinary ion beam analysis laboratory. Figure 12.7 shows $\sigma(^{12}C)$, $\sigma(^{13}C)$, and $\sigma(^{13}C)/\sigma(^{12}C)$, respectively, as a function of deuterium beam energy (Colaux et al. 2007; Kokkoris et al. 2006), when a detector angle of 135° is used. Although the cross section of both reactions, especially $^{13}C(d, p)^{14}C$, clearly increases with increasing the detection angle from 90° to 180°, we chose a detector angle of 135° in our measurements simply due to the geometry constraint in our analysis chamber. Thus, based on Figure 12.1, we chose the beam energy of 1.4 MeV and the detector angle of 135° in our experiment.

Once we chose the beam-detector parameters, the next task was to determine ^{13}C sensitivity for our experimental setup. To accomplish this, we prepared

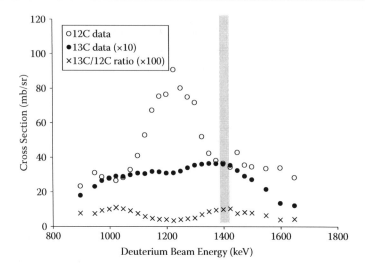

FIGURE 12.7   Reaction cross sections, σ(¹²C), σ(¹³C), and σ(¹³C)/σ(¹²C), respectively, as a function of deuterium beam energy for a fixed detector angle of 135°. (From Colaux, J. L. et al. 2007. *Nuclear Instruments and Methods B* 254:25; Kokkoris, M. et al. 2006. *Nuclear Instruments and Methods B* 249:77.)

a series of known ¹³C standard specimens by implanting 40 keV ¹³C ions into preselected pure tin foils to different fluencies ranging from $1 \times 10^{14} - 5 \times 10^{16}$ ions/cm² using a Varian DF-3000 industrial ion implanter. The ¹³C ion energy and tin target were chosen since a similar beam-sample configuration was used in our inertial confinement fusion (ICF) experiment where samples of interest were produced. The projected range of 40 keV ¹³C ions in Sn is approximately 65 nm, based on the SRIM prediction (Ziegler et al. 1985). The implantation was done using 99.2% highly enriched $^{13}CO_2$ gas in the ion source to ensure that the $(^{12}C^1H)^+$ molecular beam mass interference to $^{13}C^+$ ion beam becomes insignificant during the implantation. The deuterium beam for the (d, p) reaction was produced on an NEC 9SDH-2 Pelletron tandem ion accelerator. A large (300 mm²) surface barrier silicon detector was placed at 135° in reference to the incident beam direction and provided a large solid angle of approximately 100 msr in our setup. The detector has a thick enough depleted layer (>300 μm) to ensure that all the proton particles produced by the reactions are detected with 100% intrinsic detection efficiency. An aluminum foil (with thickness of 25.4 μm) was placed in front of the detector to filter out high flux deuterium particles backscattered from the target nuclei. Total collected charge varied from 6 μC up to 240 μC depending on the amount of ¹³C content in targets. To reduce the neutron radiation background from the slits and targets, the deuterium beam current was limited to ~30 nA with a beam spot size of ~4 mm². The target chamber has a typical vacuum of ~5×10^{-7} Torr during the analysis. Since the carbon of interest is from the near surface, the relationship between

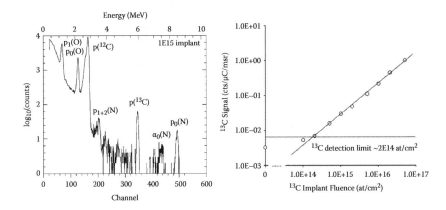

FIGURE 12.8 (a) Energy spectrum of a tin foil implanted with 40 keV ^{13}C to a fluence of 1×10^{15} ions/cm^2 (with the deuterium beam charge of 240 μC); (b) ^{13}C peak area as a function of ^{13}C implant fluence from the tin foil standard targets, showing that a ^{13}C detection limit of ~2×10^{14} atoms/cm^2 is possible if other isotope interference to the ^{13}C signal can be avoided. (From Wang, Y.Q. et al. 2010.)

the number of counts in each carbon peak, $^{12,13}A$, and the respective concentrations, $^{12,13}n$, is given by the following simple formula:

(12.2)
$$\frac{^{13}A}{^{12}A} = \frac{^{13}n}{^{12}n} \frac{^{13}\sigma}{^{12}\sigma}$$

where $^{12,13}\sigma$ are the (d, p) reaction cross sections on the respective carbon isotopes.

Figure 12.8(a) shows a typical energy spectrum from the tin foil standard that was implanted to a fluence of 1×10^{15} ions/cm^2 with a total collected deuterium beam charge of 240 μC. Except for some trace amounts of N and O, the spectrum indicates that the standard tin specimen is very pure and clean. Figure 12.8(a) shows that the proton peak from ^{13}C is well isolated and the peak area can be accurately determined using direct integration with a simple linear background subtraction. However, the proton peak from ^{12}C is noticeably nonsymmetric with interference signals embedded in the lower energy tail of the peak, perhaps due to ^{14}N(d, α_1), ^{2}H(d, p)^{3}H, etc. To determine the peak area for the ^{12}C signal properly, a Gaussian fit to the data points higher than the peak channel was used (not shown). It should be pointed out that the ^{12}C signal pileup is unlikely since its counting rate is extremely low (~8 counts per second) with the beam current of ~30 nA in our measurements. Furthermore, if the pileup from the ^{12}C signal were showing up, it should have been located at channel 330, which is still well separated from the ^{13}C proton signal (channel 350). Figure 12.8(b) shows ^{13}C peak area as a function of the ^{13}C implant fluence from the implanted tin standard targets. It shows that the ^{13}C detection limit of ~2 × 10^{14} at./cm^2 is possible provided that the other isotope interference to ^{13}C signal is negligible. For any unknown ^{13}C target, Figure 12.8(b) provides a system calibration curve from which ^{13}C content of an unknown sample can be obtained based on the measured ^{13}C peak area. The amount of ^{13}C determined by this curve will obviously include the background or residual ^{13}C contamination already present in the unknown samples.

We want to point out that instantaneous change of the $^{13}C/^{12}C$ ratio during the analysis as a result of the deuterium beam bombardment is unlikely. First, if the beam-induced chemical process such as radiolysis or surface redeposition affects carbon content in the sample, it should affect both ^{12}C and ^{13}C components in the same way; hence, the ratio of $^{13}C/^{12}C$ should not be affected. Second, the beam-induced physical process such as surface sputtering or material evaporation (if any) should also affect both ^{12}C and ^{13}C components similarly. Third, the beam-induced nuclear process will, in theory, result in ^{12}C decrease through the $^{12}C(d, p)^{13}C$ reaction and ^{13}C net increase through both $^{12}C(d, p)^{13}C$ and $^{13}C(d, p)^{14}C$ reactions. However, the extent for either ^{12}C decrease or ^{13}C increase is at least seven orders of magnitude smaller than the detection limits of these (d, p) nuclear reaction techniques. Thus, simultaneous measurement of both $^{12}C(d, p)^{13}C$ and $^{13}C(d, p)^{14}C$ reaction products should provide a reliable means for $^{13}C/^{12}C$ ratio determination.

It is obvious that the preceding statement is meaningful only if both ^{13}C and ^{12}C can be accurately measured with these reactions in the experiments. Kinematically, there are various light elements (^6Li, ^7Li, ^9Be, ^{10}B, ^{11}B, ^{14}N, ^{19}F, etc.) (Wang and Nastasi 2009) that could impose particle energy interferences to ^{13}C and/or ^{12}C proton signals in a standard range foil–silicon detector setup like ours. This interference is especially troublesome to the ^{13}C proton signal since it is typically three orders of magnitude weaker than the ^{12}C proton signal unless the specimen of interest has been purposely enriched with ^{13}C constituent. If a carbon sample of interest is a thin film, substrate interference could greatly degrade the analysis sensitivity of ^{13}C or even make $^{13}C(d, p)^{14}C$ reaction analysis totally useless if not properly chosen. We have found that common substrate materials such as silicon, aluminum, and stainless steel, as well as any N-bearing targets such as GaN, etc., all are a poor choice of substrate for ^{13}C analysis with $^{13}C(d, p)^{14}C$ reaction. One piece of good news is that oxygen, in particular ^{16}O, does not produce interference to either ^{12}C or ^{13}C in these (d, p) reaction analyses. Our experiments (data not shown) seem to indicate that any pure element, alloy, or oxide targets that consist of elements heavier than Ti should not impose significant interference to the $^{13}C(d, p)^{14}C$ proton signal when the deuterium beam energy is kept below 1.5 MeV.

We now apply this (d, p) nuclear reaction analysis technique to gauge the effectiveness of a target debris collection setup that is used to diagnose the (d, d) nuclear fusion implosion process in an ICF application. In ICF, the nature and amount of hydrodynamical mixing at the fuel–ablator interface may be diagnosed by measuring the production of ^8Li, from (t, α) reactions in the ^9Be shell of the imploded capsule (Hayes et al. 2006; Grim et al. 2008). To assay the ^8Li, it is necessary to collect the target debris following the implosion, which is typically moving at 10^7–10^8 cm/s and therefore difficult to collect directly. To address this and other challenges, a number of debris collection experiments have been carried out using CH targets at the Department of Energy's OMEGA Laser at the University of Rochester in New York. Plastic CH targets were chosen due to their relative ease of production and low cost. Further, their implosion dynamics at the OMEGA laser are similar to what is expected in the National Ignition Facility (NIF) at Lawrence Livermore National Laboratory. Unfortunately, due to the ubiquitous nature of C and the small quantities of the target mass (~40 µg), directly assaying target material using ^{12}C as the signature is too challenging. To circumvent this, 90% enriched ^{13}CH targets were manufactured, with the intent of using the change in the ratio $^{13}C/^{12}C$ from background measurements as the signature for collection of target material. Further, by measuring the concentration of ^{12}C present, we can then assay the amount of ^{13}C deposited on a collector substrate.

To collect debris coming from the implosion and ablation process, high-purity tin foils were placed at different distances away from the ICF shot epicenter as fusion debris collectors. Two tin collector foils (located 65 cm away from the epicenter) were used in this preliminary study: One foil experienced shots from a natural CH capsule and another one experienced shots from a 90% ^{13}C-enriched ^{13}CH capsule. $^{13}C/^{12}C$ ratios on these tin foils were analyzed using $^{13}C(d, p)^{14}C$ and $^{12}C(d, p)^{13}C$ reactions at the deuterium beam energy of 1.4 MeV. Extracted ^{13}C and ^{12}C contents from the collected proton energy spectra (not shown here; see Wang et al. 2010) from these shot specimens along with the tin foil reference are listed in Table 12.2. There are two important observations from these measurements: (1) ICF shots significantly removed surface carbon layer on the Sn collector foil; and (2) only ICF shots using the ^{13}CH plastic capsule produced enhanced ^{13}C signal on the Sn collector foil.

TABLE 12.2   Summary of the (d, p) NRA Data for the Various Sn Foils Analyzed

Sample	^{13}C Signal (counts/ 240 µC)	^{12}C Signal (counts/ 240 µC)	^{13}C (10^{14} at./cm²)	^{12}C (10^{16} at./cm²)	$^{13}C/^{12}C$
Natural Sn foil	243	220,363	4.60 ± 0.29	4.14 ± 0.009	0.0111 ± 0.0007
Sn-4 front (^{13}CH exposure)					
	238	154,892	4.50 ± 0.29	2.91 ± 0.007	0.0154 ± 0.0010
Sn-4 back (no ^{13}CH exposure)	290	244,848	5.48 ± 0.32	4.60 ± 0.009	0.0119 ± 0.0007
Sn-2 front (nat. CH exposure)	184	159,151	3.48 ± 0.25	2.99 ± 0.008	0.0116 ± 0.0009

In summary, we have demonstrated that $^{13}C(d, p)^{14}C$ and $^{12}C(d, p)^{13}C$, when used simultaneously, can provide reliable $^{13}C/^{12}C$ ratio measurements. Detection sensitivity of ~2E14 at./cm^2 for ^{13}C measurement is achievable provided that other isotope interferences are carefully avoided. Common substrate materials such as silicon, aluminum, and stainless steel produce unwanted interferences to ^{13}C measurements. While oxygen (in particular, ^{16}O) does not interfere with ^{13}C or ^{12}C proton signals, nitrogen isotopes (^{14}N and ^{15}N) can produce serious interference to both ^{12}C and ^{13}C measurements. Preliminary results from ICF specimens suggest that the (d, p) reaction analysis method can provide useful information to help understand the fuel–ablator mix mechanism in the ICF implosion process for fusion energy applications and may find applications in forensic and environmental research.

PROBLEMS

12.1. For the sample shown in Figure 12.1, estimate the surface depth resolutions for Nb and Cu layers when 2 and 4.5 MeV He incident beams are used. Explain why a 4.5 MeV He beam is used to resolve Nb or Cu sublayers instead of a 2.0 MeV He beam.

12.2. John was using a 3 MeV ^4He ion beam to analyze a deuterated polystyrene film. The ERD setup used included a 75° incident beam angle in reference to the sample normal, a 30° detector angle in reference to the beam direction, and a thin Al range foil. Depth resolution of 40 nm was obtained for the surface D measurement. If the beam is replaced with a 12 MeV $^{12}C^{3+}$ ion beam for ^4He plasma-treated polystyrene film, what would you expect the surface depth resolutions for D and He in the films to be? Note: The intrinsic detector resolution of 14 keV can be assumed in both ERD experiments.

12.3. To reduce the background interference and increase the He detection sensitivity, a coincidence technique is often used where both protons and alphas from the same reaction $^3He(d, p)^4He$ are detected. If detector A is placed at 150° in reference to the incident beam direction for proton detection, where should detector B be placed to detect coincidental alpha particles? If detector A at 150° is used to detect alpha particles, where should detector B be placed to detect coincidental protons?

12.4. Suppose that a carbon film was deposited on a very thin foil substrate so that deuterium backscattering events from a 1.4 MeV deuterium beam at 135° can be well separated from the reaction products without using any range foil. If the carbon film contains 2E15 ^{13}C atoms/cm^2, what would be the expected proton and alpha yields from the experiment if the measurement were done using the same experimental geometry and the maximum charge of the deuterium beam (240 μC) in the chapter?

REFERENCES

Amsel, G., and David, D. 1969. La microanalyse de l'azote par l'observation directe de réactions nucléaires applications. *Revue de Physique Appliquée* 4:383.

Amsel, G., and Davies, J. 1983. Precision standard reference targets for microanalysis by nuclear reactions. *Nuclear Instruments and Methods B* 218:177.

Bhattacharyya, D. M., Demkowicz, J., Wang, Y.-Q., Nastasi, M., and Misra, A. 2012. A transmission electron microscopy study of the effect of interfaces on bubble formation in He-implanted nano-scale Cu-Nb multilayers. *Microscopy & Microanalysis* 18:152–161.

Bonner, T. W., Conner, J. P., and Lillie, A. B. 1952. Cross section and angular distribution of the $He^3(d,p)He^4$ nuclear reaction. *Physics Reviews* 88:473.

Bulter, J. W. 1958. US Naval Research Laboratory, NRL Report 5282.

Clemens, B. M. Kung, H., and Barnett, S. A. 1999. *MRS Bulletin* 24:20.

Colaux, J. L., Thomé, T., and Terwagne, T. 2007. Cross section measurements of the reactions induced by deuteron particles on ^{13}C. *Nuclear Instruments and Methods B* 254:25.

Demkowicz, M. J., Bhattacharyya, D., Usov, I., Misra, A., Wang, Y. Q., and Nastasi, M. 2010. The effect of excess atomic volume on He bubble formation at fcc–bcc interfaces. *Applied Physics Letters* 97:161903.

Demkowicz, M. J., and Hoagland, R. G. 2009. *International Journal of Applied Mechanics* 1:421.

Demkowicz, M. J., Hoagland, R. G., and Hirth, J. P. 2008. Interface structure and radiation damage resistance in Cu-Nb multilayer nanocomposites. *Physics Reviews Letters* 100:136102.

Demkowicz, M. J., Misra, A., and Caro, A. J. 2012. The role of interface structure in controlling high helium concentrations. *Current Opinions in Solid State and Materials Science* 16 (3): 101–108.

Demkowicz, M. J., Wang, J., and Hoagland, R. G. 2008. In *Dislocations in solids,* vol. 14, ed. J. P. Hirth, 141. New York: Elsevier.

Demkowicz, M. J., Wang, Y. Q., Hoagland, R. G. et al. 2007. Mechanisms of He escape during implantation in CuNb multilayer composites. *Nuclear Instruments and Methods B* 261:524.

Doolittle, L. R. 1985. Algorithms for the rapid simulation of Rutherford backscattering spectra. *Nuclear Instruments and Methods in Physics Research B* 9:344.

Fu, E. G., and Zhang, X. 2010. Interface enabled defects reduction in helium ion irradiated Cu/V nanolayers. *Journal of Nuclear Materials,* 407:178.

Galindo, R. E., van Veen, A., Evans, J. H. et al. 2004. Protrusion formation and surface porosity development on thermally annealed helium implanted copper. *Nuclear Instruments and Methods B* 217:262.

Grim, G. P. et al. 2008. *Review of Scientific Instruments* 79:10E503-1-5.

Gurbich, A. S., and Molotdsov, S. 2004. Application of IBA techniques to silicon profiling in protective oxide films on a steel surface. *Nuclear Instruments and Methods B* 226:637.

Hayes, A. C., Jungman, G., Solem, J. C., Bradley, P. A., and Rundberg, R. S. 2006. *Modern Physics Letters* A 21:1029.

Hoagland, R. G., Mitchell, T. E., Hirth, J. P., and Kung, H. 2002. *Philosophical Magazine* A 82:643.

Höchbauer, T., Misra, A., Hattar, K., and Hoagland, R. G. 2005. *Journal of Applied Physics* 98:123516.

Jiang, W., Shutthanandan, V., Thevuthasan, S., McCready, D. E., and Weber, W. J. 2003. Oxygen analysis using energetic ion beams. *Nuclear Instruments and Methods B* 207:453.

Kokkoris, M., Misaelides, P., Kossionides, S., Zarkadas, Ch., Lagoyannis, A., Vlastou, R., Papadopoulos, C. T., and Kontos, A. 2006. A detailed study of the 12C(d, p0)13C reaction at detector angles between 135° and 170°, for the energy range $E_{d, \text{ lab}}$ = 900–2000 keV. *Nuclear Instruments and Methods B* 249:77.

Laakmann, J., Jung, P., and Uelhoff, W. 1987. *Acta Metallica* 35:2063.

Marion, J. B., and Hagedorn, F. B. 1956. *Physics Reviews* 104:1028.

Mayer, M. 1997. *SIMNRA user's guide.* Garching, Germany: Max Planck Institut fur Plasmaphysik.

Misra, A., 2006. Mechanical behavior of metallic nanolaminates. In *Nanostructure control of materials,* eds. A. J. Hill and R. H. J. Hannink, chap. 7, 146–176. Woodhead Publishing Co., UK.

Misra, A., Demkowicz, M. J., Zhang, X., and Hoagland, R. G. 2007. Radiation damage tolerance of ultra-high strength nanolayered composites. *Journal of Materials* 59:62.

Misra, A., and Hoagland, R. G. 2007. Plastic flow stability of metallic nanolaminate composites. *Journal of Materials Science* 42:1765.

Misra, A., Hoagland, R. G., and Kung, H. 2004. Thermal stability of self-supported nanolayered Cu/Nb films. *Philosophical Magazine* 84:1021.

Misra, A., and Thilly, L. 2010. Structural metals at extremes. *MRS Bulletin* 35:965.

Paszti, F. 1992. Microanalysis of He using charged particle accelerators. *Nuclear Instruments and Methods B* 66:83.

Pellegrino, S. Beck, L., and Trouslard, Ph. 2004. Differential cross-sections for nuclear reactions $^{14}N(d,p_5)^{15}N$, $^{14}N(d,p_0)^{15}N$, $^{14}N(d,\alpha_0)^{12}C$ and $^{14}N(d,\alpha_1)^{12}C$. *Nuclear Instruments and Methods B* 219–220:140.

Ruhle, M., and Wilkens, M. 1975. *Crystal Lattice Defects* 6:129.

Santamaria-Fernandez, R. Carter, D., and Hearn, R. 2008. Precise and traceable $^{13}C/^{12}C$ isotope amount ratios by multicollector ICPMS. *Analytical Chemistry* 80:5963.

Seletskaia, T., Osetsky, Y., Stoller, R. E. et al. 2005. Magnetic interactions influence the properties of helium defects in iron. *Physics Reviews Letters* 94:046403.

Serruys, Y., Ruault, M. O., Trocellier, P., Miro, S., Barbu, A., Boulanger, L., Kaietasov, O., Henry, S., Leseigneur, O., Trouslard, P., Pellegrino, S., and Vaubaillon, S. 2008. Jannus: Experimental validation at the scale of atomic modeling. *Comptes Rendus Physique* 9:437.

Tanemura, M., Ogawa, T., and Ogita, N. 1983. *Journal of Computer Physics* 51:191.

Trinkaus, H., and Ullmaier, H. 1994. High temperature embrittlement of metals due to helium: is the lifetime dominated by cavity growth or crack growth? *Journal of Nuclear Materials* 303:212–215.

Wang, J., and Misra, A. 2011. An overview of interface-dominated deformation mechanisms in metallic multilayers. *Current Opinions in Solid State and Materials Science* 15:20–28.

Wang, Y. Q., and Nastasi, M. 2009. *Handbook of modern ion beam materials analysis,* 2nd ed., chaps. 5 and 6. Warrendale, PA: Materials Research Society Publisher.

Wang, Y. Q., Zhang, J., Tesmer, J. R., Li, Y. H., Greco, R., Rundberg, R. S., Obst, A. W., Wilhelmy, J. B., and Grim, G. P. 2010. Determination of $^{13}C/^{12}C$ ratios with (d, p) nuclear reactions. *Nuclear Instruments and Methods B* 268:2010.

Wong, C. S. Muir, D. C. G., and Mabury, S. A. 2003. Measurement of $^{13}C/^{12}C$ of chloroacetic acids by gas chromatography/combustion/isotope ratio mass spectrometry. *Chemosphere* 50:903.

Woodbury, H. H., Day, R. B., and Tollestrup, A. V. 1953. Gamma Radiation from the Reaction $C^{13}(p, \gamma)N^{14}$, *Physics Reviews* 92:1199.

Yamamoto, T., Odette, G. R., Miao, P. et al. 2007. The transport and fate of helium in nanostructured ferritic alloys at fusion relevant He/dpa ratios and dpa rates. *Journal of Nuclear Materials* 399:367–370.

Zhang, K. Y., Embury, J. D., Han, K., and Misra, A. 2008. Transmission electron microscopy investigation of atomic structure of interfaces in nanoscale Cu-Nb multilayers. *Philosophical Magazine* 88:2559.

Ziegler, J. F., Biersack, J. B., and Littmark, U. 1985. *The stopping and range of ions in solids,* vol. 1. New York: Pergamon.

Zinkle, S. J. 2005. *Physics Plasmas* 12:8.

Art and Archaeology Applications

Thomas Calligaro and Jean-Claude Dran

13.1 ROLE OF IBA IN CULTURAL HERITAGE STUDIES

Interest in applying ion beam analysis (IBA) techniques to artworks or archaeological artifacts dates back to the early implementation of these techniques (Coote, Whitehead, and McCallum 1972; Eriksson, Fladda, and Johansson 1972; Gordon and Kraner 1972) and stems from their outstanding analytical capacity, unmatched for decades, and their nondestructive character. The development of external beam setups (Giuntini 2011), making this analytical tool fully non-invasive, has considerably strengthened the popularity of IBA in this field, as illustrated by the numerous facilities now involved and the sessions systematically dedicated to art and archaeology applications in IBA-related conferences. It is not our intent to cover every aspect of IBA use for the study of cultural heritage materials, since extensive reviews have already been published (Bird, Duerden, and Wilson 1983; Lahanier et al. 1986; Calligaro, Dran, and Salomon 2004), but rather to restrict the scope of this chapter to a clear description of the present state of the art when other powerful techniques have shown their usefulness in this field and increasingly challenge ion beam analysis.

Among the different IBA techniques, PIXE (particle-induced x-ray emission) has been up to now the most widely applied to cultural heritage studies, due to its high sensitivity and its easy implementation at atmospheric pressure (Dran, Calligaro, and Salomon 2000). However, this powerful technique is now facing severe competition from other noninvasive ones such as x-ray fluorescence (XRF) and Raman spectroscopy, which can be implemented with portable instruments. Moreover, x-ray absorption techniques provided at synchrotron radiation facilities, increasingly applied to heritage materials, stand as a serious challenger to PIXE as they provide the concentration of elements not only with better sensitivity but also their speciation (Creagh 2007). In this context, it seems to us timely to reassess the advantages and limitations of IBA techniques for cultural heritage studies and to point out the specific areas where, in our opinion, they are still unmatched.

This chapter will be structured as follows. In the first section we emphasize the specificity of heritage materials and briefly describe the main issues relevant to archaeology and art history that can be usefully addressed using compositional data. These questions echo the successive life phases of cultural heritage materials—namely, their fabrication, use, ageing, and actions to be taken for their sustainable conservation. In the next section we review the main qualities of IBA techniques for cultural heritage studies and show their remarkable suitability for this field. The third part of this chapter is devoted to examples of recent applications that illustrate the powerful impact and unmatched capabilities of IBA techniques in the interdisciplinary field of archaeometry and conservation science.

13.2 MAJOR ART AND ARCHAEOLOGY ISSUES

Studies of cultural heritage objects pertain to two distinct branches. The first, labeled archaeometry, addresses the question of object creation and usage, mostly relevant to archaeological artifacts, but to artworks as well. The second, entitled conservation science, deals with the preservation of cultural heritage relics by studying the impact of the environment, especially in a museum. The authentication of works, an additional field where the genuineness of out-of-context object is assessed prior to entering heritage collections, takes advantage of the detailed knowledge gained in the previous branches.

13.2.1 OBJECT CREATION

The nature and the spatial distribution of the materials in an object (e.g., composition, layered or polyphased structure) are directly linked to its history, from its creation to its use. These features can provide clues to understanding ancient manufacture technologies, artists' techniques, and object destination and usage. Chemical analysis offers an efficient way to determine the nature and amounts of ingredients used by artisans in ancient manufacture techniques (e.g., fabrication of glasses, alloys, ceramic pastes, and decorations) or to infer artists' recipes for their works (e.g., choice of pigment palette, determination of paint, and preparation layer sequence). The analysis of impurities at a trace level is particularly fruitful to determine materials' provenance, in order to answer the fundamental archaeological question of the procurement routes of raw materials and trade networks of finished objects (e.g., obsidian circulation). In addition, the surface composition of artifacts can sometimes fingerprint their usage (e.g., traces of worked materials on tools). Finally, the precise identification of the materials and their layout not only serves to document archaeology and art history, but may also provide guidelines for restoration work, helping in elaboration of the best intervention procedure with the most appropriate products.

13.2.2 CONSERVATION STATE DIAGNOSIS

For the preservation of cultural heritage, the diagnosis of conservation state is of primary importance. It aims at establishing a real checkup of the object, searching for alteration of its materials with a view to stopping its progression and recommending appropriate preservation and restoration actions. Natural ageing and alteration processes affect many manufactured materials (corrosion of metals, weathering of glass, salt efflorescence on ceramics) as well as raw materials like stones or pigments (reaction with the environment or with organic binders in paintings). The chemical analysis of what can be considered as the object's "skin" allows one to identify the alteration products and to understand the surface evolution over time (e.g., migration of cations like sodium in glasses, buildup of sulfates, chlorides, and oxide layers at the surface of metals). The clues of a long-term ageing, whenever evidenced on an object of unknown origin, may also serve as an argument for its authenticity.

13.2.3 AUTHENTICATION OF HERITAGE ITEMS

Authentication of artworks or archaeological artifacts constitutes a great challenge that most often requires the combination of knowledge of different experts (art historians, archaeologists, conservation scientists). To tell whether an object is genuine or fake, it is screened for a series of criteria of style (e.g., recognized touch of the artist, typomorphology of artifacts), manufacture (e.g., traces of modern tools in allegedly ancient items), date and eventually composition obtained using different analytical techniques (Adriaens 2005). The results are subsequently compared to those gathered on genuine objects previously studied. A typical case is the search for anachronistic materials such as modern pigments in paintings supposedly made by old masters. On the basis of the well established chronology of pigment availability (Clark 2002), a forgery or an extensive restoration is evidenced when the identified pigments were not yet or no longer in use at the time the painting is supposed to have been created. Another case concerns the attribution of an artwork to an artist or to a workshop; this can be addressed using both stylistic and compositional arguments with reference to previously studied works of the same artist/workshop. Attribution is discarded when the identified pigments, although chronologically correct, do not match the artist's palette (Tuurnala and Hautojärvi 2000). Since the authentication of the object is often conducted prior to its potential purchase without ownership rights, the use of fully nondestructive and noninvasive analytical techniques is mandatory.

While the identification of anachronistic materials can generally be gained by more conventional means, like XRF or Raman spectrometry, there remain some cases, in particular when the modern materials contain light elements (e.g., boron in glass), where IBA stands as the only nondestructive way to identify them. In this search for authenticity, dating plays a central role but, unfortunately, few materials can be directly dated (ceramics using thermoluminescence, organic remains with ^{14}C, wood with dendrochronology). The extent of surface modification induced by a long exposure to environment sets the grounds of dating methods applicable to additional materials, like the fluorination of flint (Walter, Menu, and Dran 1992) or the hydration of obsidian (Laursen and Lanford 1978) and of rock crystals (Ericson et al. 2004). However, as surface alteration depends on various uncontrolled environmental parameters (e.g., T, pH), it cannot lead to a quantitative dating method, but still can be used to distinguish between ancient and modern surfaces, thus providing a useful authentication method. An example will be given in Section 13.4 of this chapter, which is devoted to applications.

13.3   ION BEAM ANALYSIS OF CULTURAL HERITAGE RELICS

In this section, we show that the early use of IBA techniques for cultural heritage studies and their expanding popularity in this field stem from their ideal adaptation to the constraints linked to the specific features of artworks and archaeological artifacts—notably their preciousness and their highly complex and variable chemical composition.

13.3.1   SPECIFICITY OF CULTURAL HERITAGE TARGETS

Artworks and archaeological artifacts generally exhibit specific features that make them different from the usual targets studied in materials science laboratories. First, the precious and possibly unique character imposes the use of noninvasive techniques (i.e., without sampling or dismounting and, more generally, that preserve object integrity). In addition, the methods have to be nondamaging—that is, not inducing any modification of the investigated area. Many objects exhibit a large size and complex shape (e.g., paintings, sculptures), which make their handling difficult and prevent their introduction in a vacuum chamber. These features frequently result from the assemblage of various materials, possibly organic, for which a vacuum could cause detrimental degassing and dehydration. In addition, they often integrate insulating materials (glass, ceramic, stone), which, in a vacuum, would need a conductive coating for operating IBA. This is basically why most heritage items, apart from small metallic objects, cannot be analyzed in a vacuum.

At a deeper level, materials constituting the works generally exhibit a complex, granular, layered, and sometimes hybrid (organic/inorganic) character. This spatial heterogeneity calls for analytical techniques featuring high spatial resolution, laterally with a probe size at least smaller than details visible to the naked eye (a few tens of micrometers) but also in depth to take into account the stratified structure of multilayered objects (e.g., paintings, gildings, glazes on ceramic) or of altered materials. Often, sample surface is not ideally flat, but rather curved and/or rough, which might cause problems. Moreover, in the case of altered objects, the surface composition is not representing the bulk, which can be a serious limitation when operating under fully noninvasive conditions. In the worst case where this altered layer is too thick (a few tens of micrometers), the only solution is to perform the analysis on a minute sample taken in depth.

13.3.2   SUITABILITY OF IBA TECHNIQUES

We now recall the main features of IBA techniques and show how they fulfill the requirements for the analysis of cultural heritage objects. First, the non-damaging feature of IBA techniques, vividly demonstrated by their implementation in external beams, is emphasized without eluding the risks for sensitive materials. We underline the usefulness of the PIXE/PIGE (particle-induced γ-ray emission) combination, covering simultaneously a wide range of elements (almost all elements except carbon, nitrogen, and oxygen) to analyze the extremely diversified heritage materials apart from those of an organic nature. It provides the bulk concentration of presumed homogeneous samples for a wide range of elements with high sensitivity, but lacks depth information. The ranges of elements dosed with PIXE/PIGE extend respectively above and below sodium and conveniently overlap for this mobile element of great importance in glassy materials. On the other hand, we show how methods like Rutherford backscattering spectrometry (RBS), nuclear reaction analysis (NRA), and elastic recoil detection analysis (ERDA) methods, albeit less sensitive and spanning a narrower range of elements, provide useful depth distribution profiles at the surface of artifacts. RBS is well suited to the characterization of heavy element layers on top of lighter substrates (e.g., gildings) whereas elastic backscattering spectrometry (EBS with protons), NRA, and ERDA allow light element profiling. Finally we will report the unique ability of the external scanning microbeam to record nondestructively elemental maps, providing chemical images of artworks and archaeological artifacts at small (several micrometers) or large (centimeters) scale. To summarize, IBA techniques based on photon detection

(PIXE and PIGE) appear ideal to identify materials, deformulate ancient recipes, and infer the provenance of raw materials.

On the other hand, those based on particle detection (RBS and NRA) are best suited to understand the manufacture and the treatment of artifact surface (decoration, gildings, patina) and to assess the object alteration state. In conclusion, a brief comparison with competing techniques for cultural heritage shows that, whereas some can be more sensitive or easier to implement, the combination of nondestructive IBA techniques provides a unique wealth of information and keeps specific unmatched capabilities, like the nondestructive depth profiling of light elements.

13.3.2.1   NONDAMAGING CHARACTER

Due to the low energy deposited in the target, IBA is widely qualified as non-destructive. However, damage can be induced on radiation-sensitive materials (Zucchiatti and Agulló–Lopez 2012). Most mineral and metallic samples do not suffer from the analyzing beam, but organic compounds may undergo chemical bond destruction (radiolysis), leading to the breaking of polymeric molecules, gaseous decomposition, loss of water (dehydration), and possibly migration of mobile elements. Damage in glasses and glazes in the form of dark or brownish stains has been reported. For paintings, in addition to the darkening of bright pigments, the degradation of organic constituents (e.g., binders) and the blistering of the surface varnish have been observed. Beam damage may occur with integrated charges in the micro-coulomb range, and is generally more related to the deposited dose on the irradiated area (beam fluence in micro-coulombs per square millimeter) than to the beam current (power dissipation). Fortunately, dark marks induced on some inorganic materials like white lead pigments or glazes, ascribed to the formation of transient color centers, can vanish with time, moderate heating, or ultraviolet (UV) illumination (Chiari, Migliori, and Mandò 2002).

Consequently, the most sensitive objects (paintings, manuscripts, papyri) need to be analyzed with a very low beam intensity (0.1 nA) applied during a short time (100 s) and the beam spread or swept over a few millimeters-square area to lower the beam fluence. More generally, as it is very difficult to predict the behavior of the sample under the beam, special attention should be paid prior to and during analysis. Irradiation tests have to be conducted on a less important part of the object prior to the experiment and constant visual monitoring of the impact point during the experiment is mandatory. The beam dose must be kept as small as possible by improving the detection efficiency (tight geometry, large detector diameter) to compensate dose reduction.

13.3.2.2   NONINVASIVE IBA IMPLEMENTATION: EXTERNAL BEAM

A major breakthrough in the application of IBA to cultural heritage objects was the introduction, in early IBA history, of external beam setups. They were initially developed to analyze targets unable to stand vacuums, such as liquids (Deconninck 1972), and extended in the early 1980s to analyze precious historical items in a fully noninvasive way (Chen et al. 1980; Swann 1982). The external beam considerably simplifies sample positioning and changing, speeding up measurements that can reach several hundred samples per day of operation. The direct operation in air provides additional benefits. The surrounding atmosphere helps cool the irradiated area, thus reducing the risk of damage, and evacuates electric charges, thus avoiding coating the object with a conductive material.

Usually, air is replaced with helium either in a closed arrangement or, more commonly, as a gas flow to reduce energy and angular beam stragglings and to benefit from the very low x-ray attenuation of helium enabling light element measurement with PIXE. For instance, 5 cm of helium has a 95% transmission for the 1 keV sodium x-ray line.

Early external beams from the 1970s had a vacuum-to-air interface based on a 10 µm pinhole collimator or on a few micrometers thick exit foil made of various materials (e.g., Kapton®, aluminum, Havar®). The beam degradation induced by these exit foils for 3 MeV protons (Table 13.1) permitted implementing only PIXE and PIGE, which are not critically dependent on incident energy, and proton-based elastic backscattering. However, this last technique, which we label indifferently p-EBS or p-RBS, whether it follows Rutherford's law or not, exhibits a poor mass resolution (Doyle 1983). The introduction in the 1990s of 50–200 nm thick exit windows made of silicon nitride (Si_3N_4), which are exceptionally resistant to pressure and beam damage, allowed extracting helium beams in the atmosphere, enabling implementation of α-RBS and α-ERDA, and focusing the beam to around 10 µm, providing a true external nuclear microprobe (Salomon, Dran, Guillou, Moignard, Pichon,

TABLE 13.1   Specifications of Various Exit Windows Used to Extract the Beam in Air

Beam	Exit Window Material	t (µm)	ΔE (keV)	δ_E (keV)	θ (mrd)	Ø (µm)	Implemented IBA Methods
3 MeV p	Aluminum	10	230	15	21	150	PIXE PIGE p-EBS
	Kapton	8	126	9	15	500	PIXE PIGE p-EBS
	Havar	2	113	11	20	200	PIXE PIGE p-EBS d-NRA
	Si_3N_4	0.1	4	1	2.3	20	PIXE PIGE p-EBS d-NRA
	Helium gas	3000	7	2	1.4		
3 MeV He	Si_3N_4	0.1	32	3	6.1	50	α-RBS α-ERDA
	Helium gas	3000	76	4	2.4	50	

Notes: t is the foil thickness, ΔE is the energy loss, δ_E is the energy straggling, θ is the mean angular deviation, and Ø is the beam diameter a few millimeters downstream from the exit foil. All values were calculated using SRIM2003. Note that only the Si_3N_4 window allows extraction of He beams.

et al., 2008). The main difficulty with external beams remains the tricky control of the beam dose, which is fundamental for quantitative analysis. Various solutions have been developed to measure the current, using a beam chopper (Del Carmine et al. 1993); the PIXE, PIGE, or RBS emission from the exit window (Calligaro, Dran, Salomon, and Walter 2004); or the argon x-rays and ionoluminescence from the air (Lill 1999).

To ensure experimental condition reproducibility, special attention has to be paid to (1) the stability of exit window thickness and its hydrocarbon buildup, (2) the steadiness of helium flow, and (3) accurate positioning of the object. Using an optimized external beam setup based on Si_3N_4 windows and particle paths in helium kept shorter than 10 mm, IBA can be carried out with specifications almost equivalent to those achieved in a vacuum. As an example of an external nuclear microprobe, the layout of the setup of the AGLAE facility in Paris, France, is depicted in Figure 13.1. The system is equipped with two Si(Li)

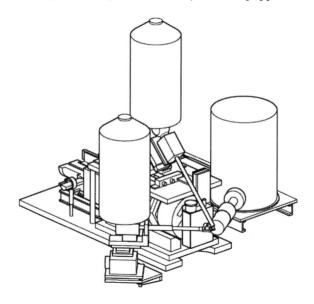

FIGURE 13.1 Layout of the AGLAE external nuclear microprobe for simultaneous PIXE/PIGE/RBS measurements. The system is fitted with two Si(Li) detectors. The first one, placed at 45° in the horizontal plane, flushed with helium gas, and protected against backscattered particles by a magnetic deflector, is dedicated to x-rays of light and major elements (10 < Z < 26). The second one, placed at 45° in the vertical plane and screened with a suitable absorber to filter out the strong emission of the main constituents, is dedicated to x-rays of heavy elements at trace level (Z > 20). An annular particle detector for RBS is centered on the beam axis and placed in vacuum in conical housing. Gamma-rays are collected using a high-purity germanium detector placed symmetrically to the first x-ray detector. Beam dose is monitored using an additional x-ray detector recording the exit window signal (not visible). (Copyright B. Moignard, C2RMF).

FIGURE 13.2   Ceramic plate with luster decoration placed in the external beam setup of AGLAE. This picture illustrates the PIXE/PIGE step of the integrated characterization of the near-surface layers of this object.

detectors for PIXE, an annular particle detector for RBS, and HPGe detector for PIGE. Note in particular the monitoring of the beam via the Si x-rays emitted by the Si_3N_4 exit window using a dedicated SDD detector, the presence of a turbo-pump to improve the vacuum at the end of the beam line, and the fine movements for an easy centering of the detector system along the beam axis (Salomon, Dran, Guillou, Moignard, and Pichon 2008). Figure 13.2 shows a close-up of the AGLAE external microbeam during the study of a ceramic item (see Section 13.4.1.).

If the noninvasive rule is strictly followed, a serious limitation arises when using IBA techniques (and also most other spectrometric methods) as they only provide the composition at the near-surface (up to a depth of a few tens of micrometers with 3 MeV protons and a few micrometers with 3 MeV helium ions), which can be quite different from the bulk. Then, as already stated, one has to rely on microsampling whenever permitted, and the nondestructive character of IBA is exploited to keep the sample intact.

13.3.2.3   MULTIELEMENTAL, QUANTITATIVE, AND HIGHLY SENSITIVE ANALYSIS: PIXE TECHNIQUE

Among the IBA techniques, PIXE is the one that best fits the requirements mentioned before and the easiest to implement at atmospheric pressure. This explains why PIXE is by far the most frequently used IBA technique for heritage studies, especially for the identification of materials and the determination of their provenance. Its main advantage is that it allows the measurement of all elements above sodium, with a sensitivity reaching at best case a few micrograms per gram. Usually a proton beam of 3 MeV with a current from 0.1 to 10 nA is employed, depending on target composition and detector geometry. The power dissipated at the beam impact is low (0.3 to 30 mW), which is fundamentally the origin of the nondamaging character of this technique. Most PIXE setups rely on the use of two x-ray detectors to acquire simultaneously the spectra of major and trace elements. The detector dedicated to the main constituents ($11 < Z < 26$) is operated without an absorber and with helium gas blown between the detector window and the sample surface. It is fitted with a magnet to deflect the trajectories of backscattered ions, which would otherwise readily damage it. The detector dedicated to trace elements ($Z > 20$), with a higher solid angle, is fitted with an adequate x-ray absorber to filter out x-rays of major elements (e.g., 50 μm aluminum foil for mineral samples). This arrangement can be optimized for most materials by choosing a combination of adequate filters and sample-to-detector distances.

Specific setups have been optimized to improve PIXE sensitivity for heritage items, in particular the ones containing heavy elements like metals (e.g., bronze, gold, silver) as major constituents. Indeed, without special caution, the background due to the pileup of the intense lines of major elements and their low energy tails would hinder the measurement of the weak lines of trace elements. This situation can be overcome using selective absorbers, usually made of single metal foils, to attenuate or even completely filter out matrix x-rays (Swann and Fleming 1990). The absorber element is chosen such that its absorption edge, with a 10-fold increase of the mass attenuation coefficient, occurs just below the main line to be filtered out. For instance, traces in gold are efficiently measured using a 75 μm copper filter that attenuates Au L-lines by a factor 10^{-7} but keeps a transmission of 30% for elements around silver (rhodium, palladium, cadmium, tin, antimony)—allowing for the latter a sensitivity of ca. 20 μg/g to be reached (Guerra 2008). Finally one may consider the use of a wavelength dispersive spectrometer, not affected by tailing and pileups of intense lines, at the expense of lower efficiency and higher cost (Kavcic 2010).

13.3.2.4   MEASUREMENT OF LIGHT ELEMENTS

Heritage materials may contain elements lighter than sodium and thus invisible with PIXE, as major (carbon, nitrogen, oxygen), minor, or trace constituents (lithium, beryllium, boron, fluorine). The nondestructive determination of such light elements is a very difficult task for most spectrometric techniques but is efficiently accomplished by IBA techniques—namely, PIGE, NRA, and ERDA. The prompt gamma-ray emission (PIGE), commonly associated with PIXE in heritage applications, usefully provides lithium, beryllium, boron, and fluorine concentrations during the same irradiation. This stems from the high γ-ray yields for these elements, yet lower than that of x-rays but partly balanced by the much higher volume and solid angle of the γ-ray detectors. In general, elements like sodium, aluminum, and silicon can be measured by both PIXE and PIGE techniques. Let us mention that PIGE allows the measurement of light elements when their x-rays are obscured by a surface layer (e.g., sodium that fingerprints the use of ultramarine pigment in paintworks when it is coated by a varnish layer; Grassi et al. 2004).

Because of the low γ-ray yields, oxygen, carbon, and nitrogen cannot be determined using PIGE with protons. Alternatively, these light elements among others can be measured using the strong γ-ray production induced by bombardment with deuterons (d-PIGE) (Szíki et al. 2004)—for instance, traces of carbon in archaeological copper alloys (Papillon and Walter 1997). The drawbacks are the short-term activation of the sample and the high background in the γ-ray spectra correlated with neutron emission. However, because of radiation protection regulations, only a few facilities are designed and authorized to carry out deuteron-based experiments.

Light elements in heritage items can also be measured using p-EBS. Indeed, scattering of protons and, to a lesser extent, of alpha particles by light nuclei frequently departs from the Rutherford law; the cross sections exhibit strong resonances and a magnitude about 10-fold higher than the Rutherford ones for 3 MeV protons. This enhanced yield compared to classical RBS has permitted the measurement of carbon, nitrogen, and oxygen on heavy element substrates like metal patina (Salomon, Dran, Guillou, Moignard, Pichon, et al. 2008) and the assessment of the organic binder proportion in paint layers (De Viguerie, Beck, and Salomon 2009) or of the amount of collagen left in prehistoric bones (Beck et al. 2011).

Finally, the determination of hydrogen content and, whenever accessible, its depth profile is of great importance in alteration studies where hydration plays a fundamental role. The specific IBA methods developed to profile this light element will be discussed in the next section. Table 13.2 contains a list of IBA techniques applicable to light element analysis in external beam mode.

13.3.2.5   DEPTH PROFILING: RBS, NRA, ERDA

A major asset of IBA techniques, invaluable in heritage studies, is the depth profiling capabilities provided by RBS, NRA, and ERDA. Early applications of these techniques to art and archaeology were carried out in vacuum and thus excluded the study of full objects, but the availability of external proton or helium ion microbeams with good energy resolution has allowed both α-RBS and d-NRA to be carried out *in situ* at atmospheric pressure on heritage targets.

RBS is ideally suited for the characterization of layers made of heavy elements on top of a substrate composed of lighter ones. This is notably the case for numerous heritage objects—for instance, gildings, lead glazes, and metallic decorations on ceramics. RBS spectra not only allow the determination of layer thicknesses, but also ordering, and assessing the roughness of layer interfaces. In external beams, RBS is often carried out in parallel with the proton beam required for PIXE (p-RBS)—consequently with a limited mass

TABLE 13.2   Light Element Analysis with External Beams

Element	Technique	Nuclear Reaction	Incident Ion	E MeV Beam	Interaction Product	E MeV Product	Applications
H	ERDA	$^1H(\alpha,\alpha)^1H$	α	3	p	1	Quartz[a]
Li	PIGE	$^7Li(p,p'\gamma)^7Li$	p	3	γ-ray	0.478	Gemstones[b]
Be	PIGE	$^9Be(p,\alpha\gamma)^6Li$	p	3	γ-ray	3.5	Gemstones[b]
Be	PIGE	$^9Be(\alpha,n\gamma)^{12}C$	α	2.5	γ-ray	4.44	Gemstones[c]
B	PIGE	$^{10}B(p,\alpha\gamma)^7Be$	p	3.85	γ-ray	0.429	Glass[d]
C	NRA	$^{12}C(d,p)^{13}C$	d	2	p	3	Cu alloys[e], gold[f]
C	d-PIGE	$^{12}C(d,p\gamma)^{13}C$	d	3	γ-ray	3.089	Cu alloys[g], pottery[h]
C	p-EBS	$^{12}C(p,p)^{12}C$	p	3	p	3	Paintings[i]
N	NRA	$^{14}N(d,p_5)^{15}N$	d	2	p	1.8	Cu alloys[e]
N	p-EBS	$^{14}N(p,p)^{14}N$	p	3	p	3	Bones[j]
N	d-PIGE	$^{14}N(d,p\gamma)^{15}N$	d	3	γ-ray	7	Pottery[h]
O	NRA	$^{16}O(d,p)^{17}O$	d	2	p	2.5	Cu alloys[e]
O	p-EBS	$^{16}O(p,p)^{16}O$	p	3	p	3	Painting[i]
O	d-PIGE	$^{16}O(d,p\gamma)^{17}O$	d	3	γ-ray	0.871	Pottery[h]

[a] Calligaro, T. et al. 2009. *Applied Physics A* 94: 871–878.
[b] Calligaro, T. et al. 2000. *Nuclear Instruments and Methods B* 161–163:769–774.
[c] Gutiérrez, P. C. et al. 2010. *Nuclear Instruments and Methods B* 268:2038–2041.
[d] Mäder, M. and C. Neelmeijer. 2004. *Nuclear Instruments and Methods B* 226:110–118.
[e] Ioannidou, E. et al. 2000. *Nuclear Instruments and Methods B* 161–163:730–736.
[f] Demortier, G. and A. Gilson. 1987. *Nuclear Instruments and Methods B* 18:286–290.
[g] Papillon, F. and P. Walter. 1997. *Nuclear Instruments and Methods B* 132:480–468.
[h] Szíki, G. A. et al. 2004. *Nuclear Instruments and Methods B* 219–220:508–513.
[i] Beck, L. et al. 2010. *Nuclear Instruments and Methods B* 268:2086–2091.
[j] Beck, L. et al. 2012. *Nuclear Instruments and Methods B* 273:203–207.

resolution for heavy elements. p-RBS is still useful to detect layered structures and its poor mass resolution is somewhat circumvented by the high resolving power of PIXE. In this integrated approach, p-RBS mainly serves to check that the sample is homogeneous, to ensure the proper processing of PIXE spectra (e.g., check for glass alterations). The use of α-RBS with superior specifications (Rutherford cross section, improved mass resolution) is attractive, but the He beam cannot be used to carry out PIXE simultaneously; x-ray production by He ions is equivalent to that of protons having a quarter of the energy and hence notably insufficient. Consequently, the analysis of the same area has to be carried out in two distinct runs: one with protons for PIXE and another with He ions for RBS. An example of such a combination of PIXE and RBS with both protons and He ions will be given in Section 13.4. This combination revealed the technology employed to produce the metallic-shine luster decoration of Islamic ceramics, which appeared to be made of layers of copper and silver nanoparticles embedded in a glassy matrix.

NRA with detection of particles, mainly with deuteron beams, can be considered as the natural counterpart of RBS for light elements, targeted at characterizing layers of light elements lying on top of a heavy element substrate. Such a situation is widespread in heritage studies, especially at the surface of metals (e.g., patina and corrosion layer on alloys). NRA is not affected by its implementation in air because the energy resolution is dominated by straggling in the few micrometers-thick polymer foil required to stop scattered beam particles. From Table 13.2, one can see that (d,p) reactions at 1 MeV have been mostly employed to characterize patinas at the surface of bronzes (Ioannidou et al. 2000).

It should be mentioned that resonant nuclear reaction analysis (r-NRA, so-called resonance scanning technique) has been used to profile hydrogen and fluorine at the surface of archaeological artifacts, especially in view of developing dating methods (Table 13.3). For instance, hydration of ancient glasses has been profiled by r-NRA with incident ^{15}N and ^{19}F beams. However, because the energy straggling in air would considerably worsen its depth resolution, r-NRA has always been carried out in vacuum. Recently, the measurement of hydration profiles has been performed at atmospheric pressure using elastic recoil spectrometry with He ions (ERDA). The design of the external beam ERDA setup, based on a tight control of the experiment geometry and of the surrounding atmosphere, is described in Section 13.4. Its operation is illustrated by the evidencing of the forgery nature of an allegedly pre-Columbian rock crystal skull (Calligaro et al. 2009).

TABLE 13.3   Gamma-Producing Resonant Nuclear Reactions Used for the Determination of Depth Profiles of Light Elements in Heritage Materials (r-NRA)

Ion γ-resonant reactions						
Element	Reaction	E, MeV	Eγ MeV	ΔR μm	R μm	Applications
H	$^{1}H(^{15}N,\alpha\gamma)^{12}C$	6.385	4.44	0.004	2–3	Quartz[a] obsidian dating[b]
H	$^{1}H(^{19}F, \alpha\gamma)^{16}O$	16.2	6–7	0.08		Glass weathering[c]
F	$^{19}F(p, \alpha\gamma)^{12}C$	0.872	6–7	0.1	1.4	Flint and tooth dating[d]
S	$^{32}S(p,p'\gamma)^{32}S$	3.094	2.23			Bronze patina[e]

Notes: E$_r$ is the energy of the resonance, Eγ is the energy of the emitted γ-ray. ΔR is the depth resolution at the sample surface and R the maximum probing depth in a silica matrix. Only the last reaction has been implemented in external beam mode.

[a] Dersch, O. and F. Rauch. 1999. *Fresenius Journal of Analytical Chemistry* 365:114–116.

[b] Laursen, T. and W. A. Lanford. 1978. *Nature* 276:153–156.

[c] Kossionides, S. et al. 2002. *Nuclear Instruments and Methods B* 195:408–413.

[d] Walter, P. et al. 1992. *Nuclear Instruments and Methods B* 64:494–498.

[e] Kalliabakos, G. et al. 2000. *Nuclear Instruments and Methods B* 170:467–473.

13.3.2.6   MULTISCALE IBA CHEMICAL IMAGING OF HETEROGENEOUS ARTIFACTS

Despite a clear success in many fields such as geology or biology, classical scanning nuclear microprobes operated in vacuum have seldom been employed in art and archaeology. This situation probably stems from the requirement to take a minute sample, which in addition could be damaged by the high beam density at the impact point (ca. nanoamperes per square micrometer). Moreover, once a microsample is picked up, many alternative techniques can be applied, notably the analytical scanning electron microprobe (SEM)-energy dispersive x-ray (EDX) routinely employed to study works of art. In spite of such limitations, in-vacuum microprobes were applied to the study of small artifacts or prepared samples (ceramic, glaze, and painting cross sections). The development of microprobes operated at atmospheric pressure brought considerable change to the situation. Based upon windowless micrometric pinhole collimators, early setups allowed the extraction only of low-intensity beams. As already mentioned, major progress occurred in the 1990s with the introduction of ultrathin silicon nitride foils (down to 0.1 μm thick) to maintain the air–vacuum interface. This new exit window made possible the focusing of highly intense beams with a diameter of a few tens of micrometers in the air, allowing the analysis of microscopic details inside an entire object (Calligaro et al. 2011). This rather large probe size compared to micrometer-size or less beams achieved in vacuum systems is not too detrimental: it matches the particle range in thick targets, providing the analysis of a volume of ca. $30 \times 30 \times 30 \ \mu m^3$. Moreover, as was pinpointed in previous review papers (Mandò 1994), most of the relevant chemical information in heritage items can be gained using this probe size, which by the way corresponds to the smallest details visible to the naked eye.

Note that, contrary to alternative chemical mapping techniques like SEM-EDX that only deliver qualitative maps, IBA can provide full and accurate composition including traces at each point of the image. The ability to build such quantitative maps is an important feature in heritage studies, as the comparison between these chemical images with the visible image of the same area of an artifact is usually very informative.

Two scanning modes are generally implemented in external nuclear microprobes. The first one, based on the fast magnetic scanning of the beam, permits collection of elemental micromaps over an area limited by the dimensions of the beam exit window (ca. $1 \ mm^2$). The benefits of this mode will be detailed in Section 13.4 through an example of chemical imaging of the various phases of lapis lazuli, a heterogeneous precious rock used for carvings and painting pigments. In the second mode, the beam is kept fixed and the target is mechanically raster scanned using high-precision motorized translation stages. The scanned areas can reach several tens of centimeters in width and height, only limited by the translation stage range. A major limitation is the long acquisition time needed to draw elemental images, frequently reaching 1 hour. This is required to collect enough counts per point in the image to derive quantitative results and to cover the entire area at a scanning speed limited to a few millimeters per second. Typical examples of large-scale mapping applications are the identification of the palette used for Egyptian polychrome papyri (Olsson et al. 2001) or the multiscale chemical imaging of Renaissance paintings (Grassi et al. 2007), which will also be developed in Section 13.4.

13.3.3 COMPARISON WITH COMPETING TECHNIQUES

It should be of interest to compare the qualities and limitations of IBA to those of a selection of analytical techniques now increasingly used in heritage investigations, which thus may appear as serious competitors to IBA. A synthetic presentation is given in Table 13.4, where only noninvasive or microdestructive techniques implemented in air are considered, thus excluding vacuum techniques such as SEM/EDX or SIMS (secondary ion mass spectrometry). One can note that, whereas, for several criteria, IBA techniques are surpassed by recently introduced analytical methods, they show no weak point and keep the lead in three domains: light element analysis, depth profiling, and large-scale chemical imaging. However, as far as the simple identification of compounds is

TABLE 13.4 Qualities and Limitations of IBA Techniques as Compared to Competing Techniques

Qualities	IBA	XRF	Synchrotron XAS/XRD	μ-Raman	LA-ICP AES/MS	LIBS
Noninvasive	+++ [a]	+++	+++	+++	+++	+++
Nondestructive	++ [b]	+++	+++	++ [c]	— [d]	— [d]
Multielemental	+++	++	+	N/A [e]	+++	+++
Quantitative	+++	++	+++	N/A [e]	++	+
High sensitivity	++ PIXE	+++	+++	N/A	+++	++
Depth profiling	+++ RBS, NRA, ERDA	—	+	++ [f]	+	+
Mapping	+++	++	+++	+++	—	—
Light elements	+++ NRA, PIGE, ERDA	—	—	N/A	+++	+++
Speciation	—	—	+++	++	—	—
Portability	—	+++	—	+++	—	++
Availability	++	+++	—	+++	++	++
User friendly	++ PIXE + NRA/RBS	+++	—	+++	++	++

[a] With external beam.
[b] Except for radiation sensitive matter.
[c] Risk of damage by laser on sensitive matter.
[d] Craters of ca. 100 μm.
[e] No chemical analysis but identification of compounds.
[f] With confocal optics.

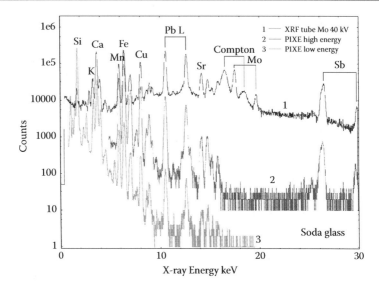

FIGURE 13.3   Comparison of XRF and PIXE spectra of a soda lime glass. The XRF spectrum was recorded using mobile equipment with a molybdenum tube run at 40 kV while the corresponding high- and low-energy PIXE spectra were acquired with 3 MeV protons. Apart from elements below silicon in the low-energy PIXE spectra and from the Mo elastic and Compton peaks in XRF, the spectra show striking similarity, in particular for trace elements (Mn, Fe, Cu, Sb, Pb, etc.).

concerned, Raman spectrometry appears as a simple alternative, which explains its growing popularity in heritage studies. As for the nondestructive determination of composition including trace elements, PIXE is severely disputed by alternative x-ray techniques. On the low end, XRF with mobile equipment features similar specifications with easier and cheaper implementation (Migliori et al. 2011). For instance, in Figure 13.3 the spectrum collected on an historical glass using mobile XRF with a molybdenum tube run at 40 kV is compared to the corresponding high- and low-energy PIXE spectra with 3 MeV protons (Calligaro 2008). Apart from elements below silicon in the low-energy PIXE spectra and from the Mo elastic and Compton peaks in XRF, the spectra show striking similarity, in particular for trace elements (Mn, Fe, Cu, Sb, Pb, etc.).

On the high end, owing to its monochromatic, polarized, and high x-ray flux, synchrotron-based XRF offers better sensitivity than PIXE and equivalent imaging capabilities. In addition, x-ray absorption spectroscopies with the synchrotron (XANES [x-ray absorption near-edge structure] and EXAFS [extended x-ray absorption fine structure]) allow the assessment of the element valence state or chemical environment, out of reach of IBA techniques (Cotte et al. 2010). This is of considerable interest for the study of materials ageing, especially when they undergo a chemical transformation with no apparent change in composition, such as the blackening of cinnabar (Cotte et al. 2006) or the degradation of inks in manuscripts (Kanngießer et al. 2004). One may wonder whether to use PIXE or XRF. The difference does not lie in detection of x-rays, which are identical for both techniques, but rather on the excitation properties, which markedly differ. For PIXE, x-ray production increases with decreasing atomic number, favoring the measurement of lighter elements such as sodium, while owing to the photoelectric process, the situation for XRF is reversed. The Z-span of dosed elements is narrower for XRF than for PIXE, so to cover an equivalent range, several acquisitions with different x-ray excitation conditions are needed (high voltage, anode nature, beam filter).

PIXE also offers a better control of the analyzed depth, which is constrained by the range of particles in the materials (less than 100 μm), whereas XRF is probing much deeper depending on the considered element (e.g., 7 mm for antimony in soda-lime glass with XRF at 50 kV). This tight depth control is notably exploited in the differential PIXE technique to profile pigment layers in paintworks. Finally, while PIXE is coming up against fierce competition from other x-ray techniques, we may recall the distinctive advantage of its simultaneous combination with p-RBS that delivers additional depth-dependent information.

Other important challengers to IBA are the micro-destructive analytical spectroscopies based on laser-ablation, which are now widely available and easily implemented. Leaving marks of ca. 50–100 μm at the surface of the objects, barely visible with the naked eye, both techniques can dose a wide range of elements with a sensitivity equivalent to PIXE for laser-induced breakdown spectroscopy (LIBS) (Giakoumaki, Melessanaki, and Anglos 2007) or several orders of magnitude better for laser ablation–inductively coupled plasma–mass spectrometry (LA-ICP-MS) (Giussani, Monticelli, and Rampazzi 2009). On the other hand, they only provide rudimentary profiles by progressive ablation of the sample and lack imaging capabilities, as it is unthinkable to cover an area of the studied object with an array of craters. Light elements can be measured with LA-ICP-MS or LIBS but the depth profiling at the micrometer level remains a distinctive feature of IBA techniques. More generally, the microdestructive character of ablation-based methods inevitably limits the number of measured points and poses the question of the significance of the microscopic analyzed area for the entire artifact.

With regard to accuracy, IBA techniques benefit from a comprehensive modeling of all the involved physical processes that has allowed the development of truly quantitative processing software. Tests have shown that the spectrum processing software for PIXE (IAEA 2003), PIGE (IAEA 1998), and NRA (Barradas et al. 2008) can provide compositional data within a few percent. XRF accuracy can be of the same order, but more pronounced matrix effects than in PIXE and its blindness to major constituents are a disadvantage. Laser-based methods are generally less accurate, mainly because the control and reproducibility of the experimental conditions during laser ablation are difficult to achieve. Micro-Raman spectrometry can hardly be used for more than simple compound identification.

13.3.4   OTHER SPECIFIC DEVELOPMENTS OF IBA TECHNIQUES TO STUDY HERITAGE MATERIALS

Heritage studies have stimulated the development of original arrangements. In addition to the previously mentioned PIXE-PIGE-RBS couplings, the implementation of light element measurement in air by p-RBS or α-ERDA, and the large-scale chemical imaging, specific setups have been designed to circumvent the limitations of IBA in particular situations. Profiling capabilities have been added to x-ray- and γ-ray-based methods, detection limits have been improved compared to PIXE for materials containing heavy elements, and the investigation of surface modification of materials relevant to cultural heritage, *in situ* and in real time, has been made possible.

13.3.4.1   DEPTH PROFILING WITH PIXE AND PIGE

In targets made of relatively thick layers of complex composition—for instance, paintworks—RBS cannot be used to determine elemental profiles. Two approaches have been developed in PIXE and PIGE modes to provide these methods with a profiling capability they are lacking or simply to verify target homogeneity. These approaches are based on the comparison of spectra recorded either at various incidence angles or at various impinging beam energies. In the first solution, the incident angle between beam and target normal is varied between 0° and 70° by steps of 10° (Weber, Martinot, et al. 2005). This approach was efficiently applied for the estimation of the decoration layer thickness and ordering on stained glass and in paintings. The second solution is the so-called differential PIXE technique. In this case, a series of PIXE spectra are collected with protons typically of 1 to 5 MeV by steps of 1 MeV (Neelmeijer et al. 2000; Mandò et al. 2005; Grassi 2009), allowing the probing of successive paint layers and the reconstruction of depth profiles for several elements by comparison of the spectra. In both methods, each PIXE spectrum generates a point in the profile.

It is worth mentioning that the combined use of PIXE and PIGE can provide a simple way to compare surface and bulk compositions. Indeed, some relatively light elements like sodium, aluminum, or silicon are measured by both techniques. Because it is not attenuated in the sample, the PIGE signal originates from a greater depth than in PIXE mode. For instance, when analyzing a soda-lime glass with 3 MeV protons, 95% of the sodium 439 keV PIGE γ-rays are emitted within a 50 μm depth in the target, whereas the 1.04 keV PIXE signal is only escaping from the first 2.6 μm. If, for a particular element, the concentrations obtained by PIXE and PIGE disagree (e.g., in glass), it can be deduced that it is nonuniformly distributed in depth and thus that the object is altered (Mäder and Neelmeijer 2004, Weber, Vanden Bemden, et al. 2005). This coupled PIXE/PIGE approach can be applied to heavier elements, such as copper, silver, and gold, for which protons induce Coulomb excitation γ-rays at 152, 310, and 279 keV, respectively. In that case, PIGE can be used to probe gold alloy composition at a greater depth than by PIXE and to evidence real gilding or copper depletion, either intentional or induced by ageing (Guerra and Calligaro 2003).

13.3.4.2 BETTER SENSITIVITY THAN PIXE: PIXE-INDUCED XRF

An exotic solution called PIXE-XRF has been developed to improve sensitivity compared to standard PIXE, providing an attractive alternative to PIXE with filters for measuring traces in metals (Demortier and Morciaux 1994). Its principle is similar to synchrotron-based XRF, as it relies on the quasi-monochromatic emission from an intermediate target bombarded with protons, which in turn is employed to induce XRF in the final target. The intermediate target composition is chosen to excite x-rays from trace elements without those from major elements of the matrix. As an example, using an arsenic intermediate target, the most energetic x-ray line emitted at 11.86 keV ($K_{\beta2}$ line) is high enough to ionize the L_{III} shell of platinum at 11.56 keV but too low to excite the L_{III} shell of gold at 11.92 keV. In this way, PIXE-XRF allowed the measurement of a few hundred micrograms per gram of platinum in archaeological gold (Guerra and Calligaro 2004).

13.3.4.3 IN SITU IBA WITH EXTERNAL BEAM IN REACTION CELLS

The availability of external beams has also permitted performance of dynamic (real time) accelerated alteration tests on materials relevant to cultural heritage, in a controlled atmosphere, in order to infer their long-term behavior under various environmental conditions. For instance, a small electrochemical cell equipped with a silicon nitride window of the same type as the previously described exit window has been built to study *in situ* the corrosion of copper alloys exposed to either gaseous or electrolyte reagent, using both PIXE and RBS (Adriaens and Dowsett 2010). Dynamic micro-RBS was also applied in an open-air, high-temperature oven to study the kinetics of copper alloy oxidation. In particular the formation of artificial patina on bronzes has been investigated at temperatures up to 360°C (Mathis et al. 2004). Finally, using a specially designed liquid cell for analyzing circulating liquids, micro-PIXE was applied to measure in real time the release of lead during the aqueous dissolution of lead-containing glass or glaze (Bouquillon et al. 2002). These experimental approaches designed for cultural heritage studies can obviously be exploited in other research fields.

13.4 EXAMPLES

13.4.1 NANOPARTICLE LUSTER DECORATION ON CERAMICS AND METAL GILDINGS: COMBINED RBS PROFILING AND PIXE BULK ANALYSIS

Luster is a ceramic decoration that appeared in the ninth century in the Middle East and spread around the Mediterranean basin in the following centuries. Luster gives a metallic shine that is due to copper and silver nanoparticles of

10–50 nm diameter distributed within a thin layer inside the glaze (glassy surface layer). The proportion and layer thickness of silver and copper nanoparticles were found to vary considerably, the former element providing a golden glint and the latter a more reddish one. Detailed deformulation of this ancient technology was gained by destructive analysis of lustered ceramic shards. In particular the nanoparticles were revealed using TEM (Figure 13.4a) and the

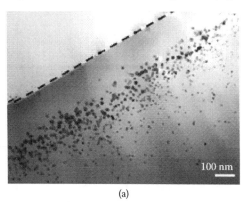

(a)

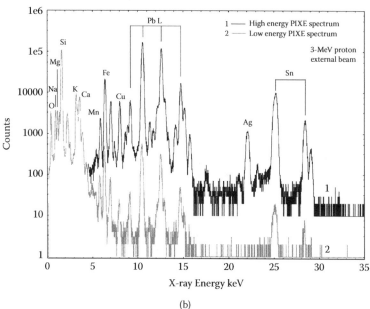

(b)

FIGURE 13.4   Characterization of nanoparticle luster decoration on ceramics. (a) TEM image of a cross section of the glaze surface showing the distribution of 20–50 nm metallic particles. (b) PIXE spectra of a lustered ceramic recorded by the low- and high-energy x-ray detectors. The high-energy spectrum reveals the presence of lead (component of glaze), tin (opacifier), and silver (luster constituent). (Continued)

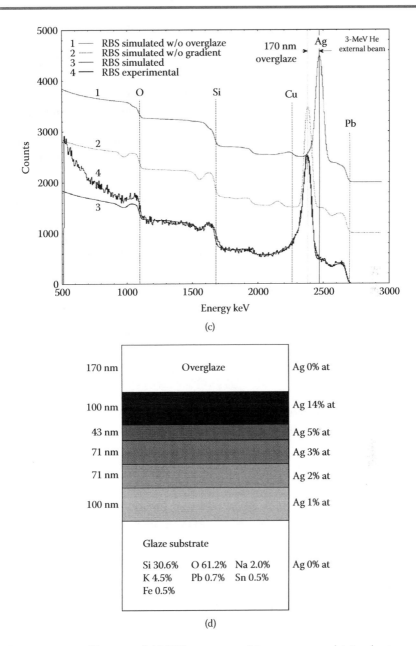

(c)

(d)

FIGURE 13.4   (Continued) (c) RBS spectrum of the same area obtained using an external beam of 3 MeV helium ions and successive simulated spectra from which the detailed multilayer structure of the glaze, shown in (d), was inferred.

speciation of copper and silver were assessed using EXAFS. However, to extend this research to the precious lustered ceramics of museum collections in order to trace the recipe evolution through ages and locations, a fully nondestructive protocol was required. The PIXE/α-RBS coupling provides an alternative nondestructive approach: PIXE delivers the detailed composition of the glaze while α-RBS gives a detailed picture of the luster structure (Polvorinos, Castaing, and Aucouturier 2006).

This is illustrated by the analysis of an early lustered ceramic from Susa, Mesopotamia, from the ninth century (Chabanne et al. 2008). In a first step, the processing of the matrix and trace PIXE spectra obtained with 3 MeV protons (Figure 13.4b) provides the composition of the glaze matrix, which appears to be of a soda type (6 wt% Na_2O) with minor lead content (6 wt% PbO) and a tin concentration (8 wt% SnO_2) that corresponds to an opacifier. The observed silver and copper lines in the PIXE spectrum of the trace elements correspond to the luster nanoparticles. In a second step, the fitting of the RBS spectra obtained with 3 MeV He using the SIMNRA simulation code permits the reconstruction of the complex luster structure. The energy position, height, and line shape of the silver or copper peak give the thickness of the overlaying glaze layer, the nanoparticle volume proportion, and the gradient of nanoparticles, which is detailed in Figure 13.4(c). First, a 100 nm layer of glaze containing 14 at.% of silver is simulated at the surface. The displacement of the copper peak in the experimental spectrum indicates the presence of a 170 nm glaze overlayer free of nanoparticles. The low-energy tailing of the copper band is finally taken into account by introducing a silver gradient extending to 400 nm in depth. The corresponding luster structure is represented in Figure 13.4(d).

13.4.2 HYDRATION DATING OF ARCHEOLOGICAL QUARTZ: PROFILING HYDROGEN DIFFUSION USING ERDA WITH EXTERNAL BEAM

In the following example, hydrogen depth profiles in two artifacts carved in rock crystal (quartz) are determined using ERDA in order to authenticate them— that is, tell if they are ancient (of pre-Columbian origin) or modern. The first object is an 11 cm height carving representing a human skull and the second one a smaller anthropomorphic mask, both acquired in the nineteenth century from the controversial French antiquity dealer E. Boban and now conserved in the Musée du Quai Branly, Paris, France. Serious doubts are indeed cast on the alleged pre-Columbian origin of the quartz skull, firstly because lapidary techniques employed to carve it seem modern. A quartz pendant from a chandelier cut and polished in AD 1740 has been used as a reference sample.

The diagnostic relies on the slow uptake of water from the environment by the freshly cut surface of quartz. The hydrogen diffusion depth increases with the square root of time and could in principle be used to estimate the manufacture date of an object. Whereas the hydration process of obsidian has already been studied in view of developing a dating method for archaeological artifacts, this approach has been extended to quartz (Ericson, Dersch, and Rauch 2004) and labeled quartz hydration dating (QHD). Its principle is to derive the time t from the diffusion profile H, in case of a constant diffusion coefficient, which is expressed as:

(13.1) $H(x, t) = (H(0, t) - Ci) \cdot erfc\{x/2(Dt) - 1/2\} + Ci$

where erfc is the complementary error function, D the diffusion coefficient, and Ci the initial bulk hydrogen concentration in quartz. In this process, the source of hydrogen is the natural thin layer of water from the environment (air, soil) naturally adsorbed on the surface of the artifact. In quartz, for historic ages, hydrogen penetrates less than 1 µm and the hydration level at the surface reaches a few atomic percent, H. The diffusion coefficient D depends upon three parameters: temperature, T, angle between the c-axis and the normal to the surface, α, and initial bulk hydrogen content, Ci.

The first hydration profiles in ancient quartz objects were obtained using NRA with ^{15}N beam. QHD is implemented here using an alternative approach—namely, ERDA with a 3 MeV external ^{4}He microbeam. The setup with a classical geometry (incident angle = exit angle = 15°) is housed in a large Plexiglas vessel flushed with helium gas (Figure 13.5a). The helium microbeam is focused on the artifact through a 100 nm thick Si_3N_4 membrane. With a beam spot size of 0.5 mm × 0.1 mm, the analyzed zone on these curved objects can be considered as flat. Recoiled hydrogen ions are collected with an annular particle detector screened with a 13 µm Kapton foil placed in an evacuated housing with the entrance sealed by a second identical membrane. Angular and energy stragglings of the incident beam and scattered ions are minimized by limiting the path length in the surrounding gas to 6 mm (3.3 keV in Si_3N_4 membrane and 5.8 keV in helium gas). An additional particle detector, fitted in the borehole of the annular detector, collects forward-scattered He ions. The spectrum acquired with this detector allows (1) simplification of geometrical alignment, (2) measurement of the orientation of the sample surface at the impact point, and (3) monitoring of the purity of the helium atmosphere inside the vessel after purging.

Positioning of the rock crystal skull in the setup is illustrated in Figure 13.5(b). The ERDA spectra recorded on the three samples (Figure 13.5c) exhibit common features: a marked peak at the surface, corresponding to the overlap of the water layer at the surface and of the first part of the diffusion profile; a decay in the 600–800 keV region with markedly different slopes for the two archaeological artifacts, the skull having a steeper slope than the mask; an intermediate slope for the reference sample; and a low-energy plateau ascribed to the intrinsic bulk hydrogen concentration Ci. Quantitative hydrogen diffusion profiles were derived from these raw data (Figure 13.5d) for the reference sample, the mask, and the skull, respectively, and the data points fitted with Equation (13.1). The hydrogen concentrations and the derived D·t products are given in Table 13.5. As expected, at large depth (>1 µm), the hydrogen levels reach a constant level Ci.

Despite uncertainties in the diffusion coefficient D, the results clearly show that the rock crystal skull is younger than the reference sample dated AD 1740.

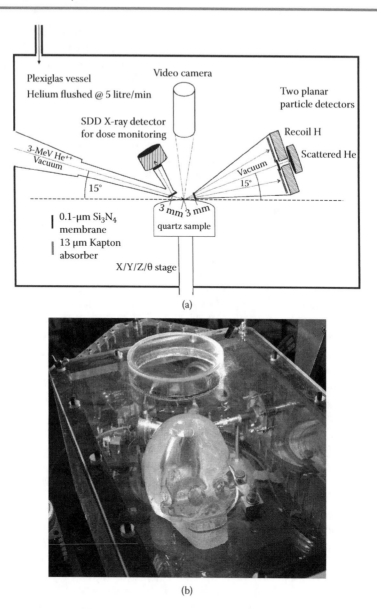

(a)

(b)

FIGURE 13.5   Determination of hydration depth profiles in rock crystal artifacts. (a) Layout of the external ERDA setup. Note the use of two particle detectors to collect, respectively, recoil hydrogen and scattered He ions from the beam (to monitor geometry). (b) Positioning of the crystal skull in the system. (Continued)

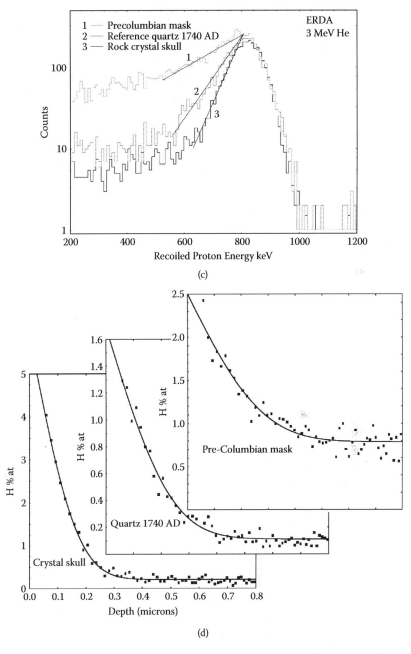

(c)

(d)

FIGURE 13.5   (Continued) (c) Raw ERDA spectra recorded on the two artifacts and the reference quartz crystal. (d) Derived quantitative hydrogen depth profiles. (Continued)

TABLE 13.5   Results Derived from the ERDA Spectra

Sample	α	H(x = 0) at.% H	C_i at.% H	D·t cm²	Manufacture date 1	Manufacture date 2
Reference	75°	1.69	0.11	1.69 10⁻¹⁰	AD 1740	
Mask	90°	2.49	0.79	2.38 10⁻¹⁰	AD 1906	AD 1278
Skull	90°	5.96	0.21	0.84 10⁻¹⁰	AD 1873	AD 1750

Notes: H(0, t) surface concentrations; D·t products and manufacture date, with opposite hypotheses on D: first, if D is assumed constant and equal to the value of the reference sample and, second, if D depends on Ci. In both cases, the rock crystal skull is too young to be pre-Columbian.

In the present case, the QHD method carried out with external ERDA has evidenced that the rock crystal skull is not pre-Columbian, but more likely a recently manufactured fake. On the other hand, the mask is found to be possibly older, depending on the model considered.

13.4.3   MICROMAPPING OF MINERAL PHASES IN LAPIS LAZULI CARVINGS

The multiscale imaging capabilities of IBA using external beam are illustrated by two examples. At small scale (1 mm or less), the scanning of the beam through the exit window permitted the characterization of the microscopic mineral phases constituting a heterogeneous rock called lapis lazuli. At a larger scale (up to several tens of centimeters), the chemical maps of selected areas of Renaissance paintworks were collected by mechanically scanning the painting under a fixed beam.

Lapis lazuli is of high interest in both archaeology and art history: Employed in Asia since the seventh millennium BC for carvings and beads, it was used in Medieval Europe in powdered form as a precious blue painting pigment called ultramarine. The chemical imaging of major and trace elements in lapis lazuli using external μ-PIXE has permitted the characterization of its mineral phases including their trace elements (Lo Giudice et al. 2009; Calligaro et al. 2011).

In the lapis lazuli, a 3 MeV proton beam penetrates around 60 μm but the effective probing depth of PIXE (depth from which 95% of the x-rays are originating) for its constituting elements is shorter, ranging from 2 μm (Na) to 40 μm (Fe). The diameter of an external microbeam is usually of ca. 30 μm due to the angular straggling of protons crossing the few millimeters gas path (air or helium) toward the sample. This coarse spatial resolution provides an image of rock at a scale comparable to classical mineralogical tools like petrographic microscopes. Most lapis lazuli mineral phases in the maps can be unambiguously identified on the basis of quantitative concentrations in major elements like Na, Al, Si, S, P, or Ca (Figure 13.6a). For example, the correlation of Na, Al, Si, and S with their relative concentrations permits the identification of the lazurite phase $Na_3Ca(Al_3Si_3O_{12})S$, responsible for the blue color. The Ca-rich

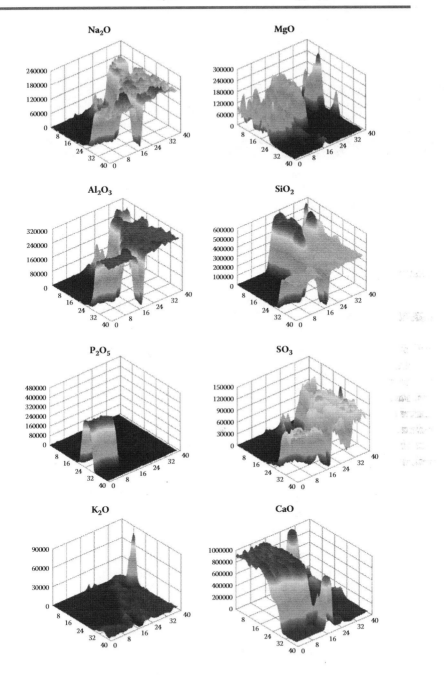

FIGURE 13.6 Characterization of mineral phases in lapis lazuli rock. (a) The quantitative maps of oxides allow the identification of the major mineral phases of the rock—for example, lazurite, calcite, and apatite. The concentration levels are given in micrograms per gram weight of oxides (1.000.000 = 100%). (Continued)

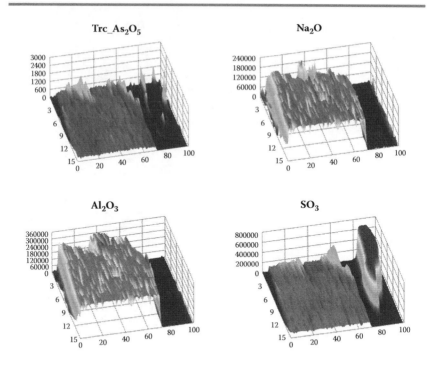

FIGURE 13.6   (Continued) (b) Maps show that, for the lapis from Tajikistan, traces of arsenic (200–400 mg/g) are homogenously distributed within the lazurite phase.

region with a low concentration of Si and Mg corresponds to calcite $CaCO_3$ and the region rich in P and Ca to apatite $Ca_5(PO_4)_3$.

The quantitative concentration maps also allow the assignment of trace elements to specific mineral phases. For instance, lapis from Tajikistan contains more arsenic than that from other sources (100 to 500 µg/g), in particular more than Afghan lapis, which it resembles in many other aspects. In lapis from Tajikistan, the correlation of arsenic with sodium, sulfur, and silicon in the recorded maps clearly shows that it is evenly distributed within the lazurite phase at a level of ca. 300 µg/g (Figure 13.6b). In contrast, in lapis from Afghanistan, arsenic concentration in lazurite was found globally below 100 µg/g and concentrated in isolated inclusions. This constitutes an interesting criterion to differentiate lapis from Tajikistan and Afghanistan.

13.4.4   LARGE-SCALE CHEMICAL IMAGING OF RENAISSANCE PAINTED MASTERWORKS

The extensive works on Renaissance paintings by the LABEC group headed by P. Mandò in Florence are particularly illustrative of the imaging capability of external beams. The following example concerns the painting *Madonna*

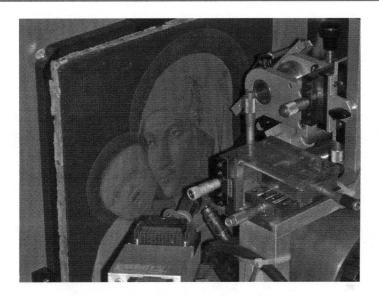

FIGURE 13.7 Chemical imaging of the Renaissance paintwork *Madonna col Bambino* by Andrea Mantegna. (a) The positioning of the painting in the LABEC external beam setup. (b) Two areas of 1.8 × 1.6 mm have been imaged in PIXE mode—namely, the pupil of the Virgin's left eye (top) and a detail of her veil (bottom) by scanning the beam through the exit window. See text for interpretation. (Continued)

col Bambino by Andrea Mantegna (1460), 43 × 31 cm², conserved in the Accademia Carrara in Bergamo, Italy. Figure 13.7(a) depicts the positioning of the painting in the LABEC external beam setup. Two areas of 1.8 × 1.6 mm have been imaged in PIXE mode—namely, the pupil of the Virgin's left eye (top) and a detail of her veil (bottom) by scanning the beam through the exit window (Figure 13.7b). For the first area, the comparison of the gold, lead, tin, and iron elemental maps with the visible image indicates that the golden glint is indeed produced by a gold-containing paint layer. The weak correlation between lead and tin maps might be ascribed to the use of the lead–tin yellow pigment. In the second area, the gold map shows the use of the same gold-containing matter, while the silicon and aluminum maps reveal that the veil detail area was obtained with ultramarine pigment. It is worth mentioning that such light element maps would hardly have been carried out by XRF.

13.5   FUTURE DIRECTIONS AND NEW CHALLENGES

Since their introduction in the early 1970s, IBA techniques have largely contributed to the advancement of various research fields and have given a strong impetus to heritage materials studies. Indeed, they have permitted the damage-free

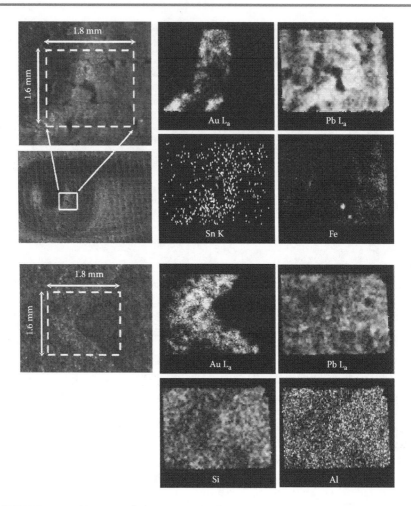

FIGURE 13.7 (Continued) (b) Two areas of 1.8 × 1.6 mm have been imaged in PIXE mode—namely, the pupil of the Virgin's left eye (top) and a detail of her veil (bottom) by scanning the beam through the exit window. See text for interpretation.

analysis of precious items with outstanding performances in terms of accuracy, sensitivity, and spatial information. The adaptation of IBA to the specificity of these materials has led to the design of increasingly sophisticated experimental approaches and setups. It should be pinpointed that, in return, this extensive experimental work has been largely beneficial to the techniques themselves with potential impact in other fields. The main progress derives from the development of external beam setups, which permit the implementation of IBA in air or in an inert gas at atmospheric pressure. This kind of setup is currently spreading in most facilities involved, not only in heritage studies but also in other fields.

Until the last decade, IBA had no valuable alternative. This is no longer true today as other spectrometric techniques have improved considerably and can in many aspects compete with IBA. However, as demonstrated in this chapter, IBA techniques remain unmatched in three domains:

- Light element analysis via the PIGE, NRA, or ERDA techniques
- Elemental depth profiling using RBS, NRA, or ERDA
- Large-scale elemental imaging mainly based on PIXE

In view of strengthening the position of IBA in art and archaeology and further convincing museum curators of the nondamaging character of these techniques, efforts must be pursued to reduce to the lowest level of the beam intensity (ca. 10 pA) and integrated charge (ca. 1 nC). To reach this goal, detection systems with much higher efficiency should be developed, such as large-sized detectors or detector arrays. (Pichon et al. 2014). Another foreseen improvement should be the systematic acquisition of elemental maps instead of single-spot analysis. This can be achieved only with a high data throughput, obtained from the previously mentioned detection systems, to obtain quantitative composition at each point. In general the study of complex heritage materials is best tackled by simultaneously using PIXE/PIGE/RBS/IBIL (ion beam-induced luminescence) and even further with a multimodal approach combining IBA with other analytical techniques such as XRD or micro-Raman spectrometry. As already mentioned, one of the most acute limitations of IBA techniques is the lack of chemical state information. High-resolution features in PIXE spectra obtained by wavelength dispersive spectrometry (satellite peaks) or IBIL could be a valuable tool to partly address this problem.

One should consider the future role of IBA in heritage studies as a member of a large pool of analytical techniques at the disposal of researchers. It now stands, after simpler and portable techniques operated in the field, such as mobile Raman spectrometry, XRF or LIBS, as a first approach that helps answer basic questions and selects the cases where the use of IBA is justified. At the end of the analytical chain, whenever the data obtained are not comprehensive enough, IBA has to be supplemented with techniques only available on larger scale facilities such as synchrotron or neutron sources. This is exemplified by the analytical strategy adopted within the CHARISMA European project where transnational access to the AGLAE accelerator-based IBA facility is offered among a set of mobile instruments named MOLAB for *in situ* diagnosis of art works (Miliani et al. 2010), along with a more restricted access to the synchrotron SOLEIL and the Budapest neutron source, for more in-depth research. In our opinion, it is within such a strategy that IBA will remain at the cutting edge of heritage study.

ACKNOWLEDGMENTS

We are highly indebted to the following colleagues, who have generously provided examples of applications that illustrate this chapter: N. Grassi and P. A. Mandò from the LABEC, Florence, Italy, for the study of paintworks, and

A. Bouquillon and M. Aucouturier from the C2RMF, Paris, for the study of lustered ceramics. We thank the AGLAE team at the C2RMF—C. Pacheco, B. Moignard, L. Pichon, and T. Guillou, not forgetting the late J. Salomon— for their skillful and efficient work over more than 20 years that has enabled a remarkable adaptation of IBA to museum objects. We acknowledge the support of M. Menu, who has coordinated the AGLAE access within two successive European programs (Eu-Artech, CHARISMA).

PROBLEMS

The following problems are aimed at justifying the use of IBA techniques and basic calculations related to examples given in Section 13.4.

13.1. RBS analysis of a gilded art work in external beam mode: experimental conditions: 3 MeV He ions, exit window 100 nm Si_3N_4; beam spot at 3 mm from the exit window; particle detector 150° to the beam direction placed behind a second 100 nm Si_3N_4 window located 3 mm from the beam spot and parallel to the detector surface; helium at atmospheric pressure. Specific gravity of helium gas 1.8 10^{-4} g/ cm³, gold 19.3 g/cm³, Si_3N_4 3.4g/ cm³. Stopping power for 3 MeV He ions in gold: 96 eV/10^{15} at./cm²; helium gas: 9.4 eV/10^{15} at./cm², Si_3N_4: 31 eV/10^{15} at./cm².

a. Calculate energy loss induced by helium gas and exit window.
b. What is the dominant contribution (gas or window)?
c. Calculate the energy position of gold front edge in RBS spectrum.
d. Estimate gilding thickness if the energy width of the gold band in the RBS spectrum is 335 keV.
e. What is the maximum gilding thickness that can be measured?

13.2. Differential PIXE to determine arrangement of paint layers: A painting is composed of two superimposed layers. The PIXE spectrum shows the mercury and lead L-lines, which are ascribed to cinnabar (HgS) and lead white ($PbCO_3$) respectively. The processing of the PIXE spectra recorded using 1.2, 2.0, and 3.9 MeV proton beams respectively gives the following apparent Hg and Pb proportions in mass:

Pigment	Element	1.2 MeV	2 MeV	3.9 MeV
Cinnabar	Hg	36%	63%	68%
Lead white	Pb	64%	37%	32%

a. Using these data, give the layer ordering. Justify your answer.
b. Would it be possible to obtain the same answer using a single spectrum recorded at one energy? (Hint: consider the L_α/L_β branching ratio.)

SOLUTIONS TO EXERCISE 13.2

a. The lead white layer is on top of the cinnabar layer. The increasing relative intensity of Hg lines with the increasing energy indicates that the beam is progressively reaching the underneath cinnabar layer.

b. Yes. For the Hg lines, the intensity ratio La over Lb is reduced by the absorption by the lead white top layer.

13.3. Sodium analysis in glass using the PIXE-PIGE combination: mass attenuation coefficient of Na x-ray lines in soda glass $\mu = 3000$ cm^2/g; specific gravity of soda glass $\rho = 2.7$ g/cm^3; stopping power of soda glass for 3 MeV protons $= 9 \cdot 10^4$ keV $\times$ cm^2/g; thick target 440 keV Na γ-ray yield:

Proton energy	1.0 MeV	1.7 MeV	2.4 MeV	3.1 MeV
Relative yield	$2.2 \cdot 10^3$	$8.3 \cdot 10^5$	$3.4 \cdot 10^6$	$9.6 \cdot 10^6$

a. Compare the probed depth (95% of the signal) for sodium in glass using PIXE and PIGE.

b. What is the maximum thickness of the Na-depleted layer produced by weathering that can be estimated in this way?

SOLUTIONS TO EXERCISE 13.3

a. For PIXE, the probed depth of low energy X-ray lines is governed by sample self absorption. 95 % of signal is collected from a depth equal to 3 radiation lengths ($3 \times 1/\mu$), that is 3.7 μm. Plotting gamma-ray yield versus proton energy shows that the yield at 1.5 MeV drops at 5% of the value at 3 MeV. The probed depth for PIGE is thus attained at 1.5 MeV, which is 62 μm according to the given stopping power.

b. The ratio Na by PIXE to Na by PIGE varies from 1 (unaltered) to 0 (totally leached). It allows measuring weathered layers ranging from zero to 3.7 μm. The sole variation of the Na by PIGE value can be used to extend this range up to 62 μm.

13.4. Selective absorbers in PIXE measurement: For the determination of trace elements in bronzes, it is common to use a cobalt absorber to attenuate copper x-ray lines: mass attenuation coefficient for Cu K-lines in cobalt $\mu = 320$ cm^2/g, for Ag-lines $\mu = 21$ cm^2/g, for Sn K-lines $\mu = 14.5$ cm^2/g; stopping power of cobalt for 3 MeV protons $= 6.4 \cdot 10^4$ keV $\times$ cm^2/g; specific gravity of cobalt $\rho = 8.9$ g/cm^3.

a. Compute the thickness of the cobalt absorber required to block backscattered protons of the 3 MeV beam.

b. Calculate the transmission for Cu, Sn, and Ag K-lines (common elements in bronzes) of this filter.

SOLUTIONS TO EXERCISE 13.4

a. Assuming the given fixed stopping power, the thickness required is 53 μm.

b. Transmission T = exp(−μx); for Cu T= 3.7 10^{-7}; for Ag K-lines T = 0.37; for Sn K-lines T = 0.51

13.5. Monitoring the integrated charge using the exit window signal: Beam monitoring in external beam mode is challenging. We consider the possible use of the PIXE and RBS signals emitted by the 100 nm Si_3N_4 exit window: Si x-ray production cross section for 3 MeV protons σ = 1000 barns; Si_3N_4 specific gravity ρ = 3.4 g/cm³.

   a. Give the PIXE count rate for the Si line using a 100 pA beam with an ideal 0.05 sr x-ray detector.

   b. Compare with RBS signal from Si with an equivalent detector.

   c. Which signal is more appropriate?

SOLUTIONS TO EXERCISE 13.5

a. Yield = thickness × rho × N_a × cross-section x solid-angle / (4 × pi × M). 100 pA corresponds to 6.25 10^8 protons/sec. The Si PIXE count rate recorded in the dose detector is 2500 counts/sec.

b. Rutherford backscattering cross-section for protons on Si is 0.04 barns(from table or formula), thus 25000 times less than PIXE cross-section.

c. Obviously, the PIXE signal is more efficient for monitoring low intensity beams.

APPENDIX

Some useful links are provided to which the reader can refer for further information. First, a comprehensive list of worldwide IBA facilities involved in the present research is given.

Laboratories Having IBA Application to Cultural Heritage as a Major Activity

Laboratory	Accelerator Type	Experimental Setup	Main Research Fields
Belgium: Liège	Cyclotron	External beam	Paintings, glasses, gemstones
France: Paris C2RMF	2 MV tandem pelletron	External microbeam	Ceramics, stones, gems, metals, glasses, enamels, glazes, manuscripts, paintings
Greece: Athens	Tandem	External beam	Metals, ceramics

Hungary: Budapest	2 MV Van de Graaff	External beam	Manuscripts, bronzes
Hungary: Debrecen	5 MV Van de Graaff	Microprobe	Paintings, gemstones, glasses, ceramics
Italy: Florence	3 MV Van de Graaff	External microbeam	Manuscripts, ceramics
Mexico: UNAM	3 MV tandem pelletron	External beam; microprobe	Stones, pigments, jewelry
Slovenia: Ljubljana	2 MV tandem tandetron	External beam; microprobe	Coins, stones
Spain: Madrid	5 MV tandem tandetron	External microbeam	Ceramics, glasses, metals, gemstones
Spain: Sevilla	3 MV tandem pelletron	External microbeam	Glasses, jewelry, metals

Note: On the basis of last decade publications.

Laboratories Occasionally Active in This Field

Laboratory	Accelerator type	Experimental Setup	Main Research Fields
Austria: Vienna	3MV tandem pelletron	External beam	Manuscripts, drawings
Belgium: Namur	2 MV tandetron	External beam	Metals
Croatia: Zagreb	1 MV tandetron 6 MV tandem VdG	Microprobe	Pigments
Finland: Turku	20 MeV cyclotron	External beam	Paintings
France: Bordeaux	4 MV singletron	External beam; microprobe	Obsidian
Germany: Berlin	65 MeV cyclotron	External beam	Metals
Germany: Rossendorf	5 MV tandem	External beam; microprobe	Bones, paintings, drawings, glasses
Italy: Lecce	3 MV tandetron	External beam; microprobe	Bone, metals
Italy: Legnaro	3 MV Van de Graaff	Microprobe	Stone, jewelry
Portugal: Sacavém	2.5 MV Van de Graaff	External beam; microprobe	Glasses
Sweden: Lund	3 MV Van de Graaff	Microprobe	Drawings
UK: Guildford	1.7 MV tandem pelletron	External beam; microprobe	Bones, glazes, manuscripts
China: Shanghai	3 MV tandem	External beam	Ceramics, manuscripts, metals
India: Bubaneshwar	3 MV tandem pelletron	Vacuum chamber	Coins
Iran: Tehran	3 MV Van de Graaff	External microbeam	Ceramics
Japan: Tokyo	2 MV tandem	External beam	Ceramics
Lebanon: Beirut	1.7 MV tandem pelletron	External beam; microprobe	Ceramics
Singapore	2.5 MV Van de Graaff	Microprobe	Bones, gemstones

Taiwan: Taipei	3 MV tandem	External beam	Coins
US: Cleveland	1.6 tandem pelletron	External beam	Ceramics
Australia: Lucas Heights	2 MV tandetron	Vacuum chamber	Pigments
South Africa: Faure	5 MV Van de Graaff	External beam	Ceramics
Chile: Santiago	3.7 Van de Graaff	Vacuum chamber	Stone

As for scientific publications, they can be found in international journals relevant to either nuclear instrumentation or cultural heritage studies. For the first case, we can mention:

Nuclear Instruments and Methods in Physics Research B
X-Ray Spectrometry
Radioanalytical and Nuclear Chemistry
Applied Physics A
Analytical and Bioanalytical Chemistry

For cultural heritage journals:

Journal of Cultural Heritage
Archaeometry
Journal of Archaeological Science

Two textbook chapters address specifically the application of PIXE and IBA to heritage, respectively:

Dran, J.-C., Calligaro, T., and Salomon, J. 2000. chapter 6, particle induced X-ray emission." In *Modern Analytical Methods in Art and Archaeology*, eds. Ciliberto and Spoto. *Chemical Analysis*, vol. 155. New York: John Wiley.

Calligaro, T., Dran, J.-C., and Salomon, J. 2004. Ion beam micronanalysis. In *Nondestructive Microanalysis of Cultural Heritage Materials*, ed. Janssens K. van Grieken. *Wilson & Wilson's Comprehensive Analytical Chemistry*, vol. 42. Amsterdam: Elsevier.

Several international IBA-related conference series are held every year. Most of them organize a session dedicated to art and archeology. Their proceedings generally appear in a special issue of *Nuclear Instruments and Methods B* and occasionally of *X-Ray Spectrometry:*

IBA: Ion Beam Analysis (every 2 years); IBA 20 was held in Brazil, April 2011.
IBA 21 was held in cambridge, UK, July 2013.
PIXE: Particle Induced X-Ray Emission and Its Applications (every 2 years); PIXE '10 was held in Guildford, UK, June 2010.
PIXE 11 was held in Gramado, Brazil, March 2013.

ICNMTA: International Nuclear Microprobe Technology and Applications (every 2 years); ICNMTA 12 was held in Leipzig, September 2010.

ICNTMA 13 was held in Lisbon, Portugal, July 2012.

ECAART: European Conference on Accelerators in Applied Research and Technology (every 3 years). ECAART 10 was held in Athens, Greece, September 2010.

ECAART 11 was held in Namur, Belgium, September 2013.

Some IBA applications also appear in art and archaeology international meeting proceedings:

ISA: International Symposium on Archaeometry (ISA). ISA32 was held in Tampa, Florida, May 2010.

ISA 39 was held in Leuwen, Belgium, May 2012.

ART: International Conference on Nondestructive Investigations and Microanalysis for the Diagnostics and Conservation of Cultural and Environmental Heritage. ART '11 was held in Florence, Italy, April 2010.

ART 14 was held in Madrid, Spain, June 2014.

Some applications can also be found in dedicated sessions in Materials Research Society (MRS) conferences, like the E-MRS spring meeting in Strasbourg, in May 2013.

Within special European programs, several networks of analytical facilities offer access to IBA for potential users in this field:

CHARISMA: Cultural Heritage Advanced Research Infrastructures: Synergy for a Multidisciplinary Approach to Conservation/Restoration 2010–2014; coordinator: B. Brunetti, U. Perugia, Italy (www.charismaproject.eu).

SPIRIT: Support of Public and Industrial Research Using Ion Beam Technology; coordinator: W. Möller, FZD-Rossendorf, Germany (www.spirit-ion.eu).

REFERENCES

Adriaens, A. 2005. Non-destructive analysis and testing of museum objects: An overview of 5 years of research. *Spectrochimica Acta Part B* 60:1503–1516.

Adriaens, A., and Dowsett, M. 2010. The coordinated use of synchrotron spectroelectro-chemistry for corrosion studies on heritage metals. *Accounts of Chemical Research* 43:927–935.

Barradas N. P., Arstila, K., Battistig, G., Bianconi, M., Dytlewski, N., Jeynes, C., Kótai, E., Lulli, G., Mayer, M., Rauhala, E., Szilágyi, E., and Thompson, M. 2008. Summary of IAEA intercomparison of IBA software. *Nuclear Instruments and Methods B* 266:1338–1342.

Beck, L., Cuif, J.-P., Pichon, L., Vaubaillon, S., Dambricourt Malassé, A., and Abel, R. L. 2012. Checking collagen preservation in archaeological bone by non-destructive studies (micro-CT and IBA). *Nuclear Instruments and Methods B* 273:203–207.

Beck, L., L. de Viguerie, Walter, Ph., Pichon, L., Gutiérrez, P. C., Salomon, J., Menu, M., and Sorieul. S. 2010. New approaches for investigating paintings by ion beam techniques. *Nuclear Instruments and Methods B* 268:2086–2091.

Bird, J. R., Duerden, P., and Wilson, D. J. 1983. Ion beam techniques in archaeology and the arts. *Nuclear Science Applications* 1:357–516.

Bouquillon, A., Dran, J.-C., Lagarde, G., Martinetto, P., Moignard, B., Salomon, J., and Walter, P. 2002. *In situ* dynamic analysis of solids or aqueous solutions undergoing chemical reactions by RBS or PIXE with external beams. *Nuclear Instruments and Methods B* 188:156–161.

Calligaro, T. 2008. PIXE in the study of archaeological and historical glass. *X-Ray Spectrometry* 37:169–177.

Calligaro, T., Coquinot, Y., Pichon, L., and Moignard, B. 2011. Advances in elemental imaging of rocks using the AGLAE external microbeam. *Nuclear Instruments and Methods B* 269:2364–2372.

Calligaro, T., Coquinot, Y., Reiche, I., Castaing, J., Salomon, J., Ferrand, G., and Le Fur, Y. 2009. Dating study of two rock crystal carvings by surface microtopography and by ion beam analyses of hydrogen. *Applied Physics A* 94:871–878.

Calligaro, T., Dran, J.-C., Poirot, J.-P., Querré, G., Salomon, J., and Zwaan, J. C. 2000. PIXE/PIGE characterization of emeralds using an external micro-beam. *Nuclear Instruments and Methods B* 161–163:769–774.

Calligaro, T., Dran, J.-C., and Salomon, J. 2004. Ion beam micronanalysis. In *Non-destructive microanalysis of cultural heritage materials*, eds. K. Janssens and R. van Grieken, chap. 5, 227–276. Wilson & Wilson's comprehensive analytical chemistry, 42, Amsterdam: Elsevier.

Calligaro, T., Dran, J.-C., Salomon, J., and Walter, Ph. 2004. Review of accelerator gadgets for art and archaeology. *Nuclear Instruments and Methods B* 226:29–37.

Chabanne, D., Aucouturier, M., Bouquillon, A., Darque-Ceretti, E., Makariou, S., Dectot, X., Faÿ-Hallé, A., and Miroudot, D. 2008. Ceramics with metallic lustre decoration. A detailed knowledge of Islamic productions from 9th century until Renaissance. http://hal.archives-ouvertes.fr/hal-00554837/fr/

Chen, J.-X., Li, H.-K., Ren, C.-G., Tang, G.-H., Wang, X.-D., Yang, F.-C., and Yao, H.-Y. 1980. PIXE research with external beam. *Nuclear Instruments and Methods B* 168:437–440.

Chiari M., Migliori, A., and Mandò, P. A. 2002. Investigation of beam-induced damage to ancient ceramics in external-PIXE measurements. *Nuclear Instruments and Methods B* 188:151–155.

Clark, R. J. H. 2002. Pigment identification by spectroscopic means: an arts/science interface. *Comptes Rendus Chimie* 5:7–20.

Coote, G. E., Whitehead, N. E., and McCallum, G. J. 1972. Rapid method of obsidian characterization by inelastic scattering of protons. *Journal of Radioanalytical Chemistry* 12:491–496.

Cotte, M., Susini, J., Dik, J., and Janssens, K. 2010. Synchrotron-based x-ray absorption spectroscopy for art conservation: Looking back and looking forward. *Accounts of Chemistry Research* 43:705–714.

Cotte, M., Susini, J., Metrich, N., Moscato, A., Gratziu, C., Bertagnini, A., and Pagano, M. 2006. Blackening of Pompeian cinnabar paintings: X-Ray microspectroscopy analysis. *Analytical Chemistry* 78:7484–7492.

Creagh, D. 2007. Synchrotron radiation and its use in art, archaeometry, and cultural heritage studies. In *Physical techniques in the study of art, archaeology and cultural heritage*, vol. 2. Oxford: Elsevier.

Deconninck, G. 1972. Quantitative analysis by (p, X) and (p, γ) reactions at low energies. *Journal of Radioanalytical Chemistry* 12:157–169.

Del Carmine, P., Lucarelli, F. Mandò, P. A., and Pecchioli, A. 1993. The external PIXE setup for the analysis of manuscripts at the Florence University. *Nuclear Instruments and Methods B* 75:480–484.

Demortier, G., and Gilson, A. 1987. Determination of traces of light elements in gold artifacts using nuclear reactions. *Nuclear Instruments and Methods B* 18:286–290.

Demortier, G., and Morciaux, Y. 1994. PIXE gadgets. *Nuclear Instruments and Methods B* 85:112–117.

Dersch, O., and Rauch, F. 1999. Water uptake of quartz investigated by means of ion-beam analysis. *Fresenius Journal of Analytical Chemistry* 365:114–116.

De Viguerie, L., Beck, L., and Salomon, J. 2009. Composition of renaissance paint layers: Simultaneous particle induced x-ray emission and backscattering spectrometry. Analytical Chemistry 81:7960–7966.

Doyle, B. L. 1983. Non-vacuum Rutherford backscattering spectrometry. *Nuclear Instruments and Methods B* 218:29–32.

Dran, J.-C., Calligaro, T., and Salomon, J. 2000. Particle induced x-ray emission. In *Modern analytical methods in art and achaeology,* vol. 155, 135–166, ed. Ciliberto and Spoto, *Chemical analysis,* New York: John Wiley.

Ericson, J. E., Dersch, O., and Rauch, F. 2004. Quartz hydration dating. *Journal of Archaeological Science* 31:883–902.

Eriksson, L., Fladda, G., and Johansson, P. A. 1972. Use of alpha-particle backscattering for study of printed and coated layers. *Journal of Radioanalytical Chemistry* 12:287–300.

Giakoumaki, A., Melessanaki, K., and Anglos. D. 2007. Laser-induced breakdown spectroscopy (LIBS) in archaeological science—Applications and prospects. *Analytical and Bioanalytical Chemistry* 387:749–760.

Giuntini, L. 2011. A review of external microbeams for ion beam analyses. *Analytical and Bioanalytical Chemistry* 40:785–793.

Giussani, B., Monticelli, D., and Rampazzi, L. 2009. Role of laser ablation—Inductively coupled plasma–mass spectrometry in cultural heritage research: A review. *Analytica Chimica Acta* 635:6–21.

Gordon, B. M., and Kraner, H. W. 1972. On the development of a system for trace element analysis in the environment by charged particle x-ray fluorescence. *Journal of Radioanalytical Chemistry* 12:181–188.

Grassi, N. 2009. Differential and scanning-mode external PIXE for the analysis of the painting Ritratto Trivulzio by Antonello da Messina. *Nuclear Instruments and Methods B* 267:825–831.

Grassi N., Giuntini, L., Mandò, P. A., and Massi, M. 2007. Advantages of scanning-mode ion beam analysis for the study of cultural heritage. *Nuclear Instruments and Methods B* 256:712–718.

Grassi N., Migliori, A., Mandò, P. A., and Calvo del Castillo, H. 2004. Identification of lapis-lazuli pigments in paint layers by PIGE measurements. *Nuclear Instruments and Methods B* 219–220:48–52.

Guerra, M. F. 2008. An overview on the ancient goldsmith's skill and the circulation of gold in the past: The role of x-ray-based techniques. *X-Ray Spectrometry* 37:317–327.

Guerra, M. F., and Calligaro, T. 2003. Gold cultural heritage objects: A review of studies of provenance and manufacturing technologies. *Measurement Science and Technology* 14:1527–1537.

———. 2004. Gold traces to trace gold. *Journal of Archaeological Science* 31:1199–1208.

Gutiérrez, P. C., Ynsa, M.-D., Climent-Font, A., and Calligaro. T. 2010. Detection of beryllium treatment of natural sapphires by NRA. *Nuclear Instruments and Methods B* 268:2038–2041.

International Atomic Energy Agency. 1998. Intercomparison of gamma ray spectrometry software packages. IAEA-TECODC-1011, Vienna. <http://www-pub.iaea.org/MTCD/publications/PDF/te_1011_prn.pdf>

———. 2003. Intercomparison of PIXE spectrometry software packages. IAEA-TECDOC-1342, Vienna, <http://www-pub.iaea.org/MTCD/publications/PDF/te_1342_web.pdf>

Ioannidou, E., Bourgarit, D., Calligaro, T., Dran, J.-C., Dubus, M., Salomon, J., and Walter, P. 2000. RBS and NRA with external beams for archaeometric applications. *Nuclear Instruments and Methods B* 161–163:730–736.

Kalliabakos, G., Kossionides, S., Misailides, P., Papadopoulos, C. T., and Vlastou, R. 2000. Determination of sulphur and copper depth distribution in patina layers using nuclear reaction techniques. *Nuclear Instruments and Methods B* 170:467–473.

Kanngießer, B., Hahn, O., Wilke, M., Nekat, B., Malzer, W., and Erko, A. 2004. Investigation of Oxidation and Migration Processes of Inorganic Compounds in Ink Corroded Manuscripts. *Spectrochimica Acta B* 59: 1511–1516.

Kavčič, M. 2010. Improved detection limits in PIXE analysis employing wavelength dispersive x-ray spectroscopy. *Nuclear Instruments and Methods B* 268:3438–3442.

Kossionides, S., Kokkoris, M., Karydas, A. G., Paradellis, T., Kordas, G., and Moraitou, G. 2002. Analysis of ancient glass using ion beams and related techniques. *Nuclear Instruments and Methods B* 195:408–413.

Lahanier, C., Amsel, G., Heitz, C., Menu, M., and Andersen, H. H. 1986. International workshop devoted to ion beam analysis in the arts and archaeology. *Nuclear Instruments and Methods B* 14:1–168.

Laursen, T., and Lanford, W. A. 1978. Hydration of obsidian. *Nature* 276:153–156.

Lill, J.-O. 1999. Charge integration in external-beam PIXE. *Nuclear Instruments and Methods B* 150:114–117.

Lo Giudice, A., Re, A., Calusi, S., Giuntini, L., Massi, M., Olivero, P., Pratesi, G., Albonico, M., and Conz, E. 2009. Multitechnique characterization of lapis lazuli for provenance study. *Analytical and Bioanalytical Chemistry* 395: 2211–2217.

Mäder, M., and Neelmeijer, C. 2004. Proton beam examination of glass—An analytical contribution for preventive conservation. *Nuclear Instruments and Methods B* 226:110–118.

Mandò, P. A. 1994. Advantages and limitations of external beams in applications to arts and archaeology, geology and environmental problems. *Nuclear Instruments and Methods B* 85:815–823.

Mandò, P. A., Fedi, M. E., Grassi, N., and Migliori, A. 2005. Differential PIXE for investigating the layer structure of paintings. *Nuclear Instruments and Methods B* 239:71–76.

Mathis, F., Salomon, J., Moignard, B., Pichon, L., Aucouturier, M., and Dran, J.-C. 2004. Real time RBS study of Cu-Sn alloy thermal oxidation by means of a 4He2+ external micro-beam. *Nuclear Instruments and Methods B* 226:147–152.

Migliori, A., Bonanni, P., Carraresi, L., Grassi, N., and Mando, P. A. 2011. A novel portable XRF spectrometer with range of detection extended to low-Z elements. *X-Ray Spectrometry* 40:107–112.

Miliani, C., Rosi, F., Brunetti, B. G., and Sgamellotti, A. 2010. *In situ* non invasive study of artworks: The MOLAB multitechnique approach. *Accounts of Chemical Research* 43:728–738

Neelmeijer, C., Brissaud, I., Calligaro, T., Demortier, G., Hautojarvi, A., Mader, M., Martinot, L., Schreiner, M., Tuurnala, T., and Weber, G. 2000. Paintings—A challenge for XRF and PIXE analysis. *X-Ray Spectrometry* 29:101–110.

Olsson, A.-M. B., Calligaro, T., Colinart, S., Dran, J. C., Lövestam, N. E. G., and Moignard, B. 2001. Micro-PIXE analysis of an ancient Egyptian papyrus: Identification of pigments used for the "Book of the Dead." *Nuclear Instruments and Methods B* 181:707–714.

Papillon, F., and Walter, P. 1997. Analytical use of the multiple gamma-rays from the 12C(d,p)13C nuclear reaction. *Nuclear Instruments and Methods B* 132:480–468.

Pichon L., Moignard, B., Lemasson, Q., Pacheco, C., and Walter, P. 2014. Development of a multi-detector and a systematic imaging system on the AGLAE external beam. *Nuclear Instruments and Methods B* 318:27–31.

Polvorinos del Rio, A., Castaing, J., and Aucouturier, M. 2006. Metallic nano-particles in lustre glazed ceramics from the 15th century in Seville studied by PIXE and RBS. *Nuclear Instruments and Methods B* 249:596–600.

Salomon, J., Dran, J.-C., Guillou, T., Moignard, B., and Pichon, L. 2008. Ion-beam analysis for cultural heritage on the AGLAE facility: Impact of PIXE/RBS combination. *Applied Physics A* 92:43–50.

Salomon, J., Dran, J.-C., Guillou, T., Moignard, B., Pichon, L., Walter, P., and Mathis, F. 2008. Present and future role of ion beam analysis in the study of cultural heritage materials: The example of the AGLAE facility. *Nuclear Instruments and Methods B* 266:2273–2278.

Swann, C. P. 1982. The study of archaeological artifacts using proton induced x-rays. *Nuclear Instruments and Methods B* 197:237–242.

Swann, C. P., and Fleming, S. J. 1990. Selective filtering in PIXE spectrometry. *Nuclear Instruments and Methods B* 49:65–69.

Szíki, G. Á., Uzonyi, I., Dobos, E., Rajta, I., Biró, K. T., Nagy, S., and Kiss, Á. 2004. A new micro-DIGE setup for the analysis of light elements. *Nuclear Instruments and Methods B* 219–220:508–513.

Tuurnala, T., and Hautojärvi, A. 2000. Original or forgery—Pigment analysis of paintings using ion beams and ionizing radiation. In *Ion beam study of art and archaeological objects,* ed. G. Demortier and A. Adriaens, 21. COST GAction, European Commission, EUR 19218.

Walter, P., Menu, M., and Dran, J.-C. 1992. Dating of archaeological flints by fluorine depth profiling: New insights in the mechanism of fluorine uptake. *Nuclear Instruments and Methods B* 64:494–498.

Weber, G., Martinot, L. Strivay, D. Garnir, H. P., and George, P. 2005. Application of PIXE and PIGE under variable ion beam incident angle to several fields of archaeometry. *X-Ray Spectrometry* 34:297–300.

Weber, G., Vanden Bemden, Y. Pirotte, M., and Gilbert, B. 2005. Study of stained glass window using PIXE–PIGE. *Nuclear Instruments and Methods B* 240:512–519.

Biomedical Applications

Harry J. Whitlow and Min-Qin Ren

14.1 INTRODUCTION

Biological materials represent one of the most important classes of materials. They make up the food chain and have been important constructional materials and energy sources since the start of life on this planet. Cells form the basis of all living matter. Despite their wide diversity, from primitive bacterial cells to complex higher organisms, their composition is remarkably similar. This is because the genetic code carried by the DNA (Alberts et al. 2004) provides a set of instructions to assemble complex biomolecules, based mainly on carbon, hydrogen, nitrogen, and oxygen that form the structural components and nanomachinery that perform functions in the cell.

Since the pioneering research in the early 1900s there has been a steady development of the life sciences toward becoming quantitative and objective. In 1953, a major leap forward came from the determination of the structure and function of DNA (deoxyribonucleic acid) by James D. Watson and Francis Crick from X-ray diffraction patterns taken by Rosalind Franklin and Raymond Gosling (Alberts et al. 2004). This gave rise to a second major step forward in the 1970s when rapid DNA sequencing became available. This facilitated the 1:1 relation between DNA sequence and amino acid sequences and other biomolecules to be established. Today DNA technology is used to tailor proteins with specific amino acid sequences and to identify individuals and their genetic origin from minute traces of matter as well as identifying and treating genetic defects. This gigantic step forward represented by the understanding of DNA and its technology is a paradigm shift just as significant as the semiconductor revolution in electronics.

The penetrating properties of MeV ions and the wide variety of different processes induced by their interaction with matter gives them unique and very useful possibilities for biomedicine. The analytical advantages of ion beam analysis (IBA) methods for biomedicine are summarized in Table 14.1. This

TABLE 14.1   Features of IBA with MeV Ions for Biomedical Research

Characteristic	Method	Advantages/Application
General and Broad Beam Analysis		
Simultaneous multielement detection	PIXE, PIGE	Capability to measure all elements
Homostatic element levels		
Detection of toxins		
Isotope detection	PIGE	^{2}H, ^{13}C, ^{15}N, ^{18}O tracer studies
		Uptake and biomolecule pathways
High sensitivity	PIXE	Trace element determination
		No preconcentration step needed
		Minimally invasive for in vivo studies
Minimal sample preparation	PIXE, RBS, PIGE, ERDA	Automated analysis
		Lower risk for inadvertent contamination
Chemically blind elemental analysis	RBS, PIXE, PIGE, ERDA	Absence of matrix effects
		True total element content
Straightforward calibration	PIXE, PIGE (internal standard)	Absolute (most quantitative)
	PIXE + RBS	Absolute with no internal standard
	RBS, ERDA	Absolute
Depth profiling	RBS, ERDA	Study of biosurface interactions
		Normalization
Nondestructive analysis	PIXE, PIGE, RBS, ERDA	Remeasurement of archived samples
MeV IBA Micro-/Nanobeam Imaging		
1D element mapping	Micro-PIXE, micro-RBS	Biosurface interaction studies in thick layers
2D element mapping	Micro-PIXE, micro-RBS	Trace element mapping
		Functionalized nanoparticle location
Energy loss measurement	STIM	Sample thickness normalization
		Subcellular imaging
Ionoluminescence	Ion beam-induced fluorescence	Subcellular imaging
		Immunohistological fluorescence marker

chapter is organized by first considering the constitution of biological matter and providing a brief introduction to the types of measurement. Sample presentation of biomedical materials is of central importance and is introduced next. Subsequently, the use of broad beam- and microbeam-based measurements is treated with emphasis on particle-induced X-ray emission (PIXE) (Johansson and Campbell 1988; Johansson, Campbell, and Malmqvist 1995; Campbell 2009), which is the most widely used IBA technique in biomedicine. Finally, some emerging MeV ion beam methods for biomedical work are discussed.

14.2 THE CONSTITUTION OF BIOLOGICAL MATTER

Biological matter contains naturally 26 or more life elements. In biochemistry *a life element is taken to mean an element whose absence will lead to death or malfunction of a biological organism.* (Watson 1972; Ren 2007; Mertz 1987) Other nonlife elements may also be found in living matter because of purposeful introduction, such as functionalized Au nanoparticles used to transport pharmaceutical preparations into cells, or ingested into organisms (e.g., from surgical prostheses and in toxins from the environment). In order to use IBA methods effectively it is important to understand how these elements are incorporated into the molecules that make up living matter (Alberts et al. 2004). A single gene from a higher organism carries the information to generate tens of thousands of the different protein chains found in their cells. These are made up of amino acids that are mainly composed of C, H, N, O, and smaller amounts of S and P. The sequence of the 20 naturally occurring amino acids in these proteins determines how they fold to give the molecule a specific shape that determines its function. The governing of amino acid sequences by the order of bases (A,T, G, and C) in the genetic code has the remarkable consequence that, with few exceptions, the composition of all living matter is quite similar (White and Whitlow 2009). Table 14.2 shows the molecular constitution of a typical cell such as an *Escherichia coli* bacterium.

Table 14.3 presents the elemental composition of living matter. The life elements can be divided into major, minor, and trace elements according to their concentrations. The major elements C, H, O, N, P, and S are the main constituents of the organic molecular units (water, amino acids, nucleic acids, saccharides, fatty acids, etc.). The minor and trace* elements serve a wide variety of functions (Alberts et al. 2004).

1. They can act as a key structural component in proteins. Examples are Fe in the oxygen-carrying protein hemoglobin (PDB[†]: 2DN1 and 2DN1) and Mg in chlorophyll (see, for example, PDB: 3PL9).

---

* Trace elements in a biomedical context are taken to mean elements with concentrations less than 100 ppm.
† PDB: denotes references to specific protein structures in the Protein Data Bank (http://home.rcsb.org/)

TABLE 14.2   Composition of a Typical Cell

Type	Total Weight %	Different Molecule Types Per Cell
H₂O	70	1
Inorganic ions	1	26
Small molecules (oglimers)		
Sugars and precursors	4	
Amino acids and precursors	0.8	200
Nucleotides and precursors	0.8	200
Fatty acids	2	50
Large molecules (macromolecules)		
Proteins	15	2000–3000
Nucleic acids	7	~ 1050
Polysaccharides	3	200

Sources: Watson, J. D. 1972. *Molecular Biology of the Gene,* 2nd ed., Philadelphia: PA: Saunders. Ren, M.-Q. 2007. PhD thesis, University of Jyväskylä.

2. Sodium, magnesium, potassium, calcium, and sulfate, chloride, nitrate, and phosphate and other life-element ions are important constituents of body fluids and the cytoplasm within cells. They regulate numerous biological functions, such as normal blood clotting (Ca), stimulation of enzyme activity (Mg), and control contraction of muscle tissue and the transmission of impulses by nerve cells (Ca and others).

Total elemental content measurements of biological material are important for process control, quality control (QC) in food (Mertz 1987), or bioenergy production applications as well as for archaeometric and art conservation

TABLE 14.3   Elemental Composition of a Typical Cell

Major Elements (10s of Atomic Percent)

Hydrogen (H), carbon (C), nitrogen (N), oxygen (O), sulfur (S), phosphorus (P)

Minor Elements (0.1–10 at.%)

Calcium (Ca), sodium (Na), chlorine (Cl), magnesium (Mg), potassium (K)

Trace Elements (less than 100 ppm)

Boron (B), copper (Cu), zinc (Zn), manganese (Mn), nickel (Ni), cobalt (Co), molybdenum (Mo), selenium (Se), chromium (Cr), iodine (I), fluorine (F), tin (Sn), silicon (Si), vanadium (V), arsenic (As)

Source: M.-Q. Ren. 2007. PhD thesis, University of Jyväskylä.

studies. Measuring the total elemental content of multifunctional elements in an entire organism is of limited physiological value. For example, calcium and phosphorus are major elements (43.4 and 13.6 at.% respectively) in calcium hydroxyapatite, $(Ca_{10}(PO_4)_6(OH)_2)$ (Markovic, Fowler, and Tung 2004), which represents the mineral content of bone. Low concentrations of Ca^{2+} ions are important signal substances (Karp 1996) that influence processes in cells (e.g., contraction, embryonic development, neurosignal transmission and apoptosis, etc.) (Alberts et al. 2004). Ca^{2+} ions exhibit large differences in concentration within cells, typically ranging from millimolarity in the extracellular fluids, endoplasmic reticulum (ER) and mitochondria down to submicromolarity levels in the cytosol.

14.2.1 TYPES OF IBA MEASUREMENTS

The simplest form of IBA measurement is determination of the total elemental content in a dry sample. A practical example of this is the level of heavy metals in biomass, which, taking *E. coli* bacteria as a prototype, has a major element composition of $CH_{1.77}O_{0.49}N_{0.24}$ (BNID 101800). The IBA measurement may be further classified according to whether the sample is thin (where the measured signal originates from the entire thickness of the sample) or thick (where the energy loss in the sample governs the thickness from which the measured signal originates). An example of a thin sample is a drop of body fluid dried on a support film. A piece of bone represents a typical thick sample. The probing beam in IBA can be focused down in a MeV ion microbeam by using specially shaped electric and magnetic fields and scanned over the sample to generate a map of compositions in one dimension (1D) or two dimensions. Two-dimensional (2D) PIXE mapping of elements can be used to study trace life elements, (Fe, Zn, Cu, etc.) in thin tissue sections (Johansson et al. 1995; Campbell 1987; Llabador and Moretto 1998; Watt and Grime 1987; Yu 2006). By combining these with angular scans or stereographic imaging, 3D mapping of the measured signal can be obtained using a standard tomographic back-projection algorithm approach as in magnetic resonance (MR) imaging. Three-dimensional (3D) images of the element distributions can also be obtained by combining the depth profiling capability of RBS with 2D lateral scanning in the MeV ion microbeam. The drawback of going to higher dimensional images is that sensitivity decreases because the signal is distributed over many pixels (voxels).

14.2.2 SAMPLE PREPARATION

The range of samples in IBA is extremely wide because of the complexity and diversity of living matter. Generally, IBA measurements are carried out under high vacuum (10^{-6} mbar or better), which implies the sample must be specially prepared for examination in vacuum (Yu 2006). Fortunately, there is a large body of experience in the biosciences on histological sample preparation that may be used directly or adapted for IBA studies (Echlin 2009; Ross and Reith 1985). In special cases where in vivo measurements are to be carried out (e.g., single-cell irradiation studies), a special environmental chamber can be used (Yu 2006).

14.2.2.1   GROSS SAMPLES

Where the total elemental content of a sample is to be determined, the sample, which can be a cell culture, organ, or entire organism, can be placed in a homogenizer (Figure 14.1), which disrupts the cell walls and membranes in the sample to form a homogenous liquid. A defined quantity of the liquid sample is then dropped on a thin substrate and allowed to dry under an aerosol-free gas flow. This approach has the advantage that it introduces only a minimum of disturbance to the elemental content of the sample. It should be borne in mind that volatile metal–organic compounds can be lost in this dehydration stage. Furthermore, in the design of experiments, it should also be considered that biological matter in its normal living state does not occur in isolation. Cells grow in cohorts and communicate by chemical and tactile signals (Alberts et al. 2004; Karp 1996). They are surrounded by extra cellular medium (ECM) and body fluids or nutrient liquid that have different chemical compositions from

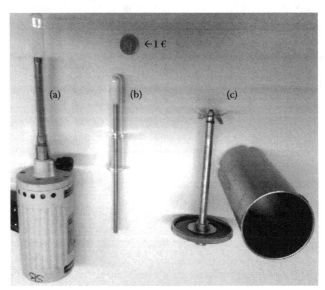

FIGURE 14.1   Homogenizer for disruption of biological materials. (a) Drive unit with homogenizer for test tube, (b) test tube homogenizer for cell cultures, (c) attachment for homogenizing tissue samples. A 1 euro coin is included to show the size.

the cells or tissue of interest. In this case, careful dissection and washing of the sample may be necessary.

14.2.2.2 CRITICAL-POINT DRYING OF CULTURED CELLS

Where delicate cultured cells are to be investigated, *critical point drying* can be used to dehydrate the cells while preserving their morphological structure (Llabador and Moretto 1998). In this technique, water is first replaced with an inert intermediate fluid that is miscible with both liquid CO_2 and water, such as methanol, ethanol, amyl acetate, or acetone. The intermediate fluid is subsequently substituted by a flow of liquid CO_2 under high pressure in a pressure bomb. Then, by sealing and heating the bomb, the CO_2 is moved 5°C–10°C above the critical point so that it remains in a supercritical state where no meniscus between gas and liquid CO_2 exists. The CO_2 is removed by isothermal release. It should be considered when using this technique that while the water is being removed, unbound minor and trace elements may be leached out by the intermediate fluid and liquid CO_2.

14.2.2.3 TISSUE SECTIONING

Where distributions of elements and biomolecules in tissues are to be investigated using a MeV ion microscope it is necessary to make thin sections of the tissue. This is carried out by *cryosectioning* because normal embedding and sectioning procedures cannot be used (Llabador and Moretto 1998) because the waxes and epoxy resins used for conventional embedding introduce changes in the elemental makeup. In cryosectioning the tissue sample is placed in an embedding medium (polyvinyl alcohol and polyethylene glycol solution) and flash-frozen in liquid N_2 to form a block containing the tissue sample. In this state it can be stored at −80°C essentially indefinitely. The sample is then sectioned in a cryomicrotome. This is similar to a normal microtome but with a cooling system so that the sample is maintained at −17°C and surrounded by a box at −23°C. The tissue sample is placed on the sample arm of the microtome. The sample arm is moved up and down by a precise motion against a fixed stainless steel knife edge, which cuts thin sections (5–20 μm) from the sample according to the feed rate. Maintaining the temperature of the box 5°C below the temperature of the sample is crucial to maintain the morphology of the thin section before it is picked up on a support coated with a supporting film.

14.2.2.4 SUPPORTING FILMS

The supporting film used to mount the specimen for thin sample analysis should not introduce spurious elemental signals. This film is in turn mounted on a metal or plastic supporting ring or an electron microscope grid, which provides mechanical stability for handling. Pioloform™ (polyvinyl butyral) and Formvar® (polyvinyl formal) are commonly used as self-supporting films for broad beam

PIXE and ion beam microscopy because they contain only C, H, and O, are easy to handle, and do not introduce any spurious heavy element signals. We have found that Pioloform can be used as a cell growth substrate, allowing cells to be grown directly on the supporting film by brief exposure to a radio frequency (rf) oxygen plasma, which is advantageous because it also sterilizes the film. Silicon nitride (Si_3N_4) is emerging as a supporting film for cell-level studies because of its low cytotoxicity and extreme surface flatness; even very thin layers (40 nm) can withstand a differential pressure of one atmosphere. However, its use for biomedical PIXE analysis is restricted because pileup from Si Kα X-ray can limit the sensitivity for heavy trace elements.

14.3 BROAD BEAM METHODS

Broad beam methods are IBA methods where a broad ion beam of 1–2 mm diameter is used to analyze the composition of samples of material. The physical basis and applications of these methods are discussed in Chapters 5–8 and here we restrict the discussion to the applications of these methods in biomedicine. The use of broad ion beams can generally be divided into two uses: analysis of biointerfaces and the analysis of homogenized gross samples to determine elemental contents.

14.3.1 BROAD BEAM STUDIES OF BIOINTERFACE INTERACTIONS

The first case of biointerface materials is the most straightforward. This covers not only surgical prosthesis materials but also the coating of components for clinical uses, food processing, and even antifouling that controls undesirable buildup of biological matter on surfaces. Today's highly successful arthroplasty (surgical replacement of joint or joint surfaces with a prosthesis) relies on careful control of the interactions between biological matter and other materials. Body fluids are highly aggressive and the mechanical wear leads to debris that may have detrimental effects such as metallosis in the body. Moreover, special coatings such as TiO_2 and diamond-like carbon (DLC) may be employed to control cell adhesion. Development of new biointerface materials requires extensive testing. Simulated body fluids are commonly used, often in conjunction with specialized test jigs to simulate wear. Later stages use animal testing where the sample is surgically implanted and retrieved by postmortem harvesting. In these cases, the changes in the prosthesis surfaces can be determined by thick-sample IBA analysis with conventional PIXE, Rutherford backscattering spectroscopy (RBS), and elastic recoil detection analysis (ERDA) methods (Chapters 5–8). These techniques can provide important quantitative information about oxidation, corrosion, and other composition changes in the surface layers and thin coatings used to enhance biocompatability. Material loss and wear can be determined by first implanting a heavy marker (e.g., Xe, Au) in the prosthesis surface and measuring the change of marker depth with RBS. This type of analysis is suitable to determine corrosion and erosion of thin wear-resistant coatings.

14.3.2 PIXE TOTAL ELEMENTAL COMPOSITION ANALYSIS IN BIOMEDICINE

In the second type of analysis the total elemental composition is measured for either an entire organism or dissected tissue samples or body fluids (e.g., amniotic fluid, cerebral–spinal fluid, serum, synovial fluid, urine, etc.). In biomedical broad beam PIXE analysis, the reduction of interfering background signals from characteristic X-ray fluorescence of analysis chamber components is of paramount importance because the X-ray signals from important life elements (Al, Cu, Fe, Cr, Mo) are small and superposed with fluorescence X-rays from chamber walls and construction materials. Figure 14.2 schematically illustrates an analysis system suited for broad beam PIXE/RBS, which is closely similar to the experimental system developed in Jyväskylä. The beam is carefully collimated to prevent spurious ions that could impinge on chamber components and prevent X-rays from the collimator generating a signal in the X-ray detector. This can be based on a metal aperture (Cu, stainless steel, Ta) followed by a high-purity graphite aperture. The sample ladder can be manufactured from pure aluminum and should have carefully chamfered edges to minimize X-ray scattering. The beam is stopped in a graphite Faraday cup, which is used to

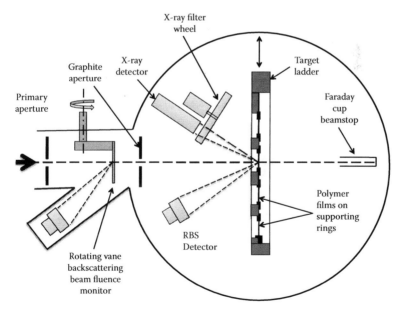

FIGURE 14.2   Schematic diagram of measurement chamber for broad beam IBA of biological samples.

measure the beam fluence and is shaped so that X-rays from the cup cannot be seen by the X-ray detector.

Figure 14.3 shows a typical PIXE spectrum from rat brain tissue on a Pioloform supporting film. This spectrum is typical of biological samples. It is seen that the dominant peaks are from the minor elements P, S, Cl, and K at low energy, while the trace life elements (Ca, Fe Cu, Zn, Br, and possibly Rb) give smaller peaks that appear at higher energies. X-rays from the most abundant major elements (C, N, O) are too low to be detected efficiently by the X-ray detector (Johansson and Campbell 1988; Johansson et al. 1995; Campbell 2009). *Pulse pileup* is where the two X-ray pulses occur so closely in time that they are not separated by the detector electronics but their amplitudes are instead summed, giving rise to a continuous background at high energies. In PIXE analysis of biological samples, the minor elements P, S, Cl, and K give strong peaks that generate a continuous pileup background that swamps the small trace element peaks unless an X-ray filter is used. X-ray filters made of thin films of Mylar and Al can be used (Johansson and Campbell 1988; Johansson et al. 1995; Campbell 2009). These have the advantage that the attenuation for each characteristic X-ray energy can be calculated from the mass attenuation coefficients. Another commonly used approach (as in Figure 14.3) is to employ a so-called *magic filter* (Johansson et al. 1995) that has pin holes so that a small fraction of the low-energy X-rays can pass unattenuated while high-energy X-rays from trace elements such as Cu, Zn, Mo, and Se pass through the filter with almost no attenuation. In this case, the sensitivity for X-rays from each element of interest must be determined experimentally.

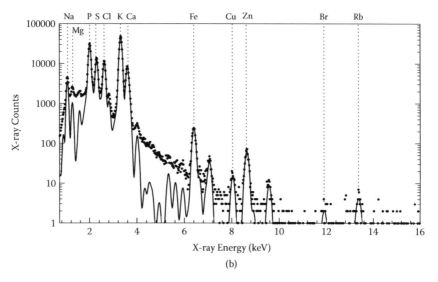

(b)

FIGURE 14.3   2 MeV H⁺ PIXE spectrum from 200 μm section of rat brain tissue. A 300 μm Perspex magic filter was used to suppress the intense low-energy X-ray lines. (From Ren, M.-Q. 2007. PhD thesis, University of Jyväskylä.)

14.3.2.1   CALIBRATION OF PIXE DATA

Calibration is an important issue for quantitative IBA work. In thin sample elemental content measurements, the most difficult part is determining how much sample is subjected to the probing ion beam. This is the case where the area of the evaporated drop of the gross sample (Section 14.2.2) is larger than the beam spot. Two general approaches are used: normalization to the RBS signal from carbon and spiking a known quantity of analyte with a solution of a reference salt. This salt should have the following properties:

1. There should be no overlap of X-ray lines with life elements.
2. There should be large excitation cross sections.
3. It should be readily available as a stable, nonvolatile water-soluble compound such as halide, nitrate, acetate, etc.
4. It should be anhydrous or have a well-defined amount of water of crystallization.
5. It should be stable and not be strongly oxidizing or reducing or have an aggressive nature.

The calibration is carried out in two stages: In the first stage, solutions containing the elements of interest at different dilutions are mixed with a defined quantity of the reference solution. Samples of the different concentrations are dropped onto the supporting films and allowed to dry in a dust-free atmosphere. Measurement of the intensity of characteristic X-ray signals from the elements of interest and the reference element then gives concentration calibration curves. Figure 14.4 illustrates a calibration curve for Ca K_β using a 10 mM ammonium monovanadate ($NH_4(VO_3)$) solution. Deviations from the straight-line dependence in the calibration curve reveal information about the

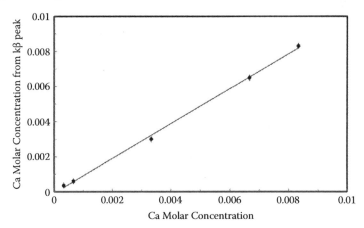

FIGURE 14.4   PIXE calibration curve for Ca Kβ against molar Ca concentration. (Unpublished data: Whitlow, H. J., Chienthavorn, O., Gilbert, L. K., Eronen, H., Norarrat, R., Sajavaara, T., and Laitinen, M.)

background level, minimum detection level, and cross talk between different elements in the sample.

In the second stage, each analyte sample is spiked by adding a known amount of the reference salt to a known amount of each homogenized sample. This mixture is then dropped onto the supporting films and allowed to dry in a dust-free environment to prevent contamination. Then, by comparing the characteristic X-ray yields from the reference salt and the elements of interest, their molarity in the liquid homogenized analyte can be determined. In the PIXE measurement seen in Figure 14.3, 5 μL 10 mM ammonium monovanadate (NH$_4$(VO$_3$)) solution was added to 300 μL sample as a concentration reference. The X-ray yields for the characteristic peaks in the PIXE spectrum can be determined using standard PIXE analysis software such as GUPIX (http://pixe.physics.uoguelph.ca/gupix/main/) or other nonlinear fitting codes. In some cases it is difficult to measure trace elements in the presence of certain major elements. An example is Ca and K where the Ca Kα peak is so close in energy to the K Kβ, which is much stronger because of the higher potassium concentrations in cells. In this case the Kβ peak of Ca can be used to provide the Ca signals. Using the databases for X-ray absorption and stopping forces included in GUPIX, quantitative composition data can be obtained from thick samples of biological matter (http://pixe.physics.uoguelph.ca/gupix/main/). The mathematical procedure for this is quite involved and outside the scope of this chapter.

14.3.2.2   INSTRUMENT AND METHOD DETECTION LIMITS

When one is dealing with extremely small concentrations, important considerations are the instrument detection limit (IDL) and the method detection limit (MDL). The IDL is often and conveniently defined as the concentration corresponding to when the signal (X-ray peak) has a significance level of 99%. For a Gaussian line shape, this corresponds to the yield of the counts in the peak exceeding three times the standard deviation σ of the counts in the background under the peak $Y_s > 3\sqrt{Y_b}$ (Figure 14.5). The background yield Y_b is usually determined by fitting an appropriate function (e.g., polynomial) to the background on either side of the peak. In qualified analytical work, it is customary to work with the MDL, which is defined as the lowest calibration standard that could be measured with the required accuracy (Wolf 1987). The MDL covers all steps in the analysis and requires the use of calibration samples for each element of interest. The MDL may be degraded by overlapping X-ray lines necessitating use of several calibration solutions. It should be noted that because the MDL specified a level of accuracy rather than a criterion for detection, MDL > IDL.

14.3.3   RBS AND ERDA IN BIOMEDICINE

RBS and ERDA are suitable for measurement of thick samples and provide depth distribution information. An example of the use of broad beam IBA

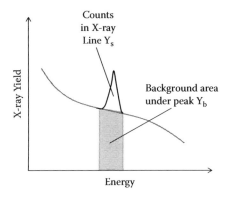

FIGURE 14.5   Illustration of the yield under the characteristic X-ray peak Y_s and the background Y_b.

methods to study biocompatible surfaces is determination of the thickness and composition of a calcium hydroxyapatite-like (HA) surface layer on glass. HA, $Ca_5(PO_4)_3$ (OH) is the mineral content of bone and is a biocompatible material that is used for a wide variety of applications such as coating prosthetics and dental implants. Figure 14.6 shows time-of-flight energy elastic recoil detection analysis (ToF-ERDA) data from a HA-like thin film that was sputter-deposited on a soda-lime glass substrate from a mixture of natural HA and red phosphorus (Sagari et al. 2007). The content of all the elements, including hydrogen, can be resolved and their composition can be quantitatively determined. This

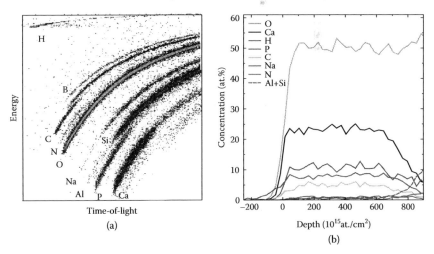

FIGURE 14.6   (a) ToF-E ERDA data from sputter deposited HA-like film on soda-lime glass obtained with a 16 MeV^{63} Cu^{7+} ion beam. (b) Quantitative depth profile obtained from the data in (a). (From Sagari, A. R. A. et al. 2007. *Nuclear Instruments and Methods B* 261:719.)

allows the composition of phosphorus in the target to be adjusted to give a correct Ca/P ratio in the sputter-deposited film.

RBS is commonly used in broad beam measurements to establish the concentration calibration for PIXE (discussed later) for simultaneously measuring the contents of the major elements C, N, and O in the sample. Then, by using a nominal composition C_3H_8 NO for the major elements in the sample, the content of heavy elements in terms of nanograms per gram of dry material can be determined from the PIXE spectrum by normalizing to an RBS spectrum measured simultaneously. It should be noted that a correction needs to be applied for the C, N, and O content of the supporting film.

14.3.4   PIGE IN BIOMEDICINE

Particle-induced gamma ray emission (PIGE) is conceptually similar to PIXE but with the difference that the X-ray detector is replaced with a γ-ray detector (e.g., a Ge detector) (Räisänen 2009). By using specific (p, γ) nuclear reactions, the content of specific isotopes, ^{7}Li, ^{10}B, ^{19}F, etc. can be measured. The advantage of PIGE is that it complements PIXE (Section 14.3.2) by allowing measurement of elements with $Z < 11$ (Na) for which PIXE is essentially blind because of X-ray absorption in the Be window of the X-ray detector. Another advantage is that PIGE is *isotope specific,* which allows the use of tracer techniques (Räisänen 2009). Rather little use of PIGE has been made in biomedicine with the notable exception of fluorine uptake in teeth. This is because lithium, beryllium boron, and fluorine are not of great physiological importance, unless in significant concentrations, where they are extremely toxic.

14.4   MICROBEAM IMAGING OF CELLS AND TISSUES

MeV ion microscopy, which is also commonly referred to as nuclear microscopy, is a powerful technique for frontline biomedical investigations. It is based on using a finely focused beam of MeV ions that is rastered over the sample. Synchronous collection of characteristic IBA signals such as characteristic X-rays and backscattered particles allows mapping in two and three dimensions of elemental concentrations to be realized. This allows concentrations in different parts of cells/tissues/organs to be determined. The MeV ion microscopy technique is analogous to scanning electron microscopy (SEM) but with the important difference that an MeV ion beam rather than electrons is used as the probing beam (Watt and Grime 1987; Grime and Watt 1984). A difficulty in focusing MeV ions beams is that they are much more rigid than electron beams because of their greater momentum. This means that cylindrical magnetic lenses, such as those used in electron microscopes, are too weak to achieve high demagnifications. Instead, magnetic quadrupole elements arranged in pairs, triplets, quadruplets, or higher order multiplets are commonly used for focusing (Grime and Watt 1984).

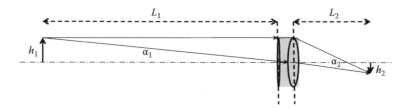

FIGURE 14.7 Demagnifying optics with a thick lens corresponding to a MeV ion microbeam. The principal planes are denoted by vertical dashed lines. Subscripts 1 and 2 denote object and image space, respectively.

Figure 14.8 shows the arrangement of pole pieces and the coil windings used to produce the magnetic excitation in a quadrupole lens element. Figure 14.9 shows a schematic of a MeV ion microprobe. The objective aperture defines the size of the beam, which is demagnified by the quadrupole triplet to produce a small spot on the target. Figure 14.7 illustrates the geometry of the demagnifying optics configuration used in MeV ion microbeams, $\alpha_1 = \alpha_2$; then, by similar

FIGURE 14.8 Pole pieces and coil windings on a magnetic quadrupole lens produced by Oxford Microbeams Ltd. for a MeV ion microbeam. (The beam tube that passes through the pole piece is not shown.)

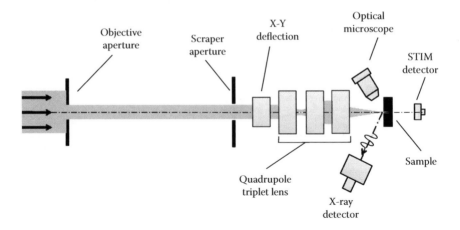

FIGURE 14.9 Schematic illustration of a MeV ion microscope with instruments for PIXE and STIM.

triangles, the demagnification is $M = h_2/h_1 = L_2/L_1$. Unlike normal spherical glass lenses, doublet and triplet quadrupole lens systems have different magnifications in the horizontal and vertical directions. A symmetrical beam spot suitable for imaging can be formed by adjusting the size of the rectangular objective aperture. A detailed discussion of lenses and aberrations in lens systems for MeV ion microscopes is outside the scope of this work; details can be found in the text by Grime and Watt (1984).

A deflection system (usually magnetic), usually placed before the lens system, is used to scan the beam spot over the sample surface. The scraper aperture spatially filters the off-axis beams to control aberrations in the beam spot on the sample. An optical microscope is often mounted in the analysis chamber to allow selection of regions of interest. A range of detectors is usually provided: An X-ray detector is used to detect X-rays from the sample for PIXE mapping. Scanning transmission ion microscopy is performed using a semiconductor detector placed behind the sample. The analysis end-station of a modern MeV ion microscope at the Center for Ion Beam Applications at the National University of Singapore is shown in Figure 14.10. This instrument is capable of point resolutions of 30 nm. Achieving such high resolutions demands extremely high mechanical stability. This requires that the entire target chamber and lens system be rigidly mounted on an optical table (Figure 14.10) with proper consideration given to thermal expansion effects and propagation of vibrations.

14.4.1 MICRO-PIXE AND -PIGE

Micro-PIXE imaging using an energy dispersive X-ray detector (Figure 14.9) is analogous to energy dispersive X-ray analysis (EDX) using a SEM. Micro-PIXE

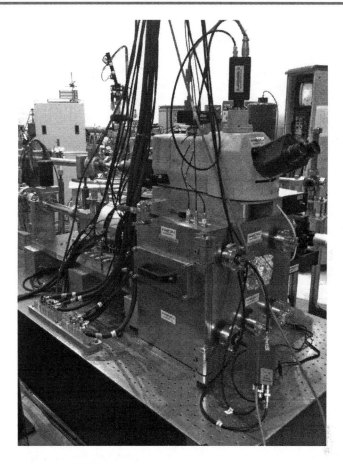

FIGURE 14.10   Second-generation MeV ion microscope at the Center for Ion Beam Applications (CIBA) at the National University of Singapore. The lens system and beam line are visible behind the sample chamber.

has the important advantage that because protons are much heavier than electrons, the dominating bremsstrahlung radiation that swamps trace element peaks in EDX measurements is almost entirely absent (Johansson and Campbell 1988; Johansson et al. 1995; Campbell 2009). The significance of this is that while EDX can only map the lateral distribution of major and some minor elements, micro-PIXE is able to map trace elements even at parts per million levels (Llabador and Moretto 1998).

Figure 14.11 shows an example of tissue-level element mapping using micro-PIXE. The images in this figure are taken from an atherosclerotic rabbit aorta (Ren et al. 2003). Not only can the correlation between S and P, which are major elements (Table 14.3), be clearly identified, but also the spatial correlation between Ca, a minor element, as well as Cu, Fe, and Zn, which are trace elements, can be clearly seen.

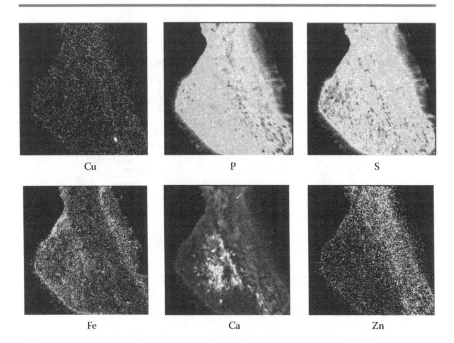

FIGURE 14.11   PIXE element maps from an atheriosclerotic rabbit aorta. The inner side of the aorta is to the left of the field. The images have been contrast expanded and the field of view is 1200 mm × 1200 mm. (After Ren, Q. et al. 2003. *Free Radical Biology and Medicine* 34:746–752.)

Micro-PIXE can be used for quantitative elemental mapping. Biological samples such as tissue sections are often inhomogeneous; therefore, it is necessary to determine the total amount of dry tissue at each point in the elemental map. The inhomogeneity implies that spiking with a reference compound will also be inhomogeneous and thus another approach must be used for calibration. This can be conveniently done using RBS (Chapter 5) to determine the dry-tissue thickness from the major element signals, as in the case of broad beam PIXE (Section 14.3.3). It should be borne in mind that RBS is blind to hydrogen and that the supporting film also contributes to the RBS spectrum and therefore corrections must be introduced to take account these factors. The data shown in Figure 14.11 were obtained in an experiment to test the hypothesis that desferrioxamine (desferal), which is an iron chelating agent, influences the development of arteriosclerosis (Ren et al. 2003). Figure 14.12 shows an optical micrograph of a section of an arteriosclerotic rabbit aorta stained with hematoxaline and eosin stain (aka H and E staining). Figure 14.13(a) illustrates schematically how maps of P and Fe from an adjacent tissue section were divided up as well as the content of Fe within the lesion and adjacent to it (Ren et al. 2003). Figure 14.13(a) shows the levels of Fe in and adjacent to the arteriosclerotic lesions for rabbits on a 1% high-cholesterol diet that were fed desferal and control specimens. It

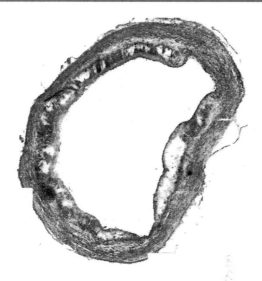

FIGURE 14.12   Hematoxylin and eosin-stained section of a rabbit aortic arch after 10 weeks on a high-cholesterol (1%) diet, showing artherosclerotic lesions extending over most of the inner wall surface (light gray region in Figure 14.13) and a small region of normal artery wall (upper right). (From Ren, M. Q. et al. 2003. *Free Radical Biology and Medicine* 34:746–752.)

is seen that at both 8 and 10 weeks the specimens fed desferal had significantly lower Fe levels, both in the lesion and adjacent to it. A full analysis revealed that enhanced Fe concentrations not only were implicated in the formation of the lesions but also that a Zn supplement reduced the Fe content.

PIGE (Räisänen 2009) at MeV ion microbeams has found little application because, as noted earlier, the life elements of physiological importance have atomic numbers greater than 11 (Na) where PIXE has good sensitivity. The main applications have been studies of the uptake of fluorine in teeth and ^{10}B from drugs used in boron neutron capture therapy.

14.4.2   SCANNING TRANSMISSION ION MICROSCOPY

In scanning transmission ion microscopy (STIM), two configurations, shown in Figure 14.14, are used. In *direct STIM* (on-axis STIM), the energy of ions and recoils that pass directly through the sample without undergoing significant deflection is measured. In *off-axis STIM*, the ions and recoils that are scattered through a forward angle, ϕ, are detected. Direct STIM measures maps of energy loss, $\Delta E = (dE/dx)\Delta x$, where (dE/dx) is the stopping force and Δx is the sample thickness. The sample thickness can be determined from the direct STIM image because biological materials have similar compositions and hence

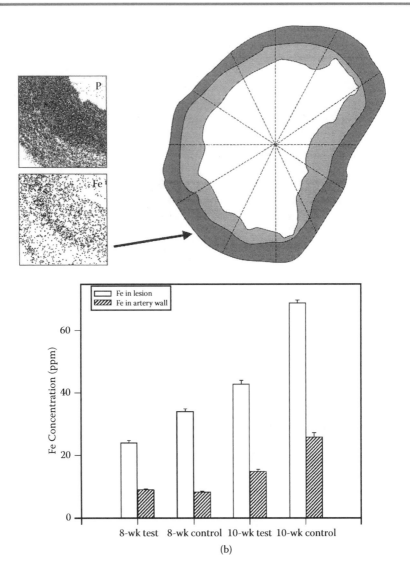

FIGURE 14.13   Analysis of Fe in rabbit aortic arch section taken adjacent to that of Figure 14.12. (top) Schematic illustration of segmentation of rabbit aortic arch cross section (Figure 14.12) with Fe and P images of a section. (bottom) Levels of Fe in aortic arches of rabbits fed a (1%) cholesterol diet with desferal (test) and without desferal (control). (From Ren, M. Q. et al. 2003. *Free Radical Biology and Medicine* 34:746–752.)

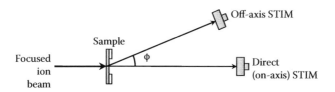

FIGURE 14.14   Schematic illustration of the geometrical configuration for direct- and off-axis STIM.

stop cross sections. Often Si p–i–n photodiodes are used as detectors for direct STIM because of their small size and low cost (Ren et al. 2007).

Figure 14.15 shows a direct STIM image taken with a 2 MeV $^4He^+$ ion beam of a cohort of human breast cancer cells. In this case the cells from the MCF-7 established cell line (American Type Culture Collection, http://www.atcc.org/) were grown directly on the Si p–i–n detector using the SiO_2 passivating oxide surface as a growth substrate (Ren et al. 2007). An alternative approach is to use a Si_3N_4 membrane as a cell-growth substrate. In this case it may be advantageous to render the surface sterile and more hydrophilic by a brief exposure to an oxygen plasma. Direct STIM is commonly used to provide structural images in conjunction with micro-PIXE and ion-induced fluorescence imaging (Ren et al. 2007; Watt et al. 2009). This is illustrated in Figure 14.15, which shows that the internal structure of the cells may be clearly seen. Notably, the nucleoli (center in the cell nucleus where the chromatin is concentrated in the

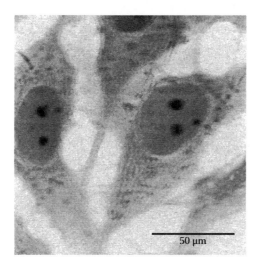

FIGURE 14.15   Proton direct-STIM energy-loss image of a cohort of human breast cancer cells. Darker contrast corresponds to larger energy losses. (Unpublished data: Ren, M. Q., Whitlow, H. J., van Kan, J. A., Bettiol, A. A., Lim, D., Gek, C. Y., Bay, B. H., Osipowicz, T., and Watt, F.)

interphase) are characterized by a very high stopping (dark contrast). The cells are well developed, with filopodia linking cells together, which provides signaling for the cohort of cells. Outside the cells a very light gray contrast that can be associated with ECM is seen.

Another application of STIM measurements is to quantify the beam fluence. Direct STIM measurements allow every ion impinging on the sample to be counted; however, the current is limited by the count rate of the p–i–n detector to 1–2×10^4 ions s^{-1}. While this is adequate for ion-induced fluorescence measurements (Section 14.4.3) and for controlling image noise by normalizing the number of ions per pixel in direct STIM imaging, it is too low to handle the ~0.1 nA (2×10^9 ions s^{-1}) currents needed for micro-PIXE. For fluence measurement in micro-PIXE, the off-axis STIM signal can be used instead because the count rate is considerably attenuated by the small scattering cross section and the finite solid angle subtended by the detector.

14.4.3 IONLUMINESCENCE MICROSCOPY

As MeV ions penetrate matter they mainly deposit energy by creating a large number of low-energy (<50 eV) electrons. These electrons are mainly created within a few nanometers of the ion track and can excite fluorescence (Townsend 2009). Because each ion excites a large number of secondary electrons, the number of fluorescence photons can be high (>1). This is contrary to the case of PIXE, where the number of high-energy (1–10 keV) excitations creating X-ray photons is very small (10^{-6}–10^{-8} per ion). In ionluminescence microscopy the fluorescence photons are counted using a photomultiplier and used to map the ion-induced fluorescence.

Optical fluorescence microscopy is a workhorse technique in biomedicine because fluorophores (fluorescence light emitting groups or nanoparticles) can be attached (functionalized) to antibodies that conjugate to specific biomolecules. This is a particularly powerful technique because it allows selective labeling of particular biomolecules, allowing their pathways to be imaged in cells and tissues using a fluorescent microscope. For example, Alexa Flour® 488 phallodin (http://products.invitrogen.com/ivgn/product/A12379) binds to F-actin filaments (Alberts et al. 2004) in the cytoskeletons of cells. Moreover, fluorophores that emit fluorescence light at different wavelengths can be conjugated with different antibodies, allowing several molecules to be imaged at the same time in different colors. The drawback of fluorescence microscopy is that the resolution is limited by the Abbe criterion to about 200 nm for typical excitation wavelengths of 300–400 nm. The situation is quite different when MeV ions are used to excite the excitation because it is confined within few nanometers of the ion track. The resolution (<100 nm) is then determined by the size of the beam spot and the multiple scattering of the ion beam inside the sample (Whitlow et al. 2009). For typical cells and tissue sections of 1–10 μm, multiple scattering gives a broadening of 10–30 nm full width at half maximum

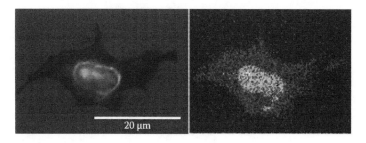

FIGURE 14.16   Image of a NA2 blastomer cell stained with Sytox green nucleic acid stain. (left) Confocal microscopy image. (right) Proton-induced ion luminescence image. The images have been mathematically enhanced by expanding contrast and median filtering with an 3 × 3 kernel. (From Watt, F. et al. 2009. *Nuclear Instruments and Methods B* 267:2113–2116.

(fwhm) for 2 MeV protons (Whitlow et al. 2009). The use of the technique is illustrated in Figure 14.16 (Watt et al. 2009), which shows a confocal microscope image and the proton-induced luminescence image of a NA2 Blastomer cell (American Type Culture Collection, http://www.atcc.org/) that has been stained with Sytox® green nucleic acid stain (Watt et al. 2009) that binds to the nucleic acid of dead cells. From the confocal microscope image, it is seen that the nucleus of the cell where DNA is concentrated has been intensely stained white; the cytoplasm takes up the stain much less readily. The sharpness of the edge of the nucleus governed by the size of the 200 nm beam spot indicates that the resolution was already better than possible for a conventional light microscope (Watt et al. 2009).

14.5   EMERGING IBA METHODS IN BIOMEDICINE

14.5.1   ANALYSIS OF ION IRRADIATION EFFECTS IN CELLS

Biomolecules are degraded by the effects of irradiation. This can lead to cell malfunction and, eventually, cell death. In particular, chromosome damage results in changes in the genetic code that lead to mutations that may be advantageous because they perpetuate evolution by creating competitive phenotypes but also detrimental because they can induce cancerous changes (Yu 2006; White and Whitlow 2009). Organisms have developed complex mechanisms to handle different forms of damage to the DNA such as single- and double-string breaks (Yu 2006). In addition, malfunctioning cells may undergo *apoptosis*, where they die in a controlled manner, releasing signal substances that trigger nearby cells to undergo apoptosis also, the so-called *bystander effect.*

In order to study these effects, a special form of MeV ion microbeam (Figure 14.17) is used for in vivo irradiation of live cells. The bottom of the petri dish used to culture the cells is a thin membrane through which the ions

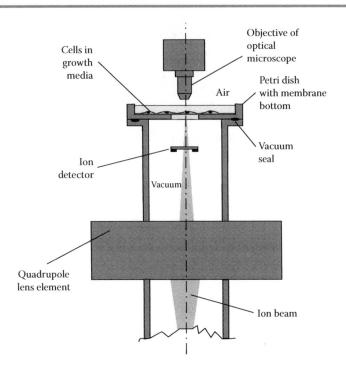

FIGURE 14.17   Schematic illustration of a vertical MeV microbeam system for single-cell irradiation studies.

pass (made, for example, of a self-supporting Mylar or Si_3N_4 film; Yu 2006). The film has to be sufficiently thin that the ions can pass through without significant angular deflection from multiple scattering, yet thick so that it is strong enough to support the atmospheric pressure difference. A detector is placed before the window to record when an ion has passed so that the beam can be stopped by an electrostatic deflector after a defined number of ions have impinged on the cell or organelle in question. Several types of detectors can be used for this purpose, such as a thin transmission Si detector, gas ionization detector, or secondary electron detectors (that register the pulse of secondary electrons emitted from the thin window when an ion traverses it). The latter are preferred because they do not introduce any perturbation of the ion energy or direction.

Prior to irradiation, the position of the cells or cell nuclei is determined using the optical microscope. This can be done manually or automatically using image recognition software. The ion beam is moved sequentially to the positions of the cell or organelle and a defined number of ions are used to irradiate this. The petri dish is then removed and the cells incubated for a period of time. Subsequently, the number and form of cell alteration are analyzed using an optical microscope.

14.5.2   ION BEAM FABRICATION OF MICROANALYTICAL TOOLS

An emerging analytical application is based on MeV ion beam lithography (MeV-IBL) techniques where ion beams are used for micromachining. These lithographic techniques include proton beam writing (PBW) with focused beams (van Kan 2009), the use of shaped beams (programmable proximity aperture lithography) (Puttaraksa et al. 2008), and ion track lithography (Trautmann 2009), where the developable tracks are used to create 10–1000 nm diameter channels through material. The MeV ion beam lithography techniques have some special characteristics that make them extremely well suited for a wide range of analytical applications in biomedicine (van Kan 2009). These include:

- The almost straight trajectories of MeV ions in matter allow high aspect-ratio structures with straight, sharply defined sidewalls to be written creating 3D structures that can form the basis of micro- and nanofluidic devices, waveguide structures, and Bragg gratings. This gives a tighter control over the geometrical size than is possible with conventional lithography. MeV-IBL facilitates writing microfluidic structures that allow volumes on a subpicoliter scale to be handled with high accuracy.
- The majority of the secondary electrons have short ranges (few nanometers) and consequently give a vanishingly small proximity exposure. This allows writing of features down to 10s of nanometers with high spatial density. This, coupled with the high aspect ratio, facilitates handling large (microliter) volumes with only small chip area consumption.)
- MeV ions interact with matter in a multitude of ways (Zhang and Whitlow 2005). This allows development of the patterns in different manners such as electrochemical etching, selective solvent dissolution, selective etching, etc. This facilitates the use of MeV-IBL for producing 3D structures in semiconductors; positive and negative tone resist polymer (PMMA, SU-8, etc.) glasses, such as SiO_2; and even metals by electrochemical deposition.
- Generally, biomedical microfluidic test devices have to be tailored to a specific application because of the wide range of sizes and concentration of the target particles (proteins, viruses, cells, etc.). The direct writing capability of the MeV-IBL allows fast and simple tailoring of the structures in software without the need to produce expensive masks.
- The 3D structures may be used as master molds for mass production of analytical tools such as microfluidic devices by soft lithography or by electrochemical deposition to make hard metal (e.g., Ni) stamps for nanoimprinting (van Kan and Ansari 2009).

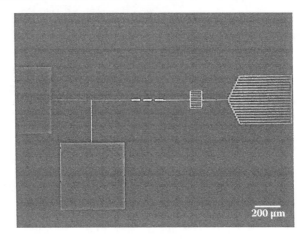

FIGURE 14.18   Optical microscope image of a prototype microfluidic analytical device for immunological use fabricated using MeV ion beam lithography with 3 MeV[4] He ions in 12 mm PMMA/Si. (Unpublished data: Wang, L. P., Gilbert, L. K., and Whitlow, H. J.)

As an example of the analytic potential of these methods, Figure 14.18 shows a prototype open-channel microfluidic device. From left to right in this figure can be seen: two fluid-drop reservoirs, a T-junction that functions as a valve, a vortex generator, an analysis volume with walls coated with antibodies, and a capillary pump. The prototype, which uses capillary force to actuate the transport of fluid, was intended for immunological testing. Magnometric detection could be used to sense the number of superparamagnetic particles functionalized with specific antibodies that are trapped in the antigen-coated walls of the analysis volume. The channel is 10 μm wide and 15 μm deep. This type of test device has the capability to combine the extreme selectivity of immunological testing (~1 in 10^{18} molecules) with the extreme sensitivity afforded by using microscale detection.

ACKNOWLEDGMENTS

Assoc. Prof. Orapin Chienthavorn and Prof. Ragnar Hellborg are thanked for their helpful advice in preparing the manuscript. HJW's work was supported by the Academy of Finland, Centre of Excellence in Nuclear and Accelerator Based Physics, ref. 213503 and grant 129999.

PROBLEMS

Here we focus on examples that deal with cell and tissue samples. The other substrate analysis is dealt with in other chapters.

14.1.   Trace metal contents in proteins: Mammalian hemoglobin (e.g., human hemoglobin a) is normally composed of four proteins, each containing one Fe atom in a heme group, and has a molecular weight of about 48 kDa. The atomic composition for dry protein is $\sim C_3H_8NO$.
   a. What is the atomic percent of Fe in a hemoglobin molecule?
   b. Estimate the molarity of Fe in normal human blood that contains $\sim 5 \times 10^6$ erythrocytes μL^{-1} of blood, each containing 35% hemoglobin and having a volume of 90 fL.

14.2.   Concentration evaluation from a spiked sample: 5 µL of 3 mM $(NH_3)_4V$ solution was added to 300 µL 0.1 mM of NaI solution. A 3 mm drop of this solution was allowed to dry on a pure polymer substrate and a PIXE analysis was performed with a 1 mm diameter beam at the center of the dried spot. The areas under the X-ray peaks for V Kα and I Lα were 300 and 184, respectively.
   a. What is the ratio of V Kα and I Lα characteristic X-rays from $(NH_3)_4V$ and NaI solutions of the same molarity?
   In the next step, 200 µg of dried seaweed was homogenized in water and made up to 20 mL. To 100 µL of this mixture, 10 µL of 3 mM $(NH_3)_4V$ solution was added. A drop of the mixture was dried and a PIXE spectrum measured as described before. The V Kα and I Kα characteristic X-ray yields were 214 and 23, respectively. The background under the I Kα peak was four counts.
   b. What is the molarity of the iodine in the homogenized seaweed mixture?
   c. What is the weight percent of iodine in the original dry sample?

14.3.   Instrument detection limit and sensitivity: The background under the I Kα peak was four counts in the previous measurement.
   a. What is the IDL for iodine in the PIXE measurement?
   b. What is the uncertainty in the result and is it statistically significant?

14.4.   Demaginification in a MeV ion microbeam: A MeV ion microbeam has D_x and D_y, the demagnifications in the horizontal and vertical planes of 35 and 110, respectively. If the contribution to the beam spot size from the aberrations is just equal to the demagnified image of the objective aperture for a spot of 1×1 µm, what size of objective aperture is needed?

14.5.   Beam-spot size, magnification, scanned size, fluence in MeV ion beam microscopy: In a MeV ion microbeam, when fixed cells are irradiated with a proton fluence of 10^{16} ions cm^{-2}, the onset of shrinkage is observed. Imaging with a 512×512 pixel image is carried out with a beam of 0.1 nA of protons. The beam spot size was larger than the pixel size and the scan rate was one raster per minute. A 1 × 1 μm beam spot is scanned over a 40 × 40 μm area on a tissue sample at one raster per minute.

a. Calculate the fluence per raster scan.

b. How many raster scans can be completed before the ion fluence exceeds the onset of shrinkage (10^{16} ions cm^{-2})?

c. The scan size is reduced to 10 × 10 μm. How large is each pixel with the same resolution?

REFERENCES

Alberts, B., Bray, D., Hopkin, K., Johnston, A., Lewis, J., Raff, M., Roberts, K., and Walter, P. 2004. *Essential cell biology,* 2nd ed., p. 554. New York: Garland Science.

Campbell, J. L. 2009. Particle-induced X-ray emission: PIXE. In *Handbook of modern ion beam analysis,* eds. Wang, Y. and Nastisi, M. 233–246. Warrendale, PA: Materials Research Society.

Echlin, P. 2009. *Handbook of sample preparation for scanning electron microscopy and X-ray microanalysis.* New York: Springer.

Grime, G. W., and Watt, F. 1984. *Beam optics of quadrupole probe-forming systems.* Bristol: Adam Hilger.

Johansson, S. A. E., and Campbell, J. L. 1988. *PIXE: A novel technique for elemental analysis.* New York: Wiley.

Johansson, S. A. E. Campbell, J. L., and Malmqvist, K. G. 1995. *Particle-induced X-ray emission spectroscopy (PIXE).* New York: Wiley.

Karp, G. 1996. *Cell and molecular biology* (582–597 DNA repair; 649–688 signaling; 695–697 cancer). New York: Wiley.

Llabador, Y., and Moretto, P. 1998. *Applications of nuclear microprobes in the life sciences.* Singapore: World Scientific.

Markovic, M., Fowler, B. O., and Tung, M. S. 2004. Preparation and comprehensive characterization of a calcium hydroxyapatite reference material. *Journal of Research at the National Institute of Standards and Technology* 109:553–568.

Mertz, W. 1987. *Trace elements in human and animal nutrition,* 5th ed., vol. 1. San Diego: Academic Press.

Puttaraksa, N., Gorelick, S., Sajavaara, T., Laitinen, M., Singkarat, S., and Whitlow, H. J. 2008. Programmable proximity aperture lithography with MeV ion beams. *Journal of Vacuum Science and Technology* B 26:1732–1739.

Räisänen, J. 2009. Particle-induced gamma emission: PIGE. In *Handbook of modern ion beam analysis,* ed. Wang, Y. and Nastisi, M. 147–172. Warrendale, PA: Materials Research Society.

Ren, M. van Kan, J. A., Bettiol, A. A., Lim, D., Chan, Y. G., Bay, B. H., Whitlow, H. J., Osipowicz, T., and Watt, F. 2007. Nano-imaging of single cells using STIM. *Nuclear Instruments and Methods B* 260:124.

Ren, M.-Q. 2007. Nuclear microscopy: Development and applications in artherosclerosis, Parkinson's disease and materials physics. PhD thesis, University of Jyväskylä.

Ren, M. Q., Watt, F., Tan, B. K. H., and Halliwell, B. 2003. Correlation of iron and zinc with lesion depth in newly formed atheriosclerotic lesions. *Free Radical Biology and Medicine* 34:746–752.

Sagari, A. R. A., Lautauret, C., Gorelick, S., Laitinen, M., Rahkila, P., Putkonen, M., Arstila, K., Sajavaara, T. Cheng, S., and Whitlow, H. J. 2007. Ion sputtering deposition of Ca-P-O films for microscopic imaging of osteoblast cells. *Nuclear Instruments and Methods B* 261:719.

Townsend, P. D. 2009. Diagnostic ion beam luminescence. In *Ion beams in nanoscience and technology,* eds. Hellborg, R. Whitlow, H. J., and Zhang, Y. 211–217. Heidelberg: Springer.

C. Trautmann. 2009. Micro- and nanoengineering with ion tracks. In *Ion beams in nanoscience and technology,* eds. Hellborg, R. Whitlow, H. J., and Zhang, Y. 369–387. Heidelberg: Springer.

van Kan, J. A. 2009. Tissue engineering and bioscience methods using proton beam writing. In *Ion beams in nanoscience and technology,* eds. Hellborg, R. Whitlow, H. J. and Zhang, Y. 315–318. Heidelberg: Springer.

van Kan, J. A., and Ansari, K. 2009. Stamps for nanoimprint lithography. In *Ion beams in nanoscience and technology,* eds. Hellborg, R. Whitlow, H. J. and Zhang, Y. 315–318. Heidelberg: Springer.

Watson, J. D. 1972. *Molecular biology of the gene,* 2nd ed., Philadelphia, PA: Saunders.

Watt, F., Betiol, A. A., van Kan, J. A., Ynsa, M. D., Minqin, R., Rajendran, R., Cui Huifang, Sheu Fwe-Shen, and Jenner, A. M. 2009. Imaging of single cells and tissues using MeV ions. *Nuclear Instruments and Methods B* 267:2113–2116.

Watt, F., and Grime, G. W. 1987. *Principles and applications of high energy ion microbeams.* Bristol: Adam Hilger.

White, D. J., and Whitlow, H. J. 2009. Nanoscale engineering in the biosciences. In *Ion beams in nanoscience and technology,* eds. Hellborg, R. Whitlow, H. J. and Zhang, Y. 3–20. Heidelberg: Springer.

Whitlow, H. J., Ren, M., van Kan, J. A., Osipowicz, T., and Watt, F. 2009. Angular and lateral spreading of ion beams in biomedical nuclear microscopy. *Nuclear Instruments and Methods B* 267:2153.

Wolf, W. R. 1987. Quality assurance for trace element analysis. In *Trace elements in human and animal nutrition,* 5th ed., vol. 1, ed. Mertz, W. 57–78. San Diego: Academic Press.

Yu, Z. 2006. *Introduction to ion beam technology.* New York: Springer.

Zhang, Y., and Whitlow, H. J. 2005. Modification of materials by MeV ion beams. In *Electrostatic accelerators fundamentals and applications,* ed. Hellborg, R. 506. Heidelberg: Springer.

IBA Software

15

Nuno P. Barradas, Matej Mayer, Miguel A. Reis, and François Schiettekatte

Once an ion beam analysis (IBA) spectrum has been acquired, the user wants to extract physical information from it, usually a depth distribution of the elemental concentrations of the different elements in the sample. Sometimes, the user simply needs the thickness of a simple, single element layer on a flat substrate. We will see in the first part of this chapter how this can be carried out by hand or with simple software. The reader is urged to attempt such an analysis in order to understand the fundamentals of spectrum conversion and simulation, so as to avoid mistakes by using more sophisticated software such as a black box. Still, the sample is frequently complex or the spectrum features unexpected impurities or a strange background, and a more sophisticated analysis is needed, most of the time using simulation software. This chapter discusses which effects are taken into account by the different software and gives a few examples and problems to illustrate how they influence the analysis.

15.1 A BRIEF INTRODUCTION TO IBA DATA ANALYSIS

Not all IBA requires quantification. Knowing which impurities are present in a given sample can be more important than their exact concentration; in channeling analysis, visual comparison between the impurity and lattice behavior is often all that is needed; in microbeam techniques, the two-dimensional (2D) patterns obtained are seldom quantified. There are many other cases in which the information required is qualitative. That said, quantitative data analysis is an essential part of IBA techniques such as Rutherford backscattering spectrometry (RBS), including elastic backscattering spectrometry (EBS; i.e., with non-Rutherford cross sections), elastic recoil detection analysis (ERDA), and nonresonant nuclear reaction analysis (NRA), or particle-induced x-ray emission (PIXE). Elemental concentrations, layer thickness values, and, more generally, depth profiles are determined from experimental data via some numerical procedure. This can be very simple, but it is often intricate and best done with software that performs the repetitive calculations required.

Many codes dedicated to quantitative data analysis of IBA data have been presented over the decades, and it is beyond the scope of this chapter to review all of them. Information can be found in an exhaustive review by Rauhala et al. (2006) on the status of IBA software, with special focus on the history and capabilities of codes. That paper is complemented by the intercomparison of seven software packages made by Barradas et al. (2007), in which specific simulations were presented by the authors of the codes, displaying the state of the art (without PIXE). A useful summary was presented by Barradas et al. (2008). PIXE software was dealt with in another extensive work published by the International Atomic Energy Agency in 2003 (Fazinic 2003) that includes several PIXE codes. The second edition of the *Handbook of Modern Ion Beam Materials Analysis,* which is geared to ion beam practitioners and accelerator laboratories, included a chapter by Barradas and Rauhala (2009) on IBA data analysis software, as well as a chapter on PIXE (Campbell 2009).

The aim of this chapter is to give the reader, including students and casual practitioners, modern and up-to-date tools to analyze data meaningfully and correctly. Compared to the references mentioned, it is more practical, based on examples and exercises, and links to the codes described.

The scope of this chapter is restricted to RBS, including EBS, ERDA, nonresonant NRA, and PIXE. Resonant NRA, such as particle-induced gamma emission (PIGE) or in general techniques where the integral yield of a reaction product is measured, is included in one of the codes but no examples are given here.

15.1.1 MANUAL DATA ANALYSIS

Data are not always analyzed using dedicated software. Before codes were written and became popular, data analysis was made manually—that is, using the principles described in Chapters 2–4 of this book and the methods explained in Chapters 5–9.

First, the data analyst needs to "read" a spectrum—that is, identify features and structures and relate them to expected (and often unexpected) properties of the sample. Then, actual calculations have to be made, using the relevant formulas and algorithms, which almost always involve approximations. The analyst must be able to understand them as well as the errors involved. The procedure is long even for an expert, and spectra with complex overlapping elemental signals can be virtually impossible to "read" and make sense of.

Nevertheless, we advise IBA practitioners, and in particular students involved in IBA experiments and data analysis, to learn how to do it on relatively simple data. It improves one's understanding of the techniques and consequently reduces the probability of making mistakes. Furthermore, manual data analysis is still often used to determine spectra features such as peak and edge positions, signal heights and integrals, or in the course of discussions to extract estimates and suggest how a measurement could be improved. The thickness of a thin layer can be calculated easily by manual analysis, which may often be faster than a full simulation.

Consider the RBS spectrum shown in Figure 15.1. A Re buffer was deposited on a Si substrate, and then Co and Re. The ^{4}He beam energy was 1 MeV, the angle of scattering 160°, and the angle of incidence 45° in the Cornell geometry (the exit angle is thus 48.4°; see Chapter 5). To make a manual analysis, the following steps can be followed:

1. Use the approximation $\dfrac{Re}{Co} \approx \dfrac{H_{Re}}{H_{Co}} \dfrac{Z^2_{Co}}{Z^2_{Re}}$, where H is the height of signals and Z the atomic number, to obtain [Re] = 12.6 at.%, [Co] = 87.4 at.%, starting from H_{Re} = 11.3 cm and H_{Co} = 9.8 cm, determined

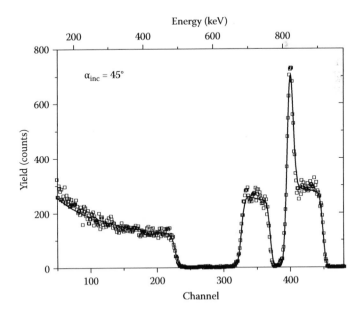

FIGURE 15.1   RBS spectrum of a sample with Co and Re deposited on a Re buffer on a Si substrate. 1 MeV ^{4}He beam detected at a 160° angle of scattering, angle of incidence 45° in the Cornell geometry. The simulation was made with RUMP.

with a ruler on an amplified print. This was a usual method when computers were not readily available.

2. Use those values and SRIM (stopping and range of ions in matter; Ziegler et al. (2009) to calculate the stopping cross-section factors $[\varepsilon]_{Re}$ = 241.0 eV/10^{15} at./cm^2, $[\varepsilon]_{Co}$ = 222.1 eV/10^{15} at./cm^2.

3. Recalculate the stoichiometry using $\dfrac{Re}{Co}$ $\dfrac{H_{Re}}{H_{Co}}$ $\dfrac{Z^2_{Co}}{Z^2_{Re}}$ $\dfrac{Re}{Co}$, obtaining [Re] = 13.6 at.%, [Co] = 86.4 at.%—not far from the initial estimate. This in turn leads to $[\varepsilon]_{Re}$ = 241.7 eV/10^{15} at./cm^2, $[\varepsilon]_{Co}$ = 222.7 eV/10^{15} at./cm^2. Steps 2 and 3 converge very rapidly.

4. Use the surface energy approximation (see Chapter 5) to determine the top CoRe film areal density, using the Co signal. The Re signal cannot be used directly, because it includes both the CoRe film and the Re buffer, which has a different stopping power. We have $Nt = \Delta E_{Co}/[\varepsilon]_{Co} = 3.6 \times 10^{17}$ at./cm^2, where ΔE_{Co} = 81 keV is the energy width of the Co signal. (This can be calculated after an energy calibration of the spectrum is obtained, for instance, from the surface Co and Re signals.)

We could continue the analysis to estimate the thickness of the Re buffer. All in all, this is a tedious procedure. If done completely manually it is not very accurate, because it depends on signal heights and widths and ignores several effects that influence the calculations (such as detection pulse pileup and plural scattering), not forgetting that the surface energy approximation is not very good.* However, it is a method that forces us to understand the measurement and the one that best allows us to estimate uncertainties.

In the case of PIXE, simple cases can also be handled manually. Take the case presented in Figure 15.2. In this case a thin CoFe film was deposited onto a Si substrate. The main purpose is to determine the Fe/Co ratio.

Use the approximation $\dfrac{Fe}{Co}$ $\dfrac{H_{Fe}}{(H_{Co}~k_b H_{Fe})}$ $\dfrac{T_{Fe}}{T_{Co}}$ $\dfrac{Co}{Fe}$, where H is the height of peaks, k_b = 0.123 is the fraction of the Fe K_β line relative to the K_α line (sitting below the Co K_α line; see Chapter 8), T is the x-ray absorption on filters used during the experiment, and σ the ionization cross section that can be obtained from a table such as that given by Cohen and Harrigan (1985). This leads to [Fe]/[Co] = 0.9, starting from H_{Fe} = 12.3 cm and H_{Co} = 14.0 cm determined with a ruler on an amplified print, and knowing that, for this spectrum, the ratio of absorptions is 0.86 and that the cross sections are 69.63 barn and 55.31 barn for Fe and Co, respectively. If peaks were far apart, the peak height (H) would have to be replaced by the peak area (approximated by the product of height and full width half maximum [H*FWHM]) because the width of the peak changes with energy. In this case the peaks Fe and Co are very close, so there is no need for this small correction.

---

* Better approximations, such as the mean energy approximation, are also not very good for very thick films.

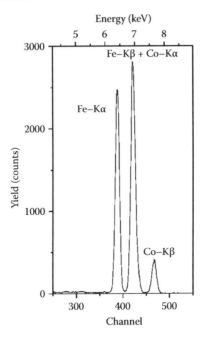

FIGURE 15.2   PIXE spectrum of a CoFe film on a Si substrate. A 2.0 MeV 1H beam was used; the angle of incidence was 22.5°.

In simple cases like this, such a very simple approach gives good and fast results. Still it is important to realize that any x-ray filters being used must be previously calibrated or be very well known in order to calculate the absorption ratios out from literature data.

15.1.2   SOFTWARE: WHAT CAN IT DO FOR YOU?

Dedicated software is the preferred way of data analysis because it can make all the repetitive calculations fast and accurately, and also because it normally includes the databases needed, such as stopping powers or x-ray absorptions.

The simplest codes for RBS and ERDA take the yield from a given element in a given channel and convert it to the concentration of that element at the corresponding depth (Børgesen et al. 1982). First versions of this so-called "direct conversion method"* use calculated Rutherford cross sections and require isolated elemental signals, but the method can be extended to non-Rutherford

---

* This method computes, thanks to the stopping power and kinematic factor, the energy interval corresponding to a depth interval and converts the number of counts in this energy interval in an element concentration using the cross section. As for the manual analysis example, the procedure is usually iterative, so, at each step, the stopping power is computed considering the element concentrations found in the previous step.

cross sections and, to some extent, superimposed signals (Jeynes et al. 2000). The results for the data from Figure 15.1 are shown in Figure 15.3. In many cases, this approach gives satisfactory results—for example, if one only needs the thickness of a single, relatively thick layer of a simple compound on a substrate. The main drawback of this method is that effects that deteriorate depth resolution remain included in the depth profile as a broadening. So the depth profiles obtained by the direct conversion method remain convoluted with the system resolution and this may lead to misinterpretations. For instance, in Figure 15.3, the Re buffer layer is retrieved as a peak with maximum concentration around 40 at.%, when in fact it is a pure Re layer, thinner than the depth resolution. Elements that have a small signal (such as light elements in RBS), are determined with a very high uncertainty, and elements that are not observed in the spectrum (such as H with RBS) are not retrieved at all. The superposition of signals and any inaccuracy in the stopping powers used lead to artifacts in the results.

This direct conversion method has been in continuous use by the heavy ion ERDA community since its development in the 1970s because this technique provides separate signals for all elements and the depth profile obtained this way is the most direct representation of the experimental data. The fact that the results include the statistical fluctuations of raw data can be seen as a drawback or as an advantage as it gives the nonexpert receiving the results an actual feeling about the uncertainty regarding, for example, the concentration of impurities and their (not necessarily uniform) distribution in a layer.

However, if one knows precisely the resolution function, more information can be extracted from a measurement by precisely fitting a properly convoluted

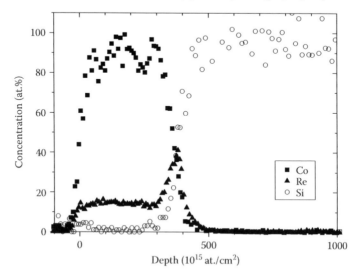

FIGURE 15.3   Depth profile obtained from the data from Figure 15.1 with the direct conversion method.

signal to the experimental spectrum. This is one of the reasons why simulation of spectra is the most widely used method of data analysis. Starting from an assumed depth profile, a spectrum is calculated and compared with the data. Then, either interactive or automated adjustment is made until a good fit is obtained. This has many advantages over the "direct conversion method." First, using simulation, superposition of signals usually found in RBS spectra is normally not a problem. A second reason, which has become increasingly important, is that effects that are practically impossible to calculate manually and difficult to include in the direct conversion method, such as plural and multiple scattering or roughness, can be implemented in the simulations. Third, a simulation allows one to examine the expected result of a measurement without actually performing it. This allows determining whether a measurement is able to provide the expected information (potential problems may be sensitivity, overlap of signals of different elements, limited mass- or depth resolution, etc.) and allows optimizing experimental parameters (such as incident ion species, beam energy, incident angle, detector solid angle) before the measurement.

The main drawback of simulations is that they usually describe a sample as slabs—that is, layers with defined thickness and flat, laterally uniform elemental concentrations. Casual practitioners will very often overestimate the validity of their assumptions or underestimate the uncertainty on the information extracted from simulations. Slabs are flat and good-looking but may not reflect reality. Beware overinterpretation!

15.1.3   THE FIRST GENERATION OF SOFTWARE—RUMP AND OTHERS

The first code for simulation of RBS spectra was presented in 1976 by Ziegler, Lever, and Hirvonen. The basic principles of simulation are still used in modern analytical codes. However, it was with the spread of PCs in the 1980s that the first generation of simulation codes became generalized. By far, the most widespread code was RUMP (Doolittle 1985, 1986), which was easy to use and included Bohr straggling, calculated the contributions of individual isotopes, supported non-Rutherford cross sections, had built-in common functions to describe depth profiles, and had some basic capability of handling layer roughness. RUMP is still in active development and it is still popular. These codes were good and reliable within their design base, which covered most applications of RBS. The simulation shown in Figure 15.1 was produced with RUMP, and it is excellent. The results were [Re] = 13.9 at.%, [Co] = 86.1 at.%, which compares very well with the manual analysis in Section 15.1.1.

Codes such as RUMP also make automatic fits to data. This is normally limited to only a few parameters at a time, and to one spectrum at a time. Early versions of the codes were dedicated to RBS only and did not include ERDA or NRA, and implemented only H and He beams. Recent versions have a window interface and include ERDA with any ion.

15.1.4   ENABLING TOOLS—SRIM, WDEPTH, AND SIGMACALC/IBANDL

IBA work relies on basic quantities that must be calculated or measured. The most important are stopping powers, energy spread, and scattering cross sections.

As seen in Chapters 4 and 5, the stopping powers determine the depth scale. SRIM (Ziegler et al. 2009) is a software that implements a semi-interpolative scheme, based on a very large database of experimental stopping powers and on different models, to provide stopping powers for any ion in any element. There are alternatives available (see Paul and Schinner 2003, 2005), but SRIM, also well known for ion implantation simulation, is almost universally used. The data analysis codes usually implement SRIM, so the user does not need to be concerned with it. Different versions of SRIM may be implemented, and that can be a user option. We cannot stress strongly enough the importance of using the best stopping powers available. In many cases, the inaccuracy of stopping powers is the main uncertainty of IBA measurements. The accuracy of SRIM-2003 for H ions is about 4.0%, for He ions 3.9%, for Li ions 4.8%, and for heavier ions (from Be to U) 5.8%, resulting in an overall accuracy of 4.68% (from www. srim.org, 2012). However, in some cases it can be off experimental values by more than 10%. In such cases, the expert interested in high accuracy will need to use experimental values or a different scheme.

The scattering cross sections are the basis for quantification of elemental content. For RBS and ERDA they are usually calculated with the Rutherford formula, including electron screening (see Chapter 3). However, at high energies strong deviations from the Rutherford formula are observed. The cross section is non-Rutherford for protons of almost any energy and for ^{4}He at energies as low as 2 MeV for some light elements. These non-Rutherford cross sections cannot be calculated a priori, and we need to rely on experiment. The largest database for cross sections of interest for IBA is IBANDL, supported by the International Atomic Energy agency (a still larger cross-section database is EXFOR, also supported by the IAEA, but it is not IBA specific). However, measuring cross sections accurately is difficult, and sometimes very large discrepancies between different measurements of the same cross section exist in the literature. Typical accuracies of experimental cross-section data are 5%–10%, but sometimes they are much worse than the stated accuracies. Gurbich (2009) developed the code SigmaCalc, which uses different theoretical models to evaluate existing data sets for a given reaction. The result is the cross section for any angle, within the energy range considered in the evaluation. The resulting accuracy is usually better than the accuracy of individual experiments. Unless there are strong reasons why a given experimental cross section is preferable in a specific case, SigmaCalc should be used wherever it is available (which includes the most usual reactions in IBA).

Energy spread determines how sharp signals are. This is important for diffusion, intermixing, and roughness studies. At the surface, it is mainly determined by the resolution of the experimental apparatus and by the energy spread of the primary beam. It normally becomes worse with depth, what is normally called "straggling" (see Chapter 4). Whenever the width of signals is important—for instance, in studies of layer mixing, diffusion, roughness, etc.—an accurate evaluation of straggling is essential. The main sources of energy spread are energy loss straggling (due to the statistical nature of energy loss), geometric broadening (due to both the angular spread of the beam and the finite dimensions of the beam and the detector), and multiple scattering (the ions undergo not one single scattering event, but rather very many, leading to trajectories that are not straight). For light elements, usually energy loss straggling determines the total straggling, while for heavier elements energy spread due to multiple scattering is often dominant. All these factors are fairly straightforward to calculate, except multiple scattering. Szilágyi, Pászti, and Amsel (1995) developed in the 1990s the code DEPTH, based on the multiple scattering theory by Amsel, Battistig, and L'Hoir (2003), which includes all these effects to calculate the depth resolution for any ion and target. The current version WDEPTH is windowed and user friendly. RESOLNRA by Mayer (2008) is another code for depth resolution calculations and optimizations. It is based on the SIMNRA simulation code and uses identical input files, thus providing a common environment for depth resolution calculations and simulations.

15.1.5   CURRENTLY POPULAR SOFTWARE—SIMNRA, NDF

In the late 1990s, two driving forces for progress came into play, both generated by the success of IBA techniques. Some laboratories became regional centers providing IBA to many partners. Better and more efficient algorithms were necessary to handle large volumes of data, including some degree of automation. At the same time, with the development of increasingly complex materials, the samples and systems to analyze became more complex. This required the introduction in the codes of physical effects that previously had been ignored or overlooked, and also the implementation of RBS, ERDA, NRA, and even PIXE within one single software package.

With these two codes NDF (Barradas et al. 1997) and SIMNRA (Mayer 1997) emerged in the same year. Both implement RBS, ERDA, and nonresonant NRA. NDF also implements resonant NRA, PIXE, and SIMS (secondary ion mass spectrometry), and NDP (neutron depth profiling) SIMNRA set out to implement advanced physics capabilities accurately, with a modern user interface that was easy to use even in complex cases involving, for instance, elastic and inelastic reactions. It is currently the most popular modern IBA code.

The main initial drive behind NDF was full automation of analysis by combining all the information from (possibly) several spectra to provide a unique solution. It then followed the goal of improving the accuracy of the analysis by developing and implementing new models. It has the most features, but it is harder to use and in the late 2000s it was cited around half the times that SIMNRA was. The main features in these two codes are the following:

- State-of-the-art calculation of depth resolution as a function of depth: NDF calls WDEPTH in run time, while SIMNRA implements directly a similar algorithm to WDEPTH. This is essential whenever energy spread has to be considered. We show in Figure 15.4 a spectrum from the same Co/Re sample already shown in Figure 15.1, but measured at an 83° angle of incidence (that is, 7° angle with the surface on the way in, 6.6° on the way out). It turns out the sample was not a homogeneous CoRe film on a Re buffer on Si, but a Si substrate/Re buffer/(Co 2 nm/Re 0.5 nm)$_{\times 15}$ multilayer. The simulation assuming only Chu straggling is clearly not suited in this case where the largest contribution to energy spread is multiple scattering. To get a "good fit" using a code that only has Bohr straggling, we could introduce an ad hoc multiplicative factor to the straggling, but then the actual width of the Re peaks would lose the precious information that it carries: It can tell us how much roughness there is in this sample. It is worth mentioning that the tool RESOLNRA provided with SIMNRA can be used to find the optimal experimental configuration to achieve the most accurate measurement of given specific parameters (such as thickness) for a given assumed sample.

- Calculation of roughness, albeit with very different methods: SIMNRA sums spectra over a distribution function of some sample characteristics, such as surface height or film thickness. It can model real surfaces. NDF either sums spectra like SIMNRA, or it can also approximate the effect of roughness as an equivalent energy broadening in spectral features for given model types of roughness. Sommation of spectra is more accurate when the actual surface is known, while the equivalent energy broadening method in NDF is appropriate for routine analysis of low to moderate levels of roughness. Both codes are restricted to some specific types of roughness. We show in Figure 15.5 the same data as in Figure 15.4, including energy spread as calculated with WDEPTH, with and without the effect of roughness as calculated with NDF. This leads to a quantitative determination of the sample roughness (in this case, around 0.7 nm). Only the surface region is shown, since it carries all the information about roughness, and, in fact, lower channels can be safely ignored. We must stress that the effect of roughness on the spectrum is nearly impossible to distinguish from the effect of layer mixing or interdiffusion, and complementary microscopy measurements are often required.

- SIMNRA comes with a large library of non-Rutherford cross sections. This allows the user to select the appropriate ones easily, which is extremely useful and strongly facilitates analysis of EBS and resonant NRA data. NDF supports the evaluated cross sections available in SigmaCalc as well as experimental cross sections, but they must be introduced via a file.

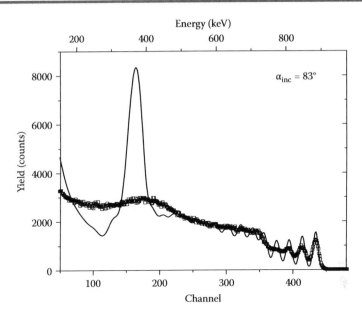

FIGURE 15.4   Spectrum collected at an 83° angle of incidence from the same sample shown in Figure 15.1. The simulation assumes Chu straggling only.

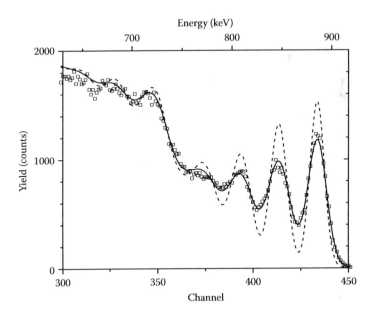

FIGURE 15.5   Same data as shown in Figure 15.3. The NDF simulations include all the sources of energy spread as calculated with WDEPTH. The dashed line is for the nominal structure. The solid line includes the effect of roughness.

- In EBS, one often uses a sharp resonance to enhance the yield from light elements. For near-surface layers both SIMNRA and NDF handle this. For deeply buried films, NDF implemented an algorithm that correctly simulates the shape and yield of resonances (Barradas et al. 2006). This is also available in SIMNRA since 2014.

- Both codes implement an analytical calculation of double scattering, which is sufficient for many applications. However, when higher order plural scattering events are important—for instance, in grazing angle geometry—or when the scattering cross section becomes very high (low energy ions, high Z, or thick materials), double scattering does not entirely account for observed backgrounds. Ratios of 2 (or somewhat larger) between experimental and calculated backgrounds can be expected (Steinbauer, Bauer, and Biersack 1990). A ratio of 10 probably means some other phenomenon is at play, such as beam ions scattering on the slits defining the beam at the entrance of the chamber. SIMNRA implements double scattering for RBS. NDF includes double scattering for ERDA as well.

- The yield at low energies is notoriously difficult to calculate, and the low-energy signal must often be disregarded in the analysis. Besides plural scattering and slit scattering, the increasing energy spread with depth leads to distortions of the calculated signal, which can be very strong, particularly at grazing angle. NDF implements an advanced algorithm that allows the calculation to be extended to lower energies (Barradas 2007).

- Both codes implement a statistical model of pileup based on the work of Wielopolski and Gardner (1977), which is applicable for low to medium count rates. NDF also implements the algorithm by Molodtsov and Gurbich (2009), which includes tail pileup and is also appropriate for higher count rates.

We can now return to the grazing angle Co/Re data. The full spectrum is shown in Figure 15.6, with the simulation including energy spread and roughness. Below channel 200, the misfit is very large, with the data much above the simulation. This is mainly due to double scattering, which is very high at grazing angle. A simulation including double scattering, pileup, and the low-energy yield calculation is also shown. Excellent qualitative agreement is obtained even if this spectrum is difficult to simulate. However, quantitatively, there are deviations at low energies, and, in fact, no extra information has been gained relative to the simulation with correct energy spread and roughness only. Except for that, if the models work well in this case, they probably also work well in simpler spectra.

- In the case of PIXE, GUPIX (Maxwell, Teesdale, and Campbell 1995) is the program most commonly used. However, it does not handle RBS spectra. Comparing, PIXE data and RBS spectra simulations for a single sample model is a feature that only NDF currently possesses, by

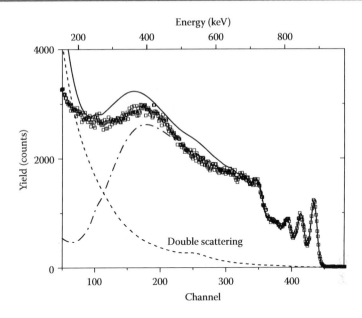

FIGURE 15.6   State-of-the-art simulation for the same data as shown in Figure 15.4, made by NDF. The contribution of double scattering is shown as a dashed line.

incorporating the open source routine LibCPIXE (Pascual-Izarra, Reis, and Barradas 2006). This is in the spirit of NDF to take into account simultaneously information from several spectra in an analysis. This is extremely useful because PIXE is very complementary to RBS in terms of elemental resolution and sensitivity.

■ NDF can do fully automated analysis. This is suitable for most spectra, and for large volumes of data it increases the efficiency of data analysis by a significant factor. It is not appropriate in cases where the important information is in small details, or where the data analysis requires specialized knowledge from the analyst. NDF can also perform a local search on all parameters (including the depth profile and roughness as well as experimental parameters such as energy calibration, collected charge, beam energy, and angle of incidence and of scattering), starting in a user-provided initial guess. SIMNRA uses the Simplex algorithm for fitting the depth profile and roughness (one layer at a time), the energy calibration, and the collected charge.

■ NDF can fit simultaneously any number of spectra collected from the same sample, using the same depth profile. There are no restrictions on the beams or techniques used, so if one sample has been measured (e.g., with ^{4}He-RBS, proton RBS, ^{4}He-ERDA, heavy ion ERDA, and PIXE), all the spectra collected can be analyzed in one run, extracting all the information present in each spectrum to retrieve one single depth profile. This is extremely powerful, particularly in complex samples where one single spectrum is not enough to obtain a full picture of the system.

■ SIMNRA is also an OLE automation server, which allows other applications to control SIMNRA, including batch work and fitting. This, unlike other SIMNRA features, is not user friendly as it requires a great deal of knowledge. Full documentation and examples are available.

■ One point to keep in mind is that fitted parameters are often correlated (e.g., two relatively thin layers containing similar amounts of the same element, and their thickness). Because of this correlation, the uncertainty on these parameters is much larger than anticipated by the user, given the resolution of the technique and the statistics collected. The best method to estimate uncertainties correctly is the so-called uncertainty budget (Jeynes and Barradas 2009), which, however, requires specialized knowledge.

15.1.6   MONTE CARLO: A "BRUTE FORCE" APPROACH

Most effects are taken correctly, if not exactly, into account by the analytical codes. One important exception is multiple scattering (i.e., the fact that ions do not fly in straight trajectories inside the target). An exact analytical theory has never been presented. The theory developed by Amsel et al. (2003), used by SIMNRA and NDF, is valid for small deflections. By including dual scattering, SIMNRA and NDF account for the effect of a second important collision in the trajectory of an ion, but in some cases, such as heavy ion RBS or ERDA, or in low-energy tails of RBS spectra, collisions at higher orders must be taken into account. This also applies to grazing incidence measurements.

Monte Carlo simulation approaches the problem by actually simulating the trajectory of ions. Hence, all the effects can be naturally taken into account, including the exact beam, sample, and detecting system configuration. This can include peculiar effects such as slit scattering, detector foil scattering, or complex detection geometries. The drawback of Monte Carlo is that at MeV energies and with the typical small solid angles used in IBA measurements, only about 10^{-9} of the incident particles are detected, so that a simulation requires typically about 10^{12} incident particles, resulting in very long computing times.

After some initial work directed to theoretical studies of plural and multiple scattering in RBS (Steinbauer et al. 1990; Eckstein and Mayer 1999), codes directed toward HI-ERDA data analysis were presented in the early 2000s (Arstila, Sajavaara, and Keinonen 2001; Johnston et al. 2000). These codes limit the number of particles that have to be calculated by a number of approximations, especially that scattered/recoiled atoms are pitched only in the detector direction, and a virtual detector (usually 100 times larger than the real one) is used, leading to acceptable computing times. These researchers were very successful in analyzing complex HI-ERDA data, reproducing the width of signals and the low-energy tails found in the data. However, the codes were difficult to use, and in practice their usage was restricted to their authors and collaborators.

This changed with the presentation of the MC code Corteo (Schiettekatte 2008), which includes a user-friendly graphical user interface. It is as easy to use as the standard analytical codes. Corteo can simulate RBS and ERDA, including with time-of-flight detector, as well as coincidence experiments. Nevertheless, MC codes introduce some new problems (namely due to the cutoff angle and the virtual detector), and the user should get familiar with these.

Despite improvements, MC calculations remain long, so this approach should be considered mainly to attack specific problems. The general features of a spectrum can be obtained in minutes, but low-probability or low-energy events require much longer calculations to obtain reliable statistics. Current versions do not handle non-Rutherford cross sections and nuclear reactions, because they often have a strong angular dependence. This may be solved in the future by taking advantage of SigmaCalc, using it to calculate the scattering cross section for the required range of angles and energies, and making interpolations, as NDF does for its double scattering routine. Overall, MC simulation is a very useful complement to analytical codes.

15.1.7   PIXE

In the case of PIXE, the simulation philosophy underlying RBS fitting codes is usually not used. The starting point is the spectrum itself, rather than a hypothetical sample structure. The peaks present are recognized and assigned to given elements by the analyst or generated automatically by peak search algorithms. The areas of each peak are then converted into a concentration. This often leads to a model of the sample that treats the sample as a single homogeneous layer. See Fazinic (2003) for a detailed description of seven codes commonly used for analyzing PIXE data.

Aiming at the integration of PIXE with the other techniques within a single code, the open source library LibCPIXE (Pascual-Izarra, Reis, et al. 2006) was developed based on DATTPIXE (Reis and Alves 1992) algorithms but using a sample oriented simulation philosophy. It is now included in NDF. LibCPIXE does not simulate energy spectra. Instead, it calculates expected X-ray yields emitted from a sample, with assumed known structure and under given experimental conditions. This can then be compared with experimental peak areas. This approach is much better tuned to deal with multilayer samples that generate complex spectra than trying to unravel the sample structure starting from the spectra.

As a consequence, LibCPIXE, as well as its ancestor DATTPIXE and its present implementation in NDF, makes use of any existing code that is able to provide area values for the main peaks in a PIXE spectra (presently, AXIL/

QXAS and GUPIX outputs are accepted as input formats to NDF) and compares these main peak areas with the simulated x-ray yield values for a sample model. In the NDF implementation, this model can have a layer structure with almost no limitations, except for the usual assumption that each layer is homogeneous.

In the revision of LibCPIXE carried out recently by Taborda, Chaves, and Reis (2011) (LibCPIXE version 2.08), both H^+ and He^{2+} PIXE data can be handled and sample matrix effects corrections include the energy loss of the impinging particles and the absorption of the exiting x-rays. Secondary fluorescence corrections are not yet implemented. Roughness and, in general, lateral inhomogeneity are also not handled by LibCPIXE so, to simulate them, intermixing layers must be created. An example of this is present in the study of roughness origin in titanate multilayer targets by Reis et al. (2010).

15.2 EXAMPLES

NDF and SIMNRA are proprietary codes available commercially. However, test versions are available and can be used to make these exercises. Corteo and QXAS/LibCPIXE are freely available. SRIM, WDEPTH, and SigmaCalc are also needed to make the exercises. The Web sites are listed in Table 15.1.

15.2.1 IAEA INTERCOMPARISON EXERCISE: NDF

Download the WiNDF installation package from http://www.ionbeamcentre.co.uk/ndf, where detailed installation instructions are given. Obtain a demo license (see Table 15.1). Install WiNDF into a directory of your choice. The

TABLE 15.1 IBA Web Resources

NDF	Data analysis (full version)	http://www.ionbeamcentre.co.uk/ndf
		A time-limited demo license is obtainable on request from Dr. Chris Jeynes at C.Jeynes@surrey.ac.uk
	Data analysis (free version)	http://www.itn.pt/facilities/lfi/ndf/uk_lfi_ndf.htm
SIMNRA	Data analysis	http://www.simnra.com/
Corteo	Data analysis	http://www.lps.umontreal.ca/~schiette
QXAS	PIXE data analysis	http://www.iaea.or.at/programmes/ripc/physics/faznic/qxas.htm
LibCPIXE	PIXE data analysis	http://cpixe.sourceforge.net/
SRIM	Stopping powers	http://www.srim.org/
SigmaCalc	Cross sections	http://www-nds.iaea.org/sigmacalc/
IBANDL	Cross sections	http://www-nds.iaea.org/ibandl/
DEPTH	Energy spread	http://www.kfki.hu/~ionhp/doc/prog/wdepth.htm

installation requires administrator privileges on the computer. C:\WINDF\ is recommended for the sake of this exercise. Download SRIM from www.srim. org. Copy the files SRModule.exe, VERSION, SCOEF03.dat and SNUC03. dat into the WiNDF directory. Download WDEPTH from http://www.kfki. hu/~ionhp/doc/prog/wdepth.htm. Install it into a directory of your choice. C:\ WDEPTH\ is recommended for the sake of this exercise.

The first exercise is to make the simulations defined in the International Atomic Energy Agency (IAEA) intercomparison of IBA software (Barradas et al. 2007). Some of them rely on WDEPTH being installed in directory C:\ WDEPTH\. If it is not installed, those simulations will be made assuming Chu straggling only, so they will look very different from what is in Barradas et al. (IAEA). If it is installed in a different directory, you need to edit the files 12.DPT, 23.DPT, and 26.DPT. In all of them the first line is C: (this indicates the path to \WDEPTH\). If you installed it, for instance, in C:\IBACodes\WDEPTH\, you need to replace that first line by C:\IBACodes.

Download the IAEA_NDF_C.zip file from http://www.ionbeamcentre. co.uk/ndf (or alternatively from http://www.itn.pt/facilities/lfi/ndf/uk_lfi_ndf. htm) and unzip it into a directory of your choice. C:\WINDF\IAEA_NDF_C\ is recommended. Start WiNDF. In the File menu, click on Open (alternatively, click on the Open Batch File button—the first on the left, looking like a yellow swiss cheese). Navigate into the C:\WINDF\IAEA_NDF\ directory. Then select the file IAEA_C.SPC. You are all set! By pressing the button Run NDF (execute existing NDF.BAT), all the calculations will be made, because all are included in the same batch file that you just opened. Calculation 11 was not made by NDF in the intercomparison exercise. Calculation 28 (channeling) is carried out, but only the random spectrum is calculated. Calculations 17 and 22 make no sense, since they include geometrical straggling but no multiple scattering. They require (in NDF) manually manipulating the WDEPTH output files to remove the contribution of multiple scattering, and then inputting the resulting depth resolution files with the DEPTH option. This is not recommended usage and therefore these calculations are not included here. Calculations 12, 25, and 26 include double scattering and will be slow.

With all the calculations done, you can explore the results. Select each calculation in the main IBA DataFurnace window: In the lower left corner, you can select the Sample # number. Sample N corresponds to calculation N. You can look at the simulation with the button Plot Fit (using file bXsf. dat). You can see a detailed log of the result with the button QED - View Results File (bs.RES).

After seeing all the results, you can explore how NDF did each one of the calculations. To inspect or change the geometry, double click on one of the Geometries in the main IBA DataFurnace window. The Geometry window will open. It has the experimental parameters. The buttons More and More+ include advanced calculation options. All the calculations include at least one option, which is SRIM stopping power. Some calculations include several additional options.

Try to change some of the options—for instance, in calculation 9 (a 3.5 MeV Li beam), set ZBL85 stopping powers in More. You do not want to redo all the simulations, only that one. Select sample 9. Then use the button ORD - Open NDF.ORD editor, to check only the box sample 9. Now when you run, only calculation 9 is made.

Now download the IAEA_NDF_E.zip file from http://www.ionbeamcentre. co.uk/ndf (or alternatively from http://www.itn.pt/facilities/lfi/ndf/uk_lfi_ndf. htm) and unzip it into a directory of your choice. C:\WINDF\IAEA_NDF_E\ is recommended. This includes the four data sets analyzed in the IAEA intercomparison exercise. Start WiNDF. Open the batch file IAEA_E.SPC. You are all set! By pressing the button Run NDF (execute existing NDF.BAT), the four data sets will be analyzed. Again, you should explore the results.

15.2.2   EBS: SIMNRA

SIMNRA is a Windows program with graphical user interface and easy-to-use menus and windows. It can be downloaded from www.simnra.com. The Windows installer will install the program automatically into a directory of your choice. The installation requires administrator privileges on the computer.

By default, SIMNRA uses Ziegler/Biersack stopping powers. These are identical to SRIM 1997 stoppings. In order to achieve a higher accuracy, it is recommended to download and install the newest version of SRIM from www.srim. org and use it for all stopping-power calculations. This is achieved by clicking Options:Preferences in SIMNRA and setting the path to the SRIM program in Directories: SRIM program directory. This has to be done only once after installation. The use of SRIM stopping powers is activated by clicking Setup: Calculation and selecting Electronic stopping power data: SRIM. This setting can be made permanent by clicking Setup: Calculation: File: Save as Default.

The analysis of a thin gold layer on silicon by 2.015 MeV protons is shown in Figure 15.7. The measurement was performed at an incident angle of 5° to the surface normal in order to avoid channeling in the single-crystalline silicon substrate; the scattering angle was 165° in Cornell geometry. All experimental conditions can be examined by clicking Setup: Experiment.

The cross section for backscattering from silicon deviates considerably from the Rutherford cross section (Gurbich 1998), while the cross section for backscattering from gold is still Rutherford (Bozoian 2009). A number of different cross-section data sets for backscattering of protons from silicon have been measured and are available in SIMNRA. The available cross-section data can be examined by clicking Reactions. For silicon the SigmaCalc cross section (Gurbich 1998) should be selected, while the cross section for gold is Rutherford.

The target consists of two layers, which can be examined by clicking Target: Target. The first layer is the gold layer and the second is silicon substrate.

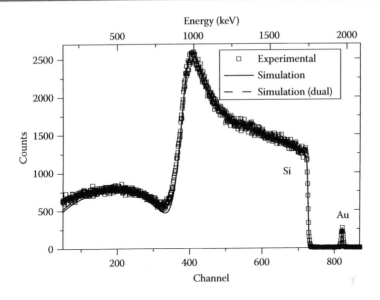

FIGURE 15.7   Backscattering spectrum of 2.015 MeV protons from a thin Au layer on Si, scattering angle 165°, analyzed with SIMNRA. Solid line: simulation using single scattering; dashed line: simulation using the dual scattering approximation.

A fit of the spectrum typically requires the following steps:

1. Create the target with the Au layer having a reasonable initial thickness and the silicon substrate as layer 2.
2. Click `Calculate: Fit Spectrum.` Check `Fit: Particles*sr` and select `From 500 To 700` as `Fitting range.` Click the `Fit` button. The value of particles*sr is now adjusted for best fit from channel 500 to 700 (i.e., to the backscattering yield from the silicon substrate).
3. Now check `Fit: Layer Composition and Thickness: Layer 1`, and select `From 812 To 830` as `Fitting range.` Because this is a thin layer, where the shape of the spectrum is not important, you can increase the accuracy by selecting `Chi2 evaluation: Integrals.` Now only the total number of counts is considered, without taking the shape of the spectrum into account. Click the `Fit` button.
4. If necessary, you can repeat steps 2 and 3. Alternatively, you can check both `Fit: Particles*sr` and `Fit: Layer Composition and Thickness: Layer 1` and then select `From 500 To 830` as `Fitting range.` In the latter case particles*sr and the composition of layer 1 are fitted simultaneously.

The complicated spectrum structure between channels 250 and 450 is a resonance in the Si backscattering yield. The deviations between experiment and simulation from channels 100 to 300 are due to plural large-angle scattering.

This can be examined by checking `Setup:Calculation:Parameter:Dual scattering`. The agreement between experiment and simulation improves, but at the cost of much longer computing time. The remaining differences are mainly due to the inaccuracy of the dual scattering approximation. This can be overcome only by Monte Carlo programs.

The discrepancies between experiment and simulation at channels below about 100 are due to a breakdown of the straggling models at very large energy losses. At energies close to zero the simulation always gets inaccurate, and the low channels should not be taken into account in any analysis.

15.2.3   NRA: SIMNRA

The energy spectrum of protons created in the $D(^3He,p)\alpha$ reaction from a thick amorphous deuterated hydrocarbon (a-C:D) film is shown in Figure 15.8. A very thick solid-state detector with a thickness of 2 mm was used in order to fully stop the high-energetic protons with maximum energies up to almost 14,000 keV. The detector had a large solid angle of about 30 msr and was covered with a nickel absorber foil in order to filter backscattered primary ^3He ions, which would otherwise cause a very high count rate at low energies. The experimental cross-section data from Alimov, Mayer, and Roth (2005) were used.

This example demonstrates a nuclear reaction with inverse kinematics, which is encountered with some light projectiles and target elements. Particles coming from the surface have the *lowest* energy, while particles from deeper in the material have *higher* energies. This is exactly opposite to RBS/EBS. The depth of origin of protons with a specific energy is marked in Figure 15.8. The depth resolution is mainly governed by geometrical spread due to the large solid angle (see Mayer et al. 2009) and exhibits the special feature that the depth resolution is very bad at the surface, but gets better deeper inside the sample. The depth resolution at the surface is 50 μm (FWHM), decreasing to 3 μm at about 30 μm depth and to 1 μm at about 36 μm depth. All information about the deuterium depth distribution within the top 25 μm is squeezed into a single Gaussian-shaped peak providing only information about the total amount of D from the near-surface layer, while the spectrum contains detailed information about the D depth profile from about 25 μm to about 37 μm. The depth profile between the surface and 25 μm as shown in Figure 15.8 (bottom) was derived by using information from additional measurements at 800, 2500, and 4000 keV incident energy (Mayer et al. 2009).

15.2.4   MONTE CARLO CALCULATIONS: CORTEO

Visit the Corteo Web site at http://www.lps.umontreal.ca/~schiette and download the package corresponding to your operating system, as well as the parameter files for the examples covered here. The program is distributed under the

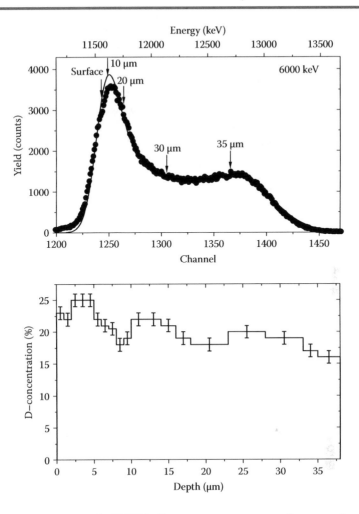

FIGURE 15.8   NRA with SIMNRA. Top: energy spectrum of protons from the D(3He,p)a nuclear reaction at 6000 keV incident energy, reaction angle 135°. Bottom: deuterium depth profile. The sample was a thick amorphous deuterated hydrocarbon film.

GNU general public license; this means that it is provided with the source code, which you can inspect, modify, and redistribute as long as you provide the source code with the modified version. Follow the installation instructions. Once everything is set up, start CorteoUI (the user interface).

As with NDF, the first example covered is the intercomparison exercise S1, for which the parameter file (once downloaded) can be loaded by clicking on the folder icon at the top of the simulation panel. The example is a standard RBS analysis with a 1.5 MeV ⁴He beam detected at a 150° scattering angle. Through this exercise, we will change a few parameters in order to illustrate the main differences of Monte Carlo simulations over analytical simulations provided by NDF and SIMNRA. Problem 15.3 at the end of the chapter is a heavy ion ERDA analysis.

First, decrease the number of simulated ions to 100,000 and run the simulation by clicking the lightning button. After a few seconds, it results in a smooth, simulated spectrum closely comparable to what is obtained with analytical codes. However, this simulation corresponds to a case where multiple collisions are "turned off." In Corteo, this is obtained by specifying a negative mean free path.

By removing the negative sign and increasing to 90° the cone angle in which ions are directed to the detector, the simulation will now include the effect of multiple and plural collisions: Ions initially directed away from the detector may reach it as a result of other collisions. But running the simulation with those parameters will produce only a few detected particles. In order to reproduce the point detector specified by simulation S1, the detector was made very small and the backscattered ions were closely directed at it (small cone angle). Detector height and width can be increased to 10 mm in the detectors panel, which is roughly the physical size of the most common detector used in RBS. Hence, the simulation now accounts for the effects due to a finite size detector. Running the simulation with these parameters produces about 100 detected ions—clearly not enough to generate a smooth simulation.

In order to improve the detection efficiency, we will have to consider a virtual detector—that is, find backscattered ions that missed the detector by a few millimeters and correct their trajectory so they hit the detector. In the detectors panel, input 10 as the virtual detector factor, both for the height and width. Another factor of 10 can be gained by increasing to 1,000,000 the number of simulated ions in the *simulation* panel, so the next simulation, albeit 10 times longer, will feature 1,000 times more counts. (Because we made the detector 100 times larger, the amplitude is 100 times higher than the first simulation. The amplitude can be multiplied by a factor by right-clicking on the corresponding legend entry and selecting operation.)

Run this simulation and then right-click left of the graph and select logarithmic Y axis from the menu. It can now be seen that, compared to the first simulation, this one features a background between the high-energy peak and the substrate, as well as some differences in the low-energy tail. These are the most obvious effects of multiple collisions, in addition to a slight broadening of the peak edges. In this case the background is small and dominated by dual collisions, so SIMNRA and NDF account very well for these effects, with faster simulations.

But whenever the ion energy per unit mass is low or the target is thick or made of heavy atoms, ions are likely to suffer more than two important collisions, and it becomes necessary to account for that. In the second example (simulation S12 of the intercomparison exercise), the spectra of elastic recoils induced by 50 MeV iodine ions hitting the same target as in the first example are simulated. Load file S12-noRot.cml and start the simulation. As for S1, this initial simulation returns the spectrum we would get if there were no multiple scattering. Again, remove the negative sign in front of the mean free path value, increase cone angle to 90° and, in the `detectors` panel, change the virtual detector factors to a width of 2 and a height of 10. (We do not want to increase the width too much so as to avoid bad trajectory correction effects.) Running this simulation shows how significant are the effects of multiple collisions in this case. The gold peak is notoriously wider and features a low-energy tail.

Heavy ion ERDA and ⁴He-ERDA for the analysis of hydrogen are also dealt with by Corteo in a very similar manner and using exactly the same methods. The recoils detected and existing stopping foils or detection systems such as E-DE detectors are given input in the same user-friendly way. The Corteo web page includes several examples that can be explored.

15.2.5   PIXE: NDF

PIXE data analysis will be considered here under the scope of simultaneous and consistent processing with the other IBA techniques (mainly EBS). This is presently only possible by using the `IBA DataFurnace` code. Most of the practical questions relating to PIXE are therefore similar to those relating to the use of this code for any other technique, as described in Section 15.2.1. The specific items relating to PIXE are the extraction of line intensities from energy spectra and the graphical output, which is not a spectrum simulation.

It is important to realize that the IBA DataFurnace does not include any routine to extract the line intensities from PIXE spectra, since this is also absent from the PIXE code library LibCPIXE. Peaks in the spectrum must therefore be first retrieved using a general fitting code such as QXAS/AXIL or even GUPIX. Why is this so? The reason is that PIXE data are integral data, which means that the information contained in the spectra is not scattered throughout all the channels of the spectra; instead, it is stored in the areas of the peaks of characteristic x-rays, which gather the contribution to that particular element of all target layers crossed by the particle beam. These areas depend on the layered structure of the target, but a single PIXE spectrum does not have information on the depth distribution of the elements. This can be obtained from the RBS data, thus making the two techniques extremely complementary to each other.

The output of IBA Datafurnace PIXE data fitting is therefore not a spectrum but a bar graph comparing fitted areas to the corresponding experimental areas extracted from the spectrum.

In order to be able to use this approach in a practical sense, a previous calibration of the PIXE system yield is necessary most of the time. Basically, this corresponds to irradiating a few homogeneous standard reference materials or pure and stable chemical stoichiometric chemical compounds.

In Pascual-Izarra, Reis, et al. (2006) and Pascual-Izarra, Barradas, and Reis (2006) several examples of PIXE analysis with NDF are presented. Here we select a MnIr thin film on a Si substrate that was measured with RBS and with PIXE. The data are shown in Figure 15.9, where the PIXE energy spectrum and

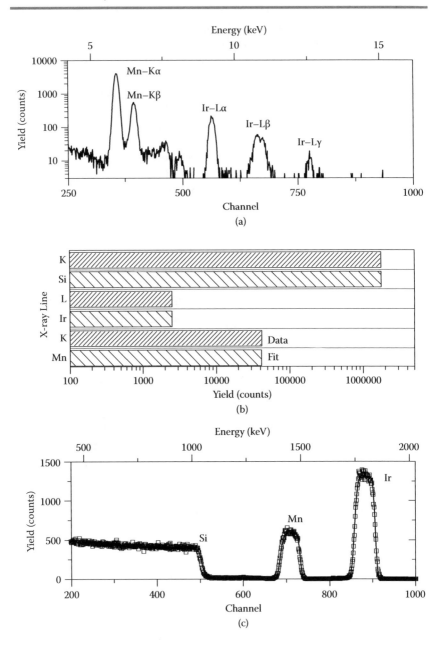

FIGURE 15.9   MnIr film on a Si substrate, analyzed by NDF. (a) PIXE energy spectrum measured with a 1.22 MeV 1H beam, (b) PIXE line intensities, and (c) RBS data collected from an MnIr thin film on Si with a 2 MeV ^{4}He beam. NDF did a simultaneous fit to the two sets of data.

the corresponding line intensities are shown, together with the RBS spectrum. First, download the PIXE_NDF_T1.zip file from http://www.ionbeamcentre. co.uk/ndf (or, alternatively, from http://www.itn.pt/facilities/lfi/ndf/uk_lfi_ndf. htm) and unzip it into a directory of your choice. C:\WINDF\PIXE_NDF_T1\ is recommended. Start WiNDF. Open the batch file PIXE_T1.SPC. You are all set! By pressing the button Run NDF (execute existing NDF.BAT), the data sets will be analyzed. Again, you should explore the results, in the same way that was done for the RBS example with NDF. Note that three samples are defined in the batch file, but only sample 1, which is a simultaneous fit to the RBS and PIXE data, is made when pressing the Run NDF (execute existing NDF.BAT) button.

15.2.6 SOFTWARE HANDLING OF CHANNELING SPECTRA

Ion channeling is dealt with in detail in Chapter 9 of this book. Channeling is a powerful tool to analyze the crystal structure of solids, including the characterization of defects and the lattice locations of impurities. While in many cases channeling data are analyzed qualitatively, often a quantitative analysis is needed. As Chapter 9 covers the other methods, here we deal only with quantitative analysis made with simulation data analysis codes.

Channeling data usually come in two different ways: as a minimum yield for a given axis or plane (more generally as an angular scan around a channel or a plane) or directly as the channeled spectrum. Many codes can analyze both types of data.

Some codes are based on analytical models, and some employ Monte Carlo methods. One thing that the codes have in common is that they are difficult to use and require extensive knowledge about the technique and about the problem at hand. Usually, new users need to be trained by the code authors or by experienced users. Also, many of these codes support only a restricted number of crystals, both in what concerns the crystalline structure and the constituent elements. Extension to other crystals often requires programming them into the code.

Very often, the Monte Carlo codes assume that the defects are located in an otherwise ideal lattice. This neglects the relaxation of the crystal lattice around the defects and leads to an overestimation of the defect density. It is possible to calculate the lattice relaxation with ab initio or molecular dynamics methods, and some Monte Carlo codes can include the results in the simulations (Balboni et al. 2002; Kovač and Hobler 2006). This requires even more specialized knowledge, and there is no package that integrates the lattice relaxation calculation with the channeling calculation in one single code.

DICADA is an analytical code developed by K. Gärtner (1997, 2005, and references therein). It analyzes channeling energy spectra to extract the depth distribution of different types of defects, including point defects randomly placed in the lattice or in specific positions and dislocation loops. It is used by different

IBA groups, and access to the source code is possible. FLUX is one of the oldest widely used Monte Carlos codes, first presented by Smulders and Boerma (1987). It can be obtained from the authors and it is used by several IBA groups, some of which have modified the source code to extend its applicability to crystalline structures not initially included.

CASSIS is another Monte Carlo code, written by Kling (1995), which makes a full 3D treatment of ion trajectories inside the crystal. It can analyze point defects and extended defects such as dislocation loops. It has been recently refurbished (Kling 2012), including a simpler user interface that requires less knowledge from the user. The source code is proprietary, but the executable one can be obtained from the author.

IMSIL and BISIC have been developed by Hobler (1995) and by Albertazzi et al. (1996), respectively. They are both 3D MC codes that support lattice relaxation and can deal with clusters of defects. Their use is normally restricted to the group where they were developed.

McChassy is an MC code presented by Nowicki et al. (2005) that can also analyze extended defects (Turos et al. 2010), and any atom displacement can be introduced. It is used by other groups as well, but it requires training by the authors.

PROBLEMS

15.1.   RBS analysis with NDF: Download the NDF_EBS.zip file from http://www.ionbeamcentre.co.uk/ndf (or, alternatively, from http://www.itn.pt/facilities/lfi/ndf/uk_lfi_ndf.htm) and unzip it into a directory of your choice. C:\WINDF\NDF_EBS\ is recommended. Start WiNDF. In the File menu, click on Open (alternatively, click on the Open Batch File button—the first on the left, looking like a yellow Swiss cheese). Navigate into the C:\WINDF\NDF_EBS\ directory. Then select the file EBS.SPC. This includes three different carbon-based samples. One spectrum was collected from each sample. You can navigate from one sample to the next in the main IBA DataFurnace window: In the lower left corner, you can select the Sample # number.

The samples are mainly carbon, but the ^{12}C:^{13}C ratio is not fixed. It also has ^2H, Be, O, and Ni. The experimental details are given together with the example, and the elements present in the sample are also defined. The problem is to use NDF to retrieve the depth profile. Hint 1: manually define a depth profile, with many layers, until you find a simulation consistent with the fit. This is going to take some time, and it will require a great deal of expertise. Hint 2: try to use the automatic fitting option Normal speed in the NDF setup window (do not forget to press Write ndf.bat and EXIT; otherwise, the changes are not saved) and this will be considerably faster. Try the `Ultra-fast speed` button and this will be even faster and almost as good. By pressing the button `Run NDF (execute existing NDF.BAT)`, fits will be made for the three samples. If you only want to fit one of the samples, use the button `ORD - Open NDF.ORD editor` to check only the box corresponding to that sample.

You can look at the fit with the button `Plot Fit (using file bXsf.dat)`. You can see the depth profile that was determined by NDF with one of the buttons `Plot fitted structure`. You can see a detailed log of the result with the button `QED - View Results File (bs.RES)`.

15.2.   NRA analysis with SIMNRA: Download SIMNRA from www. simnra.com. The Windows Installer will install the program automatically into a directory of your choice. The installation requires administrator privileges on the computer.

    a.   Calculate energy spectra of protons from the $D(^3He,p)\alpha$ nuclear reaction at different incident ^3He energies from 800 to 6000 keV.

    b.   Use the program RESOLNRA (which is part of the SIMNRA package) to calculate the depth resolution of this reaction as a function of depth. Observe the different energy spread contributions (energy-loss straggling, geometrical straggling, multiple scattering) and their contributions to the final depth resolution.

15.3.   Heavy ion ERDA analysis with Corteo: In this exercise, we determine the impurity content and distribution in a Ni layer deposited by sputtering on silicon. The data are shown in Figure 15.10.

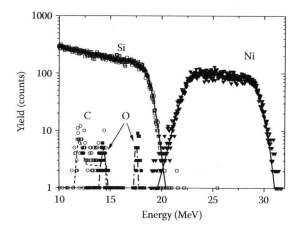

FIGURE 15.10   Corteo analysis of heavy ion ERDA data collected with a 50 MeV Ge beam from a Ni layer deposited on a Si substrate. Points: data. Lines: fit.

Download the NiSi.zip example from the Corteo Web site and uncompress it in the Corteo directory. Start Corteo and load the parameter file NiSi.cml by clicking on the folder icon at the top of the `parameters` panel. It contains a description of the sample and its analysis by elastic recoil detection (ERD) using a 50 MeV Ge beam. The beam and detector are in opposite directions, both at an angle of 15° from the surface (75° from sample normal). The corresponding spectra can be loaded in by right-clicking in the graphic area and selecting `load file`. As this is an ERD measurement, we have a separate spectrum for each element, so load each file ending with a.SPC extension in the NiSi directory. You see that the sample contains not only Ni and Si, but also some oxygen and carbon contamination.*

Then, run the simulation by clicking the lightning icon in the `Simulation` panel. The settings loaded consider only Ni on top of Si. The purpose of this exercise is to fit the oxygen and carbon spectra in order to determine where they are found and estimate their amount. Let us start with the oxygen, whose spectrum presents two peaks. We will presume that the higher energy peak consists of the native nickel oxide, NiO, on top of the Ni layer, while the lower energy peak consists of silicon native oxide, SiO_2, on top of the Si substrate. Go to the `Target` panel. The layer #0 is already configured as NiO, but with thickness set to 0. Increase the thickness and select the O in the elements, then check `Simulate this element` checkbox. Do the same for layer #3, where SiO_2 is described. Readjust the thicknesses of layers #0 and #3 until you are satisfied with the fit. By how much do you estimate the uncertainty? Also remember that it is not because the fit is OK that the target consists of well-defined, sharp-interface layers. Playing with the roughness gives an insight about the actual information that can be accessed from this measurement. As an additional exercise, find how much thickness/roughness can be included before the simulation departs from the experimental spectrum. Also remember that the thicknesses have no meaning unless the density is precisely known, which is not obvious, especially for thin layers. Only the thickness × density product is reliable.

Then, let us fit the carbon spectrum, which is slightly more challenging. The carbon concentration is low, leading to poor statistics. Still, we can say that, as for oxygen, there are peaks at the high- and low-energy ends of the spectrum, so presumably at surface and at the Ni–Si interface; however, this time, there is carbon in between, in the Ni layer. Return to the *Target* panel and add C

---

* You can edit the line color and symbol of each data set by right-clicking on its legend entry and selecting Edit from the menu.

as an element in layer #0 and specify some concentration. (Also check Simulate this element.) For the Ni–Si interface, we will presume that a layer of C was deposited on top of the SiO_2, as layer #2. Increase the thickness and check Simulate this element. Then go to layer #1 and add a small proportion of C in this Ni layer. Change the thicknesses and concentration until you are satisfied with the fit. What is your estimate of the C concentration in the Ni layer?

15.4.   PIXE analysis with NDF: Let us return to the example shown previouslyh for PIXE and RBS simultaneous analysis (Figure 15.9). The case presented was a MnIr thin film on a Si substrate. If you did not do it already, you can now download the PIXE_NDF_T1.zip file from http://www.ionbeamcentre.co.uk/ndf (or, alternatively, from http://www.itn.pt/facilities/lfi/ndf/uk_lfi_ndf.htm) and unzip it into a directory of your choice. C:\WINDF\PIXE_NDF_T1\ is recommended. Start WiNDF. Open the batch file PIXE_T1.SPC. You are all set!

By pressing the button Run NDF (execute existing NDF. BAT), the first data set corresponding to the simultaneous PIXE/RBS analysis will be fitted, as shown in the example. Now, you can explore the relative advantages of PIXE and RBS. First, use the automatic fitting option Normal speed in the NDF setup window (do not forget to press Write ndf.bat and EXIT; otherwise, the changes are not saved) to obtain a fully automated answer to the problem. Now, use the button ORD - Open NDF. ORD editor, to uncheck the box corresponding to the first case, and check the other two. The second is only the PIXE data, and the third is only the RBS. Run the fits again and explore the results.

REFERENCES

Alimov, V. Kh., Mayer, M., and Roth, J. 2005. Differential cross-section of the D(He-3, p) He-4 nuclear reaction and depth profiling of deuterium up to large depths. *Nuclear Instruments and Methods B* 234:169–175.

Amsel, G., Battistig, G., and L'Hoir, A. 2003. Small angle multiple scattering of fast ions, physics, stochastic theory and numerical calculations. *Nuclear Instruments and Methods B* 201:325–388.

Arstila, K., Sajavaara, T., and Keinonen, J. 2001. Monte Carlo simulation of multiple and plural scattering in elastic recoil detection. *Nuclear Instruments and Methods B* 174:163–172.

Balboni, S., Albertazzi, E., Bianconi, M., and Lulli, G. 2002. Atomistic modeling of ion channeling in Si with point defects: The role of lattice relaxation. *Physics Reviews B* 66:045202.

Barradas, N. P. 2007. Calculation of the low energy yield in RBS. *Nuclear Instruments and Methods B* 261:418–421.

Barradas, N. P., Alves, E., Jeynes, C., and Tosaki, M. 2006. Accurate simulation of back-scattering spectra in the presence of sharp resonance. *Nuclear Instruments and Methods B* 247:381–389.

Barradas, N. P., Arstila, K., Battistig, G., Bianconi, M., Dytlewski, N., Jeynes, C., Kótai, E., Lulli, G., Mayer, M., Rauhala, E., Szilágyi, E., and Thompson, M. 2007. International Atomic Energy Agency intercomparison of ion beam analysis software. *Nuclear Instruments and Methods B* 262:281–303.

———. 2008. Summary of "IAEA intercomparison of IBA software." *Nuclear Instruments and Methods B* 266:1338–1342.

Barradas, N. P., Jeynes, C., and Webb, R. P. 1997. Simulated annealing analysis of Rutherford backscattering data. *Applied Physics Letters* 71:291–293.

Barradas, N. P., and Rauhala, E. 2009. Data analysis software for ion beam analysis. In *Handbook of modern ion beam analysis*, 2nd ed., eds. Y. Wang and M. Nastasi, 305–345. Warrendale, PA: Materials Research Society.

Børgesen P., Behrisch, R., and Scherzer, B. M. U. 1982. Depth profiling by ion-beam spectrometry. *Applied Physics A* 27:183–195.

Bozoian, M. 2009. Actual Coulomb barriers. In *Handbook of modern ion beam analysis,* 2nd ed., vol. 2, Appendix 8, eds. Y. Wang and M. Nastasi, 113–127. Warrendale, PA: Materials Research Society.

Campbell, J. L. 2009. Particle-induced x-ray emission: PIXE. In *Handbook of modern ion beam analysis,* 2nd ed., vol. 1, eds. Y. Wang and M. Nastasi, 231–245. Warrendale, PA: Materials Research Society.

Cohen, D. D., and Harrigan, M. 1985. K- and L-shell ionization cross sections for protons and helium ions calculated in the ECPSSR theory. *Atomic Data and Nuclear Data Tables* 33:255–343.

Doolittle, L. R. 1985. Algorithms for the rapid simulation of Rutherford backscattering spectra. *Nuclear Instruments and Methods B* 9:344–351.

———. 1986. A semiautomatic algorithm for Rutherford backscattering analysis. *Nuclear Instruments and Methods B* 15:227–231.

Eckstein, W., and Mayer, M. 1999. Rutherford backscattering from layered structures beyond the single scattering model. *Nuclear Instruments and Methods B* 153:337–344.

Fazinic S., ed. 2003. Intercomparison of PIXE spectrometry software packages, IAEA-TECDOC-1342, IAEA, Vienna.

Gärtner, K. 1997. Axial dechanneling in compound crystals with point defects and defect analysis by RBS. *Nuclear Instruments and Methods B* 132:147–158.

———. 2005. Modified master equation approach of axial dechanneling in perfect compound crystals. *Nuclear Instruments and Methods B* 227:522–530.

Gurbich, A. F. 1998. Evaluation of non-Rutherford proton elastic scattering cross section for silicon. *Nuclear Instruments and Methods B* 145:578–583.

———. 2009. Evaluated elastic scattering cross sections. In *Handbook of modern ion beam analysis,* 2nd ed., vol. 2, Appendix 9, eds. Y. Wang and M. Nastasi, 129–142. Warrendale, PA: Materials Research Society.

Hobler, G. 1995. Monte Carlo simulation of two-dimensional implanted dopant distributions at mask edges. *Nuclear Instruments and Methods B* 96:155–162.

Jeynes, C., and Barradas, N. P. 2009. Pitfalls in ion beam analysis. In *Handbook of modern ion beam analysis,* 2nd ed., vol. 1, eds. Y. Wang and M. Nastasi, 347–383. Warrendale, PA: Materials Research Society.

Jeynes, C., Barradas, N. P., Wilde, J. R., and Greer, A. L. 2000. Composition of TaNiC thick films using simulated annealing analysis of elastic backscattering spectrometry data. *Nuclear Instruments and Methods B* 161–163:287–292.

Johnston, P. N., Franich, R. D., Bubb, I. F., El Bouanani, M. Cohen, D. D., Dytlewski, N., and Siegele, R. 2000. The effects of large angle plural scattering on heavy ion elastic recoil detection analysis. *Nuclear Instruments and Methods B* 161–163:314–317.

Kling, A. 1995. CASSIS—A new Monte Carlo computer program for channeling simulations of RBS, NRA and PIXE. *Nuclear Instruments and Methods B* 102:141–144.

———. 2012. Refurbishment of the CASSIS code for channeling simulations. *Nuclear Instruments and Methods B* 273:88–90.

Kovač, D., and Hobler, G. 2006. Investigation of the impact of defect models on Monte Carlo simulations of RBS/C spectra. *Nuclear Instruments and Methods B* 249:776–779.

Maxwell, J. A., Teesdale, W. J., and Campbell, J. L. 1995. The Guelph-PIXE software package-II. *Nuclear Instruments and Methods B* 95:407–421.

Mayer, M. 1997. *SIMNRA user's guide, technical report IPP9/113*. Garching: Max Planck Institut für Plasmaphysik.

———. 2008. RESOLNRA: A new program for optimizing the achievable depth resolution of ion beam analysis methods. *Nuclear Instruments and Methods B* 266:1852–1857.

Mayer, M., Gauthier, E., Sugiyama, K., and von Toussaint, U. 2009. Quantitative depth profiling of deuterium up to very large depths. *Nuclear Instruments and Methods B* 267:506–512.

Molodtsov, S., and Gurbich, A. F. 2009. Simulation of the pulse pileup effect on the pulse-height spectrum. *Nuclear Instruments and Methods B* 267:3484–3487.

Nowicki, L., Turos, A. Ratajczak, R. Stonert, A., and Garrido, F. 2005. Modern analysis of ion channeling data by Monte Carlo simulations. *Nuclear Instruments and Methods B* 240:277–282.

Pascual-Izarra, C., Barradas, N. P., and Reis, M. A. 2006. LibCPIXE: A PIXE simulation open-source library for multilayered samples. *Nuclear Instruments and Methods B* 249:820–822.

Pascual-Izarra, C., Reis, M. A., and Barradas, N. P. 2006. Simultaneous PIXE and RBS data analysis using Bayesian inference with the DataFurnace code. *Nuclear Instruments and Methods B* 249:780–783.

Paul, H., and Schinner, A. 2003. Judging the reliability of stopping power tables and programs for heavy ions. *Nuclear Instruments and Methods B* 209:252–258.

———. 2005. Judging the reliability of stopping power tables and programs for protons and alpha particles using statistical methods. *Nuclear Instruments and Methods B* 227:461–470.

Rauhala, E., Barradas, N. P., Fazinic, S., Mayer, M., Szilágyi, E., and Thompson, M. 2006. Status of IBA data analysis and simulation software. *Nuclear Instruments and Methods B* 244:436–456.

Reis, M. A., and Alves, L. C. 1992. DATTPIXE, a computer package for TTPIXE data analysis. *Nuclear Instruments and Methods B* 68:300–304.

Reis, M. A., Alves, L. C., Barradas, N. P., Chaves, P. C., Nunes, B., Taborda, A., Surendran, K. P., Wu, A., Vilarinho, P. M., and Alves, E. 2010. High resolution and differential PIXE combined with RBS, EBS and AFM analysis of magnesium titanate (MgTiO3) multilayer structures. *Nuclear Instruments and Methods B* 268:1980–1985.

Schiettekatte, F. 2008. Fast Monte Carlo for ion beam analysis simulations. *Nuclear Instruments and Methods B* 266:1880–1885.

Smulders, P. J. M., and Boerma, D. O. 1987. Computer simulation of channeling in single crystals. *Nuclear Instruments and Methods B* 29:471–489.

Steinbauer, E., Bauer, P., and Biersack, J. 1990. Monte-Carlo simulation of RBS spectra—Comparison to experimental and empirical results. *Nuclear Instruments and Methods B* 45:171–175.

Szilágyi, E., Pászti, F., and Amsel, G. 1995. Theoretical approximations for depth resolution calculations in IBA methods. *Nuclear Instruments and Methods B* 100:103–121.

Taborda, A., Chaves, P. C., and Reis, M. A. 2011. Polynomial approximation to universal ionization cross-sections of K- and L-shell induced by H and He ion beams. *X-Ray Spectrometry* 40:127–134.

Turos, A., Nowicki, L., Stonert, A., Pagowska, K., Jagielski, J., and Muecklich, A. 2010. Monte Carlo simulations of ion channeling in crystals containing extended defects. *Nuclear Instruments and Methods B* 268:1718–1722.

Wielopolski, L., and Gardner, R. P. 1977. Generalized method for correcting pulse-height spectra for peak pileup effect due to double sum pulses. *Nuclear Instruments and Methods B* 140:297–303.

Ziegler, J. F., Biersack, J. P., and Ziegler, M. D. 2009. *SRIM, the stopping and range of ions in matter*. Morrisville: Lulu Press Co.

Ziegler, J. F., Lever, R. F., and Hirvonen, J. K. 1976. Computer analysis of nuclear backscattering. In *Ion beam surface layer analysis,* vol. 1, eds. O. Meyer, G. Linker, and F. Käppeler, 163–183. New York: Plenum.

Appendix A: Crystallography

A.1 CRYSTALLOGRAPHY AND NOTATION

A crystal is composed of atoms arranged in a periodic pattern in space and is defined by a set of lattice points. The space containing this set of points can be divided into a set of cells, each identical in size, shape, and orientation; such a cell is called a *unit cell*. This cell can be described by three unit vectors—**a**, **b**, and **c**—called *crystallographic axes*, which are related to each other in terms of their lengths a, b, and c, and the angles α, β, *and* γ (see Figure A.1). Any direction in the cell can be described as a linear combination of the three axes:

(A.1) $$r = n_1\mathbf{a} + n_2\mathbf{b} + n_3\mathbf{c},$$

where n_1, n_2, and n_3 are integers.

Seven different cells are necessary to describe all possible point lattices. These define the seven crystal systems listed in Table A.1. Each corner of the unit cell of these seven systems has a lattice point; a lattice point has the same surroundings in the lattice as every other lattice point. Based on this arrangement of points, a total of 14 Bravais lattices can be produced for the seven crystal systems. Figure A.2(a) shows the face-centered cubic (fcc) with dashed circles denoting atoms in the center of the hidden (back, side, and bottom) faces. Figure A.2(b) shows the associated diamond structure of silicon and germanium, which are two interpenetrating fcc lattices. The shaded atoms are the "extra" atoms in the fcc lattice that are included in the diamond structure.

A.2 DIRECTIONS AND PLANES

The direction of any line in a lattice may be described by drawing a line through the origin parallel to the given line and then assigning the coordinates u, v, w of any point on the line. The directions [u, v, w], written in square brackets, are the indices of the direction of the line. Since this line also goes through 2u, 2v, 2w, and 3u, 3v, 3w, and so on, it is customary to convert u, v, w to a set of smallest integers.

The orientation of planes in a lattice can also be defined by a set of numbers called the *Miller indices*. We can define the Miller indices of a plane as the reciprocals of the fractional intercepts that the plane makes with the crystallographic axes.

To determine the indices for the plane P in Figure A.3, we find its intercepts with the axes along the basis vectors **a**, **b**, and **c**. Let these intercepts be x, y,

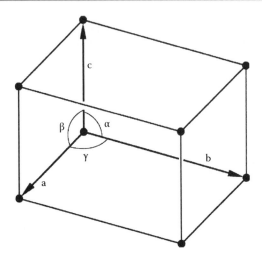

FIGURE A.1 Unit cell specified by the lengths of basis vectors a, b, and c; also by the angles between vectors.

TABLE A.1 Unit Cell, Seven Crystal Systems, and 14 Bravais Lattices

System	Axial Lengths and Angles	Bravais Lattice
Cubic	Three equal axes at right angles $a = b$ $= c, \alpha = \beta = \gamma = 90°$	Simple Body centered Face centered
Tetragonal	Three axes at right angles, two equal $a = b \neq c, \alpha = \beta = \gamma = 90°$	Simple Body centered
Orthorhombic	Three unequal axes at right angles $a \neq b \neq c, a \neq b \neq c, \alpha = \beta = \gamma \neq 90°$	Simple Body centered Base centered Face centered
Rhombohedral[a]	Three equal axes, equally inclined $a = b = c, \alpha = \beta = \gamma \neq 90°$	Simple
Hexagonal	Two equal coplanar axes at 120°, third axis at right angles $a = b \neq c$, $\alpha = \beta = 90°, \gamma = 120°$	Simple
Monoclinic	Three unequal axes, one pair not at right angles $a \neq b \neq c, \alpha = \gamma = 90° \neq \beta$	Simple Base centered
Triclinic	Three unequal axes, unequally inclined and none at right angles	Simple

[a] Also called trigonal.

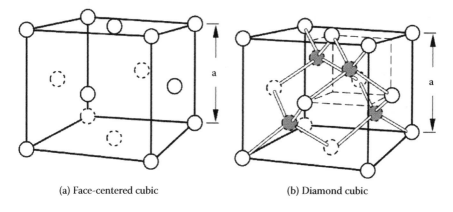

(a) Face-centered cubic   (b) Diamond cubic

FIGURE A.2 The face-centered cubic lattice and the diamond lattice of silicon with lattice parameters a.

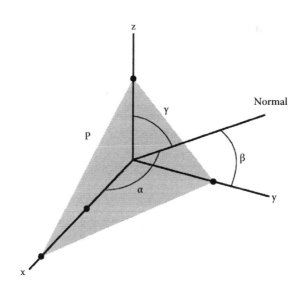

FIGURE A.3 The (122) plane and plane normal.

and z. Usually x is a fraction multiple of a, y a fraction multiple of b, and z a fraction multiple of c. The fraction triplet for these intercepts is

$$\left(\frac{x}{a}, \frac{y}{b}, \frac{z}{c}\right)$$

and we invert it to obtain the triplet

$$\left(\frac{a}{x}, \frac{b}{y}, \frac{c}{z}\right)$$

and then finally reduce this set to one composed of the smallest possible integers by multiplying by a common factor. This final set is called the Miller indices of the plane and is indicated by a set of integers (hkl). As an example consider a plan with intercepts $x = \frac{2}{3}a$, $y = 3b$, and $z = 1c$. From these intercepts we form the first triplet

$$\left[\frac{x}{a}, \frac{y}{b}, \frac{z}{c}\right] = \left(\frac{2}{3}, 3, 1\right)$$

and then invert it to form $(\frac{3}{2}, \frac{1}{3}, 1)$ and finally multiply by the lowest common denominator, which is 6, to obtain the Miller indices (926).

A.3 SPACING BETWEEN PLANES OF THE SAME MILLER INDICES

The formula for d_{hkl} is dependent on the crystal structure. For simplicity we will confine ourselves to the case in which the crystallographic axes are orthogonal. An example of various sets of planes for a two-dimensional cubic lattice is shown in Figure A.4. We can calculate the interplanar spacing by referring to Figure A.3, taking the origin as the location of another plane parallel to the one shown. The distance between these two planes, d_{hkl}, is simply the length of the normal vector between them. Let us define the angles that the normal vector makes with the a, b, and c axes as α, β, and γ, respectively. We also define the intercepts of the plane (*hkl*) with the a, b, and c axes as *x*, *y*, and *z*. From geometry and Figure A.3 we have

$$d_{hkl} = x \cos\alpha = y \cos\beta = z \cos\gamma$$

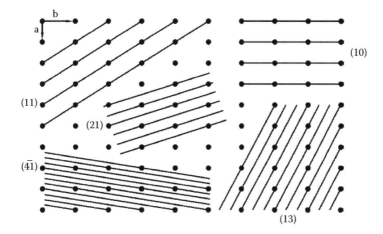

FIGURE A.4 Two-dimensional lattice showing that lines of lowest indices have the greatest spacing and the greatest density of lattice points.

where cos α, cos β, and cos γ are the directional cosines of the normal vector between the planes. Using the identity $1 = \cos^2 \alpha + \cos^2 \beta + \cos^2 \gamma$, together with the preceding equation to solve for d_{hkl} in terms of x, y, and z, gives

(A.2)
$$d_{hkl} = \frac{1}{\left(\dfrac{1}{x^2} + \dfrac{1}{y^2} + \dfrac{1}{z^2}\right)^{1/2}}.$$

From Section A.2 we know that x, y, and z are related to the Miller indices h, k, and l such that

(A.3)
$$h = n\frac{a}{x}, \quad k = n\frac{b}{y}, \quad l = n\frac{c}{z},$$

where n is the common factor used to reduce the indices to the smallest integers possible. Solving for x, y, and z from Equation (A.3) and substituting into Equation (A.2) and, considering the case of a simple cubic crystal (i.e., a = b = c), one obtains

(A.4)
$$d_{hkl} = \frac{na_c}{[h^2 + k^2 + l^2]^{1/2}}$$

which is the required formula. Thus, the interplanar distances of the (110) planes in a simple cubic crystal are $d = na_c/\sqrt{2}$, where a_c is the lattice parameter and represents the length of the unit cell cubic edge (Table A.2).

A.4 INTERATOMIC SPACING

An inspection of the periodic table of the elements indicates that the spacing between atoms of a solid (the nearest neighbor distance) varies by less than a factor of two for elements ranging between Al ($Z = 13$) and Bi ($Z = 83$). Table A.3 gives the element, atomic number, atomic density, crystal structure, nearest neighbor distance, and lattice parameters, a_c.

The nearest neighbor distance is calculated from crystallographic relations. For example, in the fcc case it represents the distance between the corner atoms in the unit cell and the atom in the face center, $\frac{\sqrt{2}}{2}a_c$. In the bcc structure it is the distance from the unit cell corners to the body center, $\frac{\sqrt{3}}{2}a_c$.

TABLE A.2 Conventions Used to Indicate Directions and Planes in Crystallographic Systems

A. Directions: line from origin to point at u, v, w:

  1. Specific directions are given in brackets. [u,v,w] .

  2. Indices uvw are the set of smallest integers. $\left[\frac{1}{2}\frac{1}{2}1\right]$ goes to [112].

  3. Negative indices are written with a bar, $\bar{u}vw$.

  4. Directions related by symmetry are given by $\langle uvw \rangle$. [111],[$\bar{1}$11], and [$\bar{1}$11] are all represented by $\langle 111 \rangle$.

B. Planes: plane that intercepts axes at $\frac{1\ \ 1\ \ 1}{h,\ k,\ l}$:

  1. Orientation is given by parentheses, (hkl).

  2. hkl are Miller indices.

  3. Negative indices are written with a bar, ($\bar{h}$kl).

  4. Planes related by symmetry are given by {h,k,l}. (100),(010), and ($\bar{1}$00) are planes of the form {100}.

C. In cubic systems: bcc, fcc, diamond:

  1. Direction [hkl] is perpendicular to plane (hkl).

  2. Interplanar spacing for cubic crystals: $d_{hkl} = a_c/\sqrt{h^2+k^2+l^2}$.

TABLE A.3   Characteristics of the Elements: Interatomic Spacing

Element	At. no. (Z)	At. Density (atom/cm³)	Crystal Structure	Lattice Parameter, a_c (nm)	Nearest Neighbor Distance (nm)
Al	13	6.02×10^{22}	fcc	0.40497	0.286
Si	14	5.0×10^{22}	Dia	0.54307	0.235
Ti[a]	22	5.66×10^{22}	bcc	0.33066	0.286
V	23	7.22×10^{22}	bcc	0.30232	0.262
Ni	28	9.14×10^{22}	fcc	0.35239	0.249
Cu	29	8.45×10^{22}	fcc	0.36148	0.256
Ge	32	4.41×10^{22}	Dia	0.56577	0.245
Nb	41	5.56×10^{22}	bcc	0.33067	0.286
Mo	42	6.42×10^{22}	bcc	0.31469	0.272
Pd	46	6.80×10^{22}	fcc	0.38908	0.275
Ag	47	5.85×10^{22}	fcc	0.40863	0.289
Au	79	5.90×10^{22}	fcc	0.40786	0.288

[a] Beta phase.

A.5   PLANE SPACINGS

The value of d, the distance between adjacent planes in the set (hkl), may be found from the following equations:

Cubic:
$$\frac{1}{d^2} = \frac{h^2 + k^2 + l^2}{a^2}$$

Tetragonal:
$$\frac{1}{d^2} = \frac{h^2 + k^2}{a^2} + \frac{l^2}{c^2}$$

Hexagonal:
$$\frac{1}{d^2} = \frac{4}{3}\left(\frac{h^2 + hk + k^2}{a^2}\right) + \frac{l^2}{c^2}$$

Rhombohedral:
$$\frac{1}{d^2} = \frac{(h^2 + k^2 + l^2)\,\sin^2\alpha + 2(hk + kl + hl)(\cos^2\alpha - \cos\alpha)}{a^2(1 - 3\cos^2\alpha + 2\cos^3\alpha)}$$

Orthorhombic:
$$\frac{1}{d^2} = \frac{h^2}{a^2} + \frac{k^2}{b^2} + \frac{l^2}{c^2}$$

Monoclinic: $\dfrac{1}{d^2} = \dfrac{1}{\sin^2 \beta}\left(\dfrac{h^2}{a^2} + \dfrac{k^2 \sin^2 \beta}{b^2} + \dfrac{l^2}{c^2} - \dfrac{2hl\cos \beta}{ac}\right)$

Triclinic: $\dfrac{1}{d^2} = \dfrac{1}{V^2}\left(S_{11}h^2 + S_{22}k^2 + S_{33}l^2 + 2S_{12}hk + 2S_{23}kl + 2S_{13}hl\right)$

In the equation for triclinic crystals, V = volume of unit cell (see following):

$$S_{11} = b^2c^2 \sin^2 \alpha$$
$$S_{22} = a^2c^2 \sin^2 \beta$$
$$S_{33} = a^2b^2 \sin^2 \gamma$$
$$S_{12} = abc^2 (\cos \alpha \cos \beta - \cos \gamma)$$
$$S_{23} = a^2bc (\cos \beta \cos \gamma - \cos \alpha)$$
$$S_{13} = ab^2c (\cos \gamma \cos \alpha - \cos \beta)$$

Appendix B: Rutherford Backscattering Spectrometry (RBS) Kinematic Factors

Tables B.1 and B.2 give the RBS kinematic factors K_{M2}, defined by Equation (2.9) in Chapter 2, for ^1H and ^4He as a projectile (M_1 = 1.0078 and 4.0026 amu) for the isotopic masses (M_2) of the elements. The kinematic factors are given as a function of scattering angle from 180° to 90° measured in the laboratory frame of reference. The first row for each element gives the average atomic weight of that element and the average kinematic factor (average K_{M2}) for that element. The subsequent rows give the isotopic masses for that element and the K_{M2} for those isotopic masses (only for light elements $Z_2 < 31$). Average K_{M2} is calculated as the weighted average of the K_{M2} values, Average K_{M2} $\sum_{M_2} \varpi_{M_2} K_{M2}$, where ϖ_{M_2} is the relative abundances for the isotopes. The data are taken from Appendix 5 of *Handbook of Modern Ion Beam Materials Analysis*, 2nd ed., eds. Y. Wang and M. Nastasi, MRS Publisher, Warrendale, PA, 2009.

TABLE B.1 Rutherford Backscattering Spectrometry Kinematic Factors for ^{1}H as a Projectile

Atomic no. (Z)	Element	Isotopic Mass (M_2) (amu)	180°	170°	160°	150°	140°	130°	120°	110°	100°	90°
3	Li	6.941	0.5568	0.5592	0.5666	0.5789	0.5960	0.6179	0.6443	0.6748	0.7090	0.7461
		6.015	0.5084	0.5110	0.5187	0.5317	0.5500	0.5733	0.6017	0.6348	0.6721	0.7130
		7.016	0.5607	0.5632	0.5705	0.5827	0.5998	0.6215	0.6477	0.6781	0.7120	0.7488
4	Be	9.012	0.6381	0.6403	0.6468	0.6576	0.6725	0.6913	0.7139	0.7397	0.7682	0.7998
5	B	10.811	0.6877	0.6897	0.6955	0.7051	0.7184	0.7352	0.7551	0.7778	0.8027	0.8293
		10.013	0.6677	0.6697	0.6758	0.6859	0.6999	0.7175	0.7385	0.7625	0.7889	0.8171
		11.009	0.6927	0.6946	0.7004	0.7099	0.7230	0.7396	0.7592	0.7816	0.8061	0.8323
6	C	12.011	0.7143	0.7161	0.7216	0.7306	0.7429	0.7585	0.7769	0.7979	0.8208	0.8452
		12.000	0.7141	0.7159	0.7214	0.7304	0.7427	0.7583	0.7768	0.7977	0.8207	0.8450
		13.003	0.7330	0.7347	0.7399	0.7484	0.7601	0.7748	0.7921	0.8118	0.8333	0.8561
7	N	14.007	0.7495	0.7512	0.7561	0.7641	0.7752	0.7891	0.8055	0.8241	0.8443	0.8658
		14.003	0.7495	0.7511	0.7560	0.7641	0.7752	0.7891	0.8055	0.8240	0.8443	0.8657
		15.000	0.7640	0.7656	0.7702	0.7779	0.7884	0.8016	0.8172	0.8347	0.8539	0.8741
8	O	15.999	0.7770	0.7785	0.7829	0.7902	0.8003	0.8128	0.8276	0.8442	0.8624	0.8815
		15.995	0.7770	0.7785	0.7829	0.7902	0.8002	0.8128	0.8275	0.8442	0.8623	0.8815
		16.999	0.7887	0.7901	0.7943	0.8013	0.8109	0.8228	0.8369	0.8527	0.8699	0.8881
		17.999	0.7992	0.8005	0.8046	0.8112	0.8204	0.8318	0.8452	0.8603	0.8767	0.8940
9	F	18.998	0.8087	0.8100	0.8138	0.8202	0.8290	0.8399	0.8527	0.8672	0.8828	0.8993
10	Ne	20.179	0.8187	0.8199	0.8236	0.8297	0.8380	0.8484	0.8606	0.8744	0.8892	0.9048
		19.992	0.8173	0.8185	0.8222	0.8284	0.8368	0.8472	0.8595	0.8733	0.8883	0.9040
		20.994	0.8252	0.8264	0.8300	0.8359	0.8439	0.8540	0.8658	0.8790	0.8934	0.9084
		21.991	0.8324	0.8336	0.8370	0.8427	0.8505	0.8601	0.8715	0.8842	0.8979	0.9124

11	Na	22.990	0.8391	0.8402	0.8435	0.8490	0.8565	0.8658	0.8767	0.8889	0.9022	0.9160
12	Mg	24.305	0.8470	0.8481	0.8512	0.8565	0.8636	0.8725	0.8829	0.8945	0.9071	0.9203
		23.985	0.8452	0.8463	0.8495	0.8548	0.8620	0.8710	0.8815	0.8933	0.9060	0.9194
		24.986	0.8509	0.8520	0.8551	0.8602	0.8671	0.8758	0.8860	0.8973	0.9096	0.9225
		25.983	0.8562	0.8572	0.8602	0.8652	0.8719	0.8803	0.8901	0.9011	0.9129	0.9253
13	Al	26.982	0.8512	0.8621	0.8650	0.8698	0.8763	0.8845	0.8939	0.9046	0.9160	0.9280
14	Si	28.086	0.8662	0.8672	0.8700	0.8746	0.8809	0.8887	0.8979	0.9081	0.9192	0.9307
		27.977	0.8658	0.8667	0.8695	0.8742	0.8805	0.8883	0.8975	0.9078	0.9189	0.9305
		28.976	0.8701	0.8710	0.8737	0.8782	0.8844	0.8920	0.9009	0.9108	0.9216	0.9328
		29.974	0.8741	0.8750	0.8777	0.8820	0.8880	0.8954	0.9040	0.9137	0.9241	0.9349
15	P	30.974	0.8779	0.8788	0.8814	0.8856	0.8914	0.8986	0.9070	0.9163	0.9264	0.9370
16	S	32.064	0.8818	0.8826	0.8852	0.8893	0.8949	0.9018	0.9100	0.9191	0.9288	0.9390
		31.972	0.8815	0.8823	0.8849	0.8890	0.8946	0.9016	0.9097	0.9188	0.9287	0.9389
		32.972	0.8849	0.8857	0.8882	0.8922	0.8976	0.9044	0.9124	0.9212	0.9307	0.9407
		33.968	0.8881	0.8889	0.8912	0.8952	0.9005	0.9071	0.9148	0.9234	0.9327	0.9424
		35.967	0.8939	0.8947	0.8970	0.9007	0.9057	0.9120	0.9194	0.9275	0.9363	0.9455
17	Cl	35.453	0.8924	0.8932	0.8955	0.8993	0.9044	0.9108	0.9182	0.9265	0.9354	0.9447
		34.969	0.8911	0.8919	0.8942	0.8980	0.9032	0.9096	0.9171	0.9255	0.9346	0.9440
		36.966	0.8967	0.8974	0.8996	0.9032	0.9082	0.9143	0.9214	0.9294	0.9380	0.9469
18	Ar	39.948	0.9040	0.9047	0.9067	0.9101	0.9147	0.9204	0.9271	0.9345	0.9425	0.9508
		35.967	0.8939	0.8947	0.8970	0.9007	0.9057	0.9120	0.9194	0.9275	0.9363	0.9455
		37.963	0.8992	0.9000	0.9021	0.9057	0.9105	0.9165	0.9234	0.9312	0.9396	0.9483
		39.962	0.9040	0.9047	0.9068	0.9102	0.9148	0.9205	0.9271	0.9345	0.9425	0.9508
19	K	39.098	0.9020	0.9027	0.9048	0.9083	0.9129	0.9188	0.9256	0.9331	0.9413	0.9497
		38.964	0.9017	0.9024	0.9045	0.9080	0.9127	0.9185	0.9253	0.9329	0.9411	0.9496

(Continued)

TABLE B.1 Rutherford Backscattering Spectrometry Kinematic Factors for ^{1}H as a Projectile (*Continued*)

Atomic no. (Z)	Element	Isotopic Mass (M_2) (amu)	180°	170°	160°	150°	140°	130°	120°	110°	100°	90°
20	Ca	40.000	0.9041	0.9048	0.9069	0.9102	0.9148	0.9205	0.9272	0.9346	0.9426	0.9508
		40.962	0.9063	0.9069	0.9089	0.9123	0.9167	0.9223	0.9288	0.9361	0.9439	0.9520
		40.078	0.9043	0.9050	0.9070	0.9104	0.9150	0.9207	0.9273	0.9347	0.9427	0.9509
		39.963	0.9040	0.9047	0.9068	0.9102	0.9148	0.9205	0.9271	0.9345	0.9425	0.9508
		41.959	0.9084	0.9090	0.9110	0.9142	0.9186	0.9241	0.9305	0.9376	0.9452	0.9531
		42.959	0.9104	0.9111	0.9130	0.9162	0.9205	0.9258	0.9320	0.9390	0.9464	0.9542
		43.956	0.9124	0.9130	0.9149	0.9180	0.9222	0.9274	0.9335	0.9403	0.9476	0.9552
		45.954	0.9160	0.9166	0.9184	0.9214	0.9255	0.9305	0.9363	0.9428	0.9498	0.9571
		47.952	0.9194	0.9199	0.9217	0.9246	0.9284	0.9333	0.9389	0.9451	0.9519	0.9588
21	Sc	44.956	0.9142	0.9148	0.9167	0.9197	0.9239	0.9290	0.9349	0.9416	0.9487	0.9561
22	Ti	47.878	0.9192	0.9198	0.9216	0.9244	0.9283	0.9332	0.9388	0.9450	0.9518	0.9588
		45.953	0.9160	0.9166	0.9184	0.9214	0.9254	0.9305	0.9363	0.9428	0.9498	0.9571
		46.952	0.9177	0.9183	0.9201	0.9230	0.9270	0.9319	0.9376	0.9440	0.9509	0.9580
		47.948	0.9194	0.9199	0.9217	0.9245	0.9284	0.9333	0.9389	0.9451	0.9519	0.9588
		48.948	0.9209	0.9215	0.9232	0.9260	0.9298	0.9346	0.9401	0.9462	0.9528	0.9597
		49.945	0.9224	0.9230	0.9247	0.9274	0.9312	0.9358	0.9413	0.9473	0.9537	0.9604
23	V	50.942	0.9239	0.9245	0.9261	0.9288	0.9325	0.9371	0.9424	0.9483	0.9546	0.9612
		49.947	0.9225	0.9230	0.9247	0.9275	0.9312	0.9358	0.9413	0.9473	0.9537	0.9604
24	Cr	50.944	0.9239	0.9245	0.9261	0.9288	0.9325	0.9371	0.9424	0.9483	0.9546	0.9612
		51.996	0.9254	0.9259	0.9275	0.9302	0.9338	0.9383	0.9435	0.9493	0.9555	0.9620
		49.946	0.9225	0.9230	0.9247	0.9275	0.9312	0.9358	0.9413	0.9473	0.9537	0.9604

Z	Element	Mass										
25	Mn	51.940	0.9253	0.9259	0.9275	0.9301	0.9338	0.9382	0.9434	0.9492	0.9555	0.9619
		52.941	0.9267	0.9272	0.9288	0.9314	0.9350	0.9394	0.9445	0.9502	0.9563	0.9626
		53.939	0.9280	0.9285	0.9301	0.9326	0.9361	0.9405	0.9455	0.9511	0.9571	0.9633
26	Fe	54.938	0.9292	0.9298	0.9313	0.9338	0.9373	0.9415	0.9464	0.9519	0.9578	0.9640
		55.847	0.9303	0.9309	0.9324	0.9349	0.9382	0.9424	0.9473	0.9527	0.9585	0.9645
		53.940	0.9280	0.9285	0.9301	0.9326	0.9361	0.9405	0.9455	0.9511	0.9571	0.9633
		55.935	0.9305	0.9310	0.9325	0.9350	0.9383	0.9425	0.9474	0.9528	0.9586	0.9646
		56.935	0.9316	0.9321	0.9336	0.9361	0.9394	0.9435	0.9483	0.9536	0.9593	0.9652
		57.933	0.9328	0.9333	0.9347	0.9371	0.9404	0.9444	0.9491	0.9544	0.9600	0.9658
27	Co	58.933	0.9339	0.9344	0.9358	0.9382	0.9414	0.9454	0.9500	0.9551	0.9607	0.9664
28	Ni	58.688	0.9336	0.9341	0.9355	0.9379	0.9411	0.9451	0.9498	0.9549	0.9605	0.9662
		57.935	0.9328	0.9333	0.9347	0.9371	0.9404	0.9444	0.9491	0.9544	0.9600	0.9658
		59.931	0.9349	0.9354	0.9368	0.9392	0.9423	0.9462	0.9508	0.9559	0.9613	0.9669
		60.931	0.9360	0.9364	0.9378	0.9401	0.9432	0.9471	0.9516	0.9566	0.9619	0.9675
		61.928	0.9370	0.9374	0.9388	0.9411	0.9441	0.9479	0.9523	0.9573	0.9625	0.9680
		63.928	0.9389	0.9393	0.9407	0.9429	0.9458	0.9495	0.9538	0.9586	0.9637	0.9690
29	Cu	63.546	0.9385	0.9390	0.9403	0.9425	0.9455	0.9492	0.9535	0.9583	0.9634	0.9688
		62.930	0.9379	0.9384	0.9398	0.9420	0.9450	0.9487	0.9531	0.9579	0.9631	0.9685
		64.928	0.9398	0.9402	0.9416	0.9437	0.9466	0.9503	0.9545	0.9592	0.9642	0.9694
30	Zn	65.396	0.9402	0.9406	0.9419	0.9441	0.9470	0.9506	0.9548	0.9595	0.9644	0.9696
		63.929	0.9389	0.9393	0.9407	0.9429	0.9458	0.9495	0.9538	0.9586	0.9637	0.9690
		65.926	0.9407	0.9411	0.9424	0.9445	0.9474	0.9510	0.9552	0.9598	0.9648	0.9699
		66.927	0.9415	0.9420	0.9433	0.9453	0.9482	0.9517	0.9558	0.9604	0.9653	0.9703
		67.925	0.9424	0.9428	0.9441	0.9461	0.9489	0.9524	0.9565	0.9610	0.9658	0.9708
		69.925	0.9440	0.9444	0.9456	0.9476	0.9504	0.9537	0.9577	0.9621	0.9667	0.9716

(Continued)

TABLE B.1   Rutherford Backscattering Spectrometry Kinematic Factors for ^1H as a Projectile (Continued)

Atomic no. (Z)	Element	Isotopic Mass (M_2) (amu)	180°	170°	160°	150°	140°	130°	120°	110°	100°	90°
31	Ga	69.723	0.9438	0.9442	0.9455	0.9475	0.9502	0.9536	0.9576	0.9619	0.9666	0.9715
32	Ge	72.632	0.9460	0.9464	0.9476	0.9495	0.9521	0.9554	0.9592	0.9634	0.9679	0.9726
33	As	74.922	0.9476	0.9480	0.9492	0.9510	0.9536	0.9568	0.9604	0.9645	0.9689	0.9735
34	Se	78.993	0.9502	0.9506	0.9517	0.9535	0.9559	0.9589	0.9624	0.9663	0.9705	0.9748
35	Br	79.904	0.9508	0.9512	0.9522	0.9540	0.9564	0.9594	0.9629	0.9667	0.9708	0.9751
36	Kr	83.800	0.9530	0.9534	0.9544	0.9561	0.9584	0.9612	0.9646	0.9682	0.9722	0.9762
37	Rb	85.468	0.9539	0.9543	0.9553	0.9569	0.9592	0.9620	0.9652	0.9688	0.9727	0.9767
38	Sr	87.617	0.9550	0.9554	0.9564	0.9580	0.9602	0.9629	0.9661	0.9696	0.9734	0.9773
39	Y	88.906	0.9557	0.9560	0.9570	0.9586	0.9608	0.9634	0.9666	0.9700	0.9737	0.9776
40	Zr	91.221	0.9568	0.9571	0.9580	0.9596	0.9617	0.9643	0.9674	0.9708	0.9744	0.9781
41	Nb	92.906	0.9575	0.9579	0.9588	0.9603	0.9624	0.9650	0.9680	0.9713	0.9749	0.9785
42	Mo	95.931	0.9588	0.9591	0.9600	0.9615	0.9636	0.9661	0.9690	0.9722	0.9756	0.9792
43	Tc	98.000	0.9597	0.9600	0.9609	0.9623	0.9643	0.9668	0.9696	0.9728	0.9761	0.9796
44	Ru	101.019	0.9609	0.9612	0.9620	0.9634	0.9654	0.9677	0.9705	0.9736	0.9768	0.9802
45	Rh	102.906	0.9616	0.9619	0.9627	0.9641	0.9660	0.9683	0.9710	0.9741	0.9773	0.9806
46	Pd	106.415	0.9628	0.9631	0.9639	0.9653	0.9671	0.9694	0.9720	0.9749	0.9780	0.9812
47	Ag	107.868	0.9633	0.9636	0.9644	0.9657	0.9675	0.9698	0.9724	0.9752	0.9783	0.9815
48	Cd	112.412	0.9648	0.9650	0.9658	0.9671	0.9688	0.9710	0.9735	0.9762	0.9792	0.9822
49	In	114.818	0.9655	0.9658	0.9665	0.9678	0.9695	0.9716	0.9740	0.9767	0.9796	0.9826
50	Sn	118.613	0.9666	0.9668	0.9676	0.9688	0.9704	0.9725	0.9748	0.9774	0.9802	0.9831
51	Sb	121.758	0.9674	0.9677	0.9684	0.9696	0.9712	0.9732	0.9755	0.9780	0.9808	0.9836

52	Te	127.586	0.9689	0.9691	0.9698	0.9709	0.9725	0.9744	0.9766	0.9790	0.9816	0.9843
53	I	126.905	0.9687	0.9690	0.9697	0.9708	0.9723	0.9742	0.9765	0.9789	0.9815	0.9842
54	Xe	131.293	0.9698	0.9700	0.9707	0.9718	0.9732	0.9751	0.9772	0.9796	0.9821	0.9848
55	Cs	132.905	0.9701	0.9703	0.9710	0.9721	0.9736	0.9754	0.9775	0.9799	0.9824	0.9849
56	Ba	137.327	0.9711	0.9713	0.9719	0.9730	0.9744	0.9762	0.9782	0.9805	0.9829	0.9854
57	La	138.905	0.9714	0.9716	0.9722	0.9733	0.9747	0.9764	0.9785	0.9807	0.9831	0.9856
58	Ce	140.115	0.9716	0.9719	0.9725	0.9735	0.9749	0.9766	0.9787	0.9809	0.9833	0.9857
59	Pr	140.908	0.9718	0.9720	0.9726	0.9737	0.9751	0.9768	0.9788	0.9810	0.9834	0.9858
60	Nd	144.242	0.9724	0.9726	0.9733	0.9743	0.9756	0.9773	0.9793	0.9814	0.9837	0.9861
61	Pm	145.000	0.9726	0.9728	0.9734	0.9744	0.9757	0.9774	0.9794	0.9815	0.9838	0.9862
62	Sm	150.360	0.9735	0.9737	0.9743	0.9753	0.9766	0.9782	0.9801	0.9822	0.9844	0.9867
63	Eu	151.965	0.9738	0.9740	0.9746	0.9756	0.9768	0.9784	0.9803	0.9824	0.9846	0.9868
64	Gd	157.252	0.9747	0.9749	0.9754	0.9764	0.9776	0.9792	0.9810	0.9829	0.9851	0.9873
65	Tb	158.925	0.9750	0.9751	0.9757	0.9766	0.9779	0.9794	0.9812	0.9831	0.9852	0.9874
66	Dy	162.498	0.9755	0.9757	0.9762	0.9771	0.9783	0.9798	0.9816	0.9835	0.9855	0.9877
67	Ho	164.930	0.9759	0.9760	0.9766	0.9775	0.9786	0.9801	0.9818	0.9837	0.9858	0.9879
68	Er	167.256	0.9762	0.9764	0.9769	0.9778	0.9789	0.9804	0.9821	0.9840	0.9860	0.9880
69	Tm	168.934	0.9764	0.9766	0.9771	0.9780	0.9791	0.9806	0.9823	0.9841	0.9861	0.9881
70	Yb	173.034	0.9770	0.9771	0.9777	0.9785	0.9796	0.9810	0.9827	0.9845	0.9864	0.9884
71	Lu	174.967	0.9772	0.9774	0.9779	0.9787	0.9799	0.9813	0.9829	0.9847	0.9866	0.9885
72	Hf	178.490	0.9777	0.9778	0.9783	0.9791	0.9803	0.9816	0.9832	0.9850	0.9868	0.9888
73	Ta	180.948	0.9780	0.9781	0.9786	0.9794	0.9805	0.9819	0.9834	0.9852	0.9870	0.9889
74	W	183.849	0.9783	0.9785	0.9790	0.9797	0.9808	0.9821	0.9837	0.9854	0.9872	0.9891
75	Re	186.207	0.9786	0.9787	0.9792	0.9800	0.9811	0.9824	0.9839	0.9856	0.9874	0.9892
77	Ir	192.216	0.9792	0.9794	0.9799	0.9806	0.9817	0.9829	0.9844	0.9860	0.9878	0.9896

(Continued)

TABLE B.1 Rutherford Backscattering Spectrometry Kinematic Factors for ^{1}H as a Projectile *(Continued)*

Atomic no. (Z)	Element	Isotopic Mass (M_2) (amu)	180°	170°	160°	150°	140°	130°	120°	110°	100°	90°
78	Pt	195.080	0.9795	0.9797	0.9802	0.9809	0.9819	0.9832	0.9846	0.9862	0.9879	0.9897
79	Au	196.967	0.9797	0.9799	0.9803	0.9811	0.9821	0.9833	0.9848	0.9864	0.9881	0.9898
80	Hg	200.588	0.9801	0.9803	0.9807	0.9814	0.9824	0.9836	0.9850	0.9866	0.9883	0.9900
81	Tl	204.383	0.9805	0.9806	0.9811	0.9818	0.9827	0.9839	0.9853	0.9869	0.9885	0.9902
82	Pb	207.217	0.9807	0.9809	0.9813	0.9820	0.9830	0.9841	0.9855	0.9870	0.9886	0.9903
83	Bi	208.980	0.9809	0.9810	0.9815	0.9822	0.9831	0.9843	0.9856	0.9871	0.9887	0.9904
84	Po	208.982	0.9809	0.9810	0.9815	0.9822	0.9831	0.9843	0.9856	0.9871	0.9887	0.9904
85	At	210.000	0.9810	0.9811	0.9816	0.9822	0.9832	0.9844	0.9857	0.9872	0.9888	0.9904
86	Rn	222.018	0.9820	0.9821	0.9825	0.9832	0.9841	0.9852	0.9865	0.9879	0.9894	0.9910
87	Fr	223.000	0.9821	0.9822	0.9826	0.9833	0.9842	0.9853	0.9865	0.9879	0.9894	0.9910
88	Ra	226.025	0.9823	0.9825	0.9829	0.9835	0.9844	0.9855	0.9867	0.9881	0.9896	0.9911
89	Ac	227.000	0.9824	0.9825	0.9829	0.9836	0.9844	0.9855	0.9868	0.9882	0.9896	0.9912
90	Th	232.038	0.9828	0.9829	0.9833	0.9839	0.9848	0.9858	0.9871	0.9884	0.9899	0.9914
91	Pa	231.036	0.9827	0.9828	0.9832	0.9839	0.9847	0.9858	0.9870	0.9884	0.9898	0.9913
92	U	238.018	0.9832	0.9833	0.9837	0.9843	0.9852	0.9862	0.9874	0.9887	0.9901	0.9916
93	Np	237.048	0.9831	0.9833	0.9836	0.9843	0.9851	0.9861	0.9873	0.9887	0.9901	0.9915
94	Pu	244.064	0.9836	0.9837	0.9841	0.9847	0.9855	0.9865	0.9877	0.9890	0.9904	0.9918
95	Am	243.061	0.9836	0.9837	0.9840	0.9846	0.9855	0.9865	0.9876	0.9889	0.9903	0.9917

TABLE B.2 Rutherford Backscattering Spectrometry Kinematic Factors for ⁴He as a Projectile

Atomic no. (Z)	Element	Isotopic Mass (M_2) (amu)	180°	170°	160°	150°	140°	130°	120°	110°	100°	90°
3	Li	6.941	0.0722	0.0735	0.0774	0.0845	0.0955	0.1115	0.1343	0.1662	0.2099	0.2680
		6.015	0.0404	0.0412	0.0438	0.0485	0.0560	0.0675	0.0846	0.1102	0.1476	0.2009
		7.016	0.0748	0.0761	0.0802	0.0874	0.0987	0.1151	0.1384	0.1708	0.2150	0.2735
4	Be	9.012	0.1482	0.1502	0.1564	0.1671	0.1832	0.2056	0.2356	0.2747	0.3241	0.3849
5	B	10.811	0.2111	0.2135	0.2208	0.2333	0.2516	0.2766	0.3090	0.3499	0.3999	0.4592
		10.013	0.1839	0.1861	0.1930	0.2049	0.2225	0.2465	0.2782	0.3186	0.3686	0.4288
		11.009	0.2178	0.2203	0.2277	0.2403	0.2589	0.2841	0.3167	0.3577	0.4076	0.4667
6	C	12.011	0.2501	0.2526	0.2604	0.2736	0.2928	0.3187	0.3518	0.3929	0.4423	0.5001
		12.000	0.2498	0.2523	0.2600	0.2733	0.2925	0.3183	0.3515	0.3926	0.4420	0.4998
		13.003	0.2801	0.2828	0.2908	0.3044	0.3240	0.3502	0.3835	0.4244	0.4731	0.5293
7	N	14.007	0.3086	0.3113	0.3194	0.3333	0.3531	0.3795	0.4127	0.4532	0.5009	0.5555
		14.003	0.3085	0.3112	0.3193	0.3332	0.3530	0.3794	0.4126	0.4531	0.5008	0.5554
		15.000	0.3349	0.3377	0.3459	0.3599	0.3798	0.4062	0.4392	0.4790	0.5257	0.5787
8	O	15.999	0.3597	0.3625	0.3708	0.3848	0.4047	0.4309	0.4635	0.5027	0.5483	0.5998
		15.995	0.3596	0.3624	0.3707	0.3847	0.4046	0.4308	0.4634	0.5026	0.5482	0.5997
		16.999	0.3830	0.3857	0.3940	0.4080	0.4278	0.4538	0.4860	0.5244	0.5689	0.6188
		17.999	0.4047	0.4074	0.4157	0.4296	0.4493	0.4750	0.5067	0.5443	0.5877	0.6362

(Continued)

TABLE B.2 **Rutherford Backscattering Spectrometry Kinematic Factors for ^{4}He as a Projectile (Continued)**

Atomic no. (Z)	Element	Isotopic Mass (M_2) (amu)	180°	170°	160°	150°	140°	130°	120°	110°	100°	90°
9	F	18.998	0.4251	0.4278	0.4360	0.4498	0.4693	0.4946	0.5258	0.5627	0.6050	0.6520
10	Ne	20.179	0.4473	0.4500	0.4582	0.4718	0.4910	0.5159	0.5464	0.5824	0.6234	0.6688
		19.992	0.4441	0.4468	0.4549	0.4686	0.4879	0.5128	0.5434	0.5795	0.6207	0.6664
		20.994	0.4621	0.4647	0.4728	0.4863	0.5053	0.5299	0.5599	0.5952	0.6354	0.6797
		21.991	0.4789	0.4816	0.4896	0.5029	0.5216	0.5458	0.5752	0.6098	0.6490	0.6920
11	Na	22.990	0.4948	0.4974	0.5053	0.5185	0.5370	0.5607	0.5896	0.6233	0.6615	0.7034
12	Mg	24.305	0.5142	0.5168	0.5245	0.5375	0.5555	0.5787	0.6069	0.6397	0.6766	0.7171
		23.985	0.5098	0.5123	0.5201	0.5331	0.5513	0.5746	0.6029	0.6360	0.6732	0.7140
		24.986	0.5240	0.5265	0.5342	0.5470	0.5649	0.5878	0.6155	0.6478	0.6842	0.7238
		25.983	0.5373	0.5399	0.5474	0.5600	0.5776	0.6001	0.6273	0.6589	0.6944	0.7330
13	Al	26.982	0.5500	0.5525	0.5600	0.5724	0.5897	0.6117	0.6384	0.6693	0.7040	0.7416
14	Si	28.086	0.5632	0.5657	0.5730	0.5852	0.6022	0.6238	0.6499	0.6801	0.7139	0.7505
		27.977	0.5620	0.5645	0.5718	0.5840	0.6010	0.6227	0.6489	0.6791	0.7130	0.7497
		28.976	0.5734	0.5759	0.5831	0.5951	0.6118	0.6331	0.6588	0.6884	0.7215	0.7573
		29.974	0.5843	0.5867	0.5938	0.6056	0.6220	0.6430	0.6681	0.6971	0.7294	0.7644
15	P	30.974	0.5946	0.5970	0.6040	0.6156	0.6318	0.6523	0.6770	0.7054	0.7370	0.7711
16	S	32.064	0.6053	0.6076	0.6145	0.6259	0.6418	0.6619	0.6861	0.7139	0.7448	0.7780
		31.972	0.6045	0.6068	0.6137	0.6251	0.6410	0.6612	0.6854	0.7132	0.7441	0.7775
		32.972	0.6139	0.6161	0.6229	0.6342	0.6498	0.6696	0.6934	0.7206	0.7509	0.7835
		33.968	0.6228	0.6250	0.6317	0.6428	0.6582	0.6776	0.7009	0.7277	0.7573	0.7892
		35.967	0.6395	0.6417	0.6482	0.6589	0.6738	0.6926	0.7150	0.7408	0.7692	0.7997

17	Cl	35.453	0.6353	0.6374	0.6440	0.6548	0.6698	0.6887	0.7114	0.7374	0.7662	0.7970
		34.969	0.6314	0.6336	0.6402	0.6511	0.6662	0.6853	0.7082	0.7344	0.7634	0.7946
		36.966	0.6474	0.6495	0.6559	0.6665	0.6811	0.6996	0.7216	0.7468	0.7747	0.8046
18	Ar	39.948	0.6689	0.6709	0.6770	0.6871	0.7010	0.7186	0.7395	0.7634	0.7897	0.8179
		35.967	0.6396	0.6417	0.6482	0.6589	0.6738	0.6926	0.7150	0.7408	0.7692	0.7997
		37.963	0.6549	0.6570	0.6633	0.6737	0.6880	0.7062	0.7279	0.7526	0.7800	0.8092
		39.962	0.6690	0.6710	0.6771	0.6872	0.7011	0.7187	0.7396	0.7635	0.7898	0.8179
19	K	39.098	0.6630	0.6651	0.6712	0.6815	0.6956	0.7134	0.7346	0.7589	0.7857	0.8142
		38.964	0.6621	0.6642	0.6703	0.6806	0.6947	0.7126	0.7339	0.7582	0.7850	0.8137
		40.000	0.6692	0.6713	0.6774	0.6874	0.7014	0.7189	0.7398	0.7637	0.7900	0.8181
		40.962	0.6756	0.6776	0.6836	0.6936	0.7073	0.7246	0.7451	0.7686	0.7944	0.8220
20	Ca	40.078	0.6697	0.6717	0.6778	0.6879	0.7018	0.7193	0.7402	0.7641	0.7903	0.8184
		39.963	0.6690	0.6710	0.6771	0.6872	0.7011	0.7187	0.7396	0.7635	0.7898	0.8179
		41.959	0.6820	0.6840	0.6899	0.6997	0.7132	0.7302	0.7504	0.7734	0.7988	0.8258
		42.959	0.6881	0.6901	0.6959	0.7055	0.7188	0.7356	0.7555	0.7781	0.8030	0.8295
		43.956	0.6940	0.6959	0.7017	0.7112	0.7243	0.7407	0.7603	0.7826	0.8070	0.8331
		45.954	0.7052	0.7071	0.7126	0.7218	0.7345	0.7505	0.7695	0.7910	0.8146	0.8398
		47.952	0.7156	0.7174	0.7228	0.7318	0.7441	0.7596	0.7780	0.7988	0.8217	0.8459
21	Sc	44.956	0.6997	0.7016	0.7073	0.7166	0.7295	0.7457	0.7650	0.7869	0.8109	0.8365
22	Ti	47.878	0.7152	0.7170	0.7224	0.7314	0.7437	0.7592	0.7776	0.7985	0.8214	0.8457
		45.953	0.7052	0.7071	0.7126	0.7218	0.7345	0.7505	0.7695	0.7910	0.8146	0.8398
		46.952	0.7105	0.7123	0.7178	0.7269	0.7394	0.7551	0.7738	0.7950	0.8182	0.8429
		47.948	0.7156	0.7174	0.7228	0.7318	0.7441	0.7596	0.7780	0.7988	0.8217	0.8459
		48.948	0.7205	0.7223	0.7276	0.7365	0.7486	0.7639	0.7820	0.8025	0.8250	0.8488
		49.945	0.7252	0.7270	0.7323	0.7410	0.7530	0.7680	0.7858	0.8060	0.8282	0.8516

(Continued)

TABLE B.2   Rutherford Backscattering Spectrometry Kinematic Factors for ^{4}He as a Projectile (Continued)

Atomic no. (Z)	Element	Isotopic Mass (M_2) (amu)	180°	170°	160°	150°	140°	130°	120°	110°	100°	90°
23	V	50.942	0.7298	0.7316	0.7368	0.7454	0.7572	0.7720	0.7896	0.8095	0.8312	0.8543
		49.947	0.7253	0.7270	0.7323	0.7410	0.7530	0.7680	0.7858	0.8061	0.8282	0.8516
		50.944	0.7298	0.7316	0.7368	0.7454	0.7572	0.7720	0.7896	0.8095	0.8312	0.8543
24	Cr	51.996	0.7345	0.7362	0.7414	0.7498	0.7615	0.7761	0.7934	0.8129	0.8344	0.8570
		49.946	0.7252	0.7270	0.7323	0.7410	0.7530	0.7680	0.7858	0.8061	0.8282	0.8516
		51.940	0.7343	0.7360	0.7411	0.7496	0.7613	0.7759	0.7932	0.8128	0.8342	0.8569
		52.941	0.7386	0.7403	0.7454	0.7537	0.7652	0.7796	0.7967	0.8160	0.8371	0.8594
		53.939	0.7428	0.7444	0.7494	0.7577	0.7690	0.7832	0.8001	0.8191	0.8399	0.8618
25	Mn	54.938	0.7468	0.7485	0.7534	0.7615	0.7727	0.7867	0.8033	0.8221	0.8425	0.8642
26	Fe	55.847	0.7504	0.7520	0.7569	0.7649	0.7760	0.7898	0.8062	0.8247	0.8449	0.8662
		53.940	0.7428	0.7444	0.7494	0.7577	0.7690	0.7832	0.8001	0.8191	0.8399	0.8618
		55.935	0.7507	0.7524	0.7572	0.7653	0.7763	0.7901	0.8065	0.8250	0.8451	0.8664
		56.935	0.7545	0.7561	0.7610	0.7689	0.7798	0.7934	0.8095	0.8278	0.8476	0.8686
		57.933	0.7582	0.7598	0.7646	0.7724	0.7831	0.7966	0.8125	0.8305	0.8501	0.8708
27	Co	58.933	0.7618	0.7634	0.7681	0.7758	0.7864	0.7997	0.8154	0.8331	0.8524	0.8728
28	Ni	58.688	0.7608	0.7624	0.7671	0.7749	0.7855	0.7989	0.8146	0.8324	0.8518	0.8723
		57.935	0.7582	0.7598	0.7646	0.7724	0.7831	0.7966	0.8125	0.8305	0.8501	0.8708
		59.931	0.7653	0.7668	0.7714	0.7791	0.7896	0.8027	0.8182	0.8356	0.8547	0.8748
		60.931	0.7686	0.7702	0.7747	0.7823	0.7926	0.8056	0.8209	0.8381	0.8569	0.8767
		61.928	0.7719	0.7734	0.7779	0.7854	0.7956	0.8084	0.8235	0.8405	0.8590	0.8786
		63.928	0.7782	0.7797	0.7841	0.7914	0.8013	0.8138	0.8285	0.8451	0.8631	0.8822

Z	Element	Mass										
29	Cu	63.546	0.7770	0.7785	0.7829	0.7902	0.8002	0.8128	0.8276	0.8442	0.8624	0.8815
		62.930	0.7751	0.7766	0.7811	0.7884	0.7985	0.8112	0.8260	0.8428	0.8611	0.8804
		64.928	0.7812	0.7827	0.7870	0.7942	0.8041	0.8164	0.8309	0.8473	0.8651	0.8839
30	Zn	65.396	0.7825	0.7840	0.7883	0.7955	0.8053	0.8175	0.8320	0.8482	0.8659	0.8846
		63.929	0.7782	0.7797	0.7841	0.7914	0.8013	0.8138	0.8285	0.8451	0.8631	0.8822
		65.926	0.7842	0.7856	0.7899	0.7970	0.8068	0.8189	0.8333	0.8494	0.8670	0.8855
		66.927	0.7870	0.7884	0.7927	0.7997	0.8094	0.8214	0.8356	0.8515	0.8689	0.8871
		67.925	0.7898	0.7912	0.7954	0.8024	0.8119	0.8238	0.8378	0.8535	0.8707	0.8887
		69.925	0.7952	0.7965	0.8007	0.8075	0.8168	0.8284	0.8420	0.8574	0.8741	0.8917
31	Ga	69.723	0.7946	0.7960	0.8001	0.8069	0.8162	0.8279	0.8416	0.8570	0.8738	0.8914
32	Ge	72.632	0.8019	0.8032	0.8073	0.8138	0.8229	0.8341	0.8474	0.8623	0.8785	0.8955
33	As	74.922	0.8074	0.8087	0.8126	0.8191	0.8279	0.8389	0.8518	0.8663	0.8820	0.8986
34	Se	78.993	0.8163	0.8176	0.8213	0.8275	0.8359	0.8464	0.8588	0.8727	0.8877	0.9035
35	Br	79.904	0.8183	0.8195	0.8232	0.8293	0.8377	0.8481	0.8603	0.8741	0.8890	0.9046
36	Kr	83.800	0.8259	0.8271	0.8307	0.8366	0.8446	0.8546	0.8664	0.8796	0.8938	0.9088
37	Rb	85.468	0.8290	0.8302	0.8337	0.8395	0.8474	0.8573	0.8688	0.8818	0.8958	0.9105
38	Sr	87.617	0.8329	0.8340	0.8375	0.8431	0.8509	0.8605	0.8718	0.8845	0.8982	0.9126
39	Y	88.906	0.8351	0.8362	0.8396	0.8452	0.8529	0.8624	0.8736	0.8861	0.8996	0.9138
40	Zr	91.221	0.8389	0.8400	0.8433	0.8488	0.8563	0.8656	0.8765	0.8888	0.9020	0.9159
41	Nb	92.906	0.8416	0.8427	0.8460	0.8514	0.8588	0.8679	0.8787	0.8907	0.9038	0.9174
42	Mo	95.931	0.8461	0.8472	0.8504	0.8556	0.8628	0.8718	0.8822	0.8939	0.9066	0.9199
43	Tc	98.000	0.8492	0.8503	0.8534	0.8585	0.8656	0.8743	0.8846	0.8961	0.9085	0.9215
44	Ru	101.019	0.8533	0.8543	0.8574	0.8624	0.8693	0.8778	0.8878	0.8990	0.9111	0.9237
45	Rh	102.906	0.8558	0.8569	0.8599	0.8648	0.8716	0.8800	0.8898	0.9008	0.9127	0.9251
46	Pd	106.415	0.8602	0.8612	0.8641	0.8689	0.8755	0.8837	0.8932	0.9039	0.9154	0.9275

(Continued)

TABLE B.2 Rutherford Backscattering Spectrometry Kinematic Factors for ^{4}He as a Projectile (Continued)

Atomic no. (Z)	Element	Isotopic Mass (M_2) (amu)	180°	170°	160°	150°	140°	130°	120°	110°	100°	90°
47	Ag	107.868	0.8620	0.8630	0.8659	0.8706	0.8771	0.8852	0.8946	0.9052	0.9165	0.9284
48	Cd	112.412	0.8672	0.8681	0.8709	0.8755	0.8817	0.8895	0.8986	0.9088	0.9198	0.9312
49	In	114.818	0.8698	0.8707	0.8735	0.8780	0.8841	0.8917	0.9007	0.9106	0.9214	0.9326
50	Sn	118.613	0.8737	0.8745	0.8772	0.8816	0.8876	0.8950	0.9037	0.9133	0.9238	0.9347
51	Sb	121.758	0.8767	0.8776	0.8802	0.8845	0.8903	0.8976	0.9060	0.9155	0.9257	0.9363
52	Te	127.586	0.8820	0.8828	0.8853	0.8894	0.8950	0.9020	0.9101	0.9192	0.9290	0.9391
53	I	126.905	0.8814	0.8823	0.8848	0.8889	0.8945	0.9015	0.9097	0.9188	0.9286	0.9388
54	Xe	131.293	0.8851	0.8860	0.8884	0.8924	0.8979	0.9046	0.9125	0.9214	0.9309	0.9408
55	Cs	132.905	0.8865	0.8873	0.8897	0.8937	0.8991	0.9058	0.9136	0.9223	0.9317	0.9415
56	Ba	137.327	0.8899	0.8907	0.8931	0.8969	0.9021	0.9086	0.9162	0.9247	0.9338	0.9434
57	La	138.905	0.8911	0.8919	0.8942	0.8980	0.9032	0.9096	0.9172	0.9256	0.9346	0.9440
58	Ce	140.115	0.8920	0.8928	0.8951	0.8988	0.9040	0.9104	0.9178	0.9262	0.9351	0.9445
59	Pr	140.908	0.8926	0.8933	0.8956	0.8994	0.9045	0.9109	0.9183	0.9266	0.9355	0.9448
60	Nd	144.242	0.8949	0.8956	0.8979	0.9016	0.9066	0.9128	0.9201	0.9282	0.9369	0.9460
61	Pm	145.000	0.8954	0.8962	0.8984	0.9021	0.9071	0.9133	0.9205	0.9286	0.9372	0.9463
62	Sm	150.360	0.8989	0.8997	0.9018	0.9054	0.9102	0.9162	0.9232	0.9310	0.9394	0.9481
63	Eu	151.965	0.9000	0.9007	0.9028	0.9064	0.9111	0.9171	0.9240	0.9317	0.9400	0.9487
64	Gd	157.252	0.9032	0.9039	0.9059	0.9093	0.9140	0.9197	0.9265	0.9339	0.9420	0.9504
65	Tb	158.925	0.9041	0.9048	0.9069	0.9103	0.9149	0.9206	0.9272	0.9346	0.9426	0.9509
66	Dy	162.498	0.9061	0.9068	0.9088	0.9121	0.9167	0.9222	0.9287	0.9360	0.9438	0.9519
67	Ho	164.930	0.9075	0.9081	0.9101	0.9134	0.9178	0.9233	0.9298	0.9369	0.9446	0.9526

68	Er	167.256	0.9087	0.9094	0.9113	0.9145	0.9189	0.9244	0.9307	0.9378	0.9454	0.9533
69	Tm	168.934	0.9096	0.9102	0.9122	0.9154	0.9197	0.9251	0.9314	0.9384	0.9459	0.9537
70	Yb	173.034	0.9116	0.9122	0.9141	0.9173	0.9215	0.9268	0.9329	0.9398	0.9471	0.9548
71	Lu	174.967	0.9125	0.9132	0.9151	0.9182	0.9224	0.9276	0.9337	0.9404	0.9477	0.9553
72	Hf	178.490	0.9142	0.9148	0.9167	0.9197	0.9238	0.9290	0.9349	0.9416	0.9487	0.9561
73	Ta	180.948	0.9153	0.9159	0.9178	0.9207	0.9248	0.9299	0.9358	0.9423	0.9494	0.9567
74	W	183.849	0.9166	0.9172	0.9190	0.9219	0.9260	0.9310	0.9368	0.9432	0.9502	0.9574
75	Re	186.207	0.9176	0.9182	0.9200	0.9229	0.9269	0.9318	0.9375	0.9439	0.9508	0.9579
76	Os	190.240	0.9193	0.9199	0.9216	0.9245	0.9284	0.9332	0.9388	0.9451	0.9518	0.9588
77	Ir	192.216	0.9201	0.9207	0.9224	0.9252	0.9291	0.9339	0.9394	0.9456	0.9523	0.9592
78	Pt	195.080	0.9212	0.9218	0.9235	0.9263	0.9301	0.9348	0.9403	0.9464	0.9530	0.9598
79	Au	196.967	0.9219	0.9225	0.9242	0.9270	0.9307	0.9354	0.9408	0.9469	0.9534	0.9602
80	Hg	200.588	0.9233	0.9238	0.9255	0.9282	0.9319	0.9365	0.9419	0.9478	0.9542	0.9609
81	Tl	204.383	0.9246	0.9252	0.9268	0.9295	0.9332	0.9377	0.9429	0.9488	0.9551	0.9616
82	Pb	207.217	0.9256	0.9262	0.9278	0.9304	0.9340	0.9385	0.9437	0.9495	0.9557	0.9621
83	Bi	208.980	0.9262	0.9268	0.9284	0.9310	0.9346	0.9390	0.9442	0.9499	0.9560	0.9624
84	Po	208.982	0.9262	0.9268	0.9284	0.9310	0.9346	0.9390	0.9442	0.9499	0.9560	0.9624
85	At	210.000	0.9266	0.9271	0.9287	0.9313	0.9349	0.9393	0.9444	0.9501	0.9562	0.9626
86	Rn	222.018	0.9304	0.9309	0.9324	0.9349	0.9383	0.9425	0.9473	0.9528	0.9586	0.9646
87	Fr	223.000	0.9307	0.9312	0.9327	0.9352	0.9386	0.9427	0.9476	0.9530	0.9587	0.9647
88	Ra	226.025	0.9316	0.9321	0.9336	0.9360	0.9394	0.9435	0.9483	0.9536	0.9593	0.9652
89	Ac	227.000	0.9319	0.9324	0.9339	0.9363	0.9396	0.9437	0.9485	0.9538	0.9595	0.9653
90	Th	232.038	0.9333	0.9338	0.9353	0.9376	0.9409	0.9449	0.9496	0.9548	0.9603	0.9661

(Continued)

TABLE B.2   Rutherford Backscattering Spectrometry Kinematic Factors for ⁴He as a Projectile (Continued)

Atomic no. (Z)	Element	Isotopic Mass (M₂) (amu)	180°	170°	160°	150°	140°	130°	120°	110°	100°	90°
91	Pa	231.036	0.9330	0.9335	0.9350	0.9374	0.9406	0.9447	0.9493	0.9546	0.9601	0.9659
92	U	238.018	0.9349	0.9354	0.9368	0.9392	0.9423	0.9462	0.9508	0.9559	0.9613	0.9669
93	Np	237.048	0.9347	0.9352	0.9366	0.9389	0.9421	0.9460	0.9506	0.9557	0.9611	0.9668
94	Pu	244.064	0.9365	0.9370	0.9384	0.9406	0.9437	0.9475	0.9520	0.9569	0.9622	0.9677
95	Am	243.061	0.9362	0.9367	0.9381	0.9404	0.9435	0.9473	0.9518	0.9568	0.9621	0.9676

Appendix C: Rutherford Cross Sections for 1 MeV ^{1}H and ^{4}He Ion Beams

The equation used to calculate the Rutherford cross sections is Equation (3.22) in Chapter 3. The mass for the scattering element used in the calculation is an average over the isotopic masses. For Rutherford cross sections at other energies, denoted by E, the values in the tables should be divided by a value of E^2 (where E takes units of MeV). In some cases, the Rutherford cross sections must be corrected for electron screening to obtain accurate scattering cross sections. The corrections, called F-factors, can be calculated from Equation (3.26) and are found in Table C.1, where E_{cm}, the center-of-mass kinetic energy in kiloelectron volts, is replaced by the laboratory energy, E_{lab}, with negligible error. Although the F-factor is relatively insensitive to angles normally used for backscattering, it is valid only for scattering angles greater than ~90°. Tables C.2 and C.3 list the calculated Rutherford cross sections for 1 MeV protons and ^{4}He ions at various backward angles. The data are taken from Appendix 6 of *Handbook of Modern Ion Beam Materials Analysis*, 2nd ed., eds. Y. Wang and M. Nastasi, MRS Publisher, Warrendale, PA, 2009.

TABLE C.1 Electron Screening F-Factors

E/Z_1 (MeV)	Z_2										
	8	14	20	29	32	39	47	56	73	79	92
0.100	0.992	0.983	0.973	0.956	0.950	0.935	0.917	0.895	0.851	0.834	0.796
0.200	0.996	0.992	0.987	0.978	0.975	0.968	0.958	0.948	0.925	0.917	0.898
0.300	0.997	0.994	0.991	0.985	0.983	0.978	0.972	0.965	0.950	0.945	0.932
0.400	0.998	0.996	0.993	0.989	0.988	0.984	0.979	0.974	0.963	0.958	0.949
0.500	0.998	0.997	0.995	0.991	0.990	0.987	0.983	0.979	0.970	0.967	0.959
0.600	0.999	0.997	0.996	0.993	0.992	0.989	0.986	0.983	0.975	0.972	0.966
0.700	0.999	0.998	0.996	0.994	0.993	0.991	0.988	0.985	0.979	0.976	0.971
0.800	0.999	0.998	0.997	0.995	0.994	0.992	0.990	0.987	0.981	0.979	0.975
0.900	0.999	0.998	0.997	0.995	0.994	0.993	0.991	0.988	0.983	0.982	0.977
1.000	0.999	0.998	0.997	0.996	0.995	0.994	0.992	0.990	0.985	0.983	0.980
1.100	0.999	0.998	0.998	0.996	0.995	0.994	0.992	0.990	0.986	0.985	0.981
1.200	0.999	0.999	0.998	0.996	0.996	0.995	0.993	0.991	0.988	0.986	0.983
1.300	0.999	0.999	0.998	0.997	0.996	0.995	0.994	0.992	0.989	0.987	0.984
1.400	0.999	0.999	0.998	0.997	0.996	0.995	0.994	0.993	0.989	0.988	0.985
1.500	0.999	0.999	0.998	0.997	0.997	0.996	0.994	0.993	0.990	0.989	0.986
1.600	1.000	0.999	0.998	0.997	0.997	0.996	0.995	0.993	0.991	0.990	0.987
1.700	1.000	0.999	0.998	0.997	0.997	0.996	0.995	0.994	0.991	0.990	0.988
1.800	1.000	0.999	0.999	0.998	0.997	0.996	0.995	0.994	0.992	0.991	0.989
1.900	1.000	0.999	0.999	0.998	0.997	0.997	0.996	0.994	0.992	0.991	0.989
2.000	1.000	0.999	0.999	0.998	0.998	0.997	0.996	0.995	0.993	0.992	0.990
2.100	1.000	0.999	0.999	0.998	0.998	0.997	0.996	0.995	0.993	0.992	0.990
2.200	1.000	0.999	0.999	0.998	0.998	0.997	0.996	0.995	0.993	0.992	0.991
2.300	1.000	0.999	0.999	0.998	0.998	0.997	0.996	0.995	0.994	0.993	0.991
2.400	1.000	0.999	0.999	0.998	0.998	0.997	0.997	0.996	0.994	0.993	0.992
2.500	1.000	0.999	0.999	0.998	0.998	0.997	0.997	0.996	0.994	0.993	0.992
2.600	1.000	0.999	0.999	0.998	0.998	0.998	0.997	0.996	0.994	0.994	0.992
2.700	1.000	0.999	0.999	0.998	0.998	0.998	0.997	0.996	0.994	0.994	0.992
2.800	1.000	0.999	0.999	0.998	0.998	0.998	0.997	0.996	0.995	0.994	0.993
2.900	1.000	0.999	0.999	0.998	0.998	0.998	0.997	0.996	0.995	0.994	0.993
3.000	1.000	0.999	0.999	0.999	0.998	0.998	0.997	0.997	0.995	0.994	0.993

Notes: E is the lab energy of the incident ion divided by its atomic number. Z_1 and Z_2 are the incident and target atoms' atomic numbers. The factors are only valid for scattering angles greater than 90°.

TABLE C.2 Rutherford Scattering Cross Sections (Barns/Steradian) of the Elements for 1 MeV ¹H (Protons)

Element	Atomic no. (Z_2)	Mass (M_2)	Scattering Angle (°)

Element	Atomic no. (Z_2)	Mass (M_2)	30	60	90	120	135	140	150	160	165	170	175
He	2	4.003	1.154E+00	8.223E−02	2.005E−02	8.550E−03	6.459E−03	5.995E−03	5.307E−03	4.868E−03	4.723E−03	4.623E−03	4.564E−03
Li	3	6.941	2.597E+00	1.860E−01	4.613E−02	2.023E−02	1.551E−02	1.446E−02	1.290E−02	1.190E−02	1.158E−02	1.135E−02	1.121E−02
Be	4	9.012	4.617E+00	3.310E−01	8.237E−02	3.632E−02	2.792E−02	2.606E−02	2.329E−02	2.151E−02	2.093E−02	2.052E−02	2.029E−02
B	5	10.811	7.215E+00	5.175E−01	1.289E−01	5.700E−02	4.388E−02	4.096E−02	3.663E−02	3.386E−02	3.295E−02	3.231E−02	3.194E−02
C	6	12.011	1.039E+01	7.453E−01	1.858E−01	8.223E−02	6.334E−02	5.914E−02	5.290E−02	4.891E−02	4.760E−02	4.669E−02	4.615E−02
N	7	14.007	1.414E+01	1.015E+00	2.532E−01	1.122E−01	8.645E−02	8.073E−02	7.224E−02	6.681E−02	6.502E−02	6.378E−02	6.305E−02
O	8	15.999	1.847E+01	1.326E+00	3.309E−01	1.467E−01	1.131E−01	1.056E−01	9.456E−02	8.746E−02	8.513E−02	8.350E−02	8.255E−02
F	9	18.998	2.338E+01	1.678E+00	4.190E−01	1.859E−01	1.434E−01	1.339E−01	1.199E−01	1.109E−01	1.080E−01	1.059E−01	1.047E−01
Ne	10	20.180	2.886E+01	2.072E+00	5.174E−01	2.296E−01	1.771E−01	1.655E−01	1.481E−01	1.370E−01	1.334E−01	1.309E−01	1.294E−01
Na	11	22.990	3.492E+01	2.507E+00	6.262E−01	2.780E−01	2.145E−01	2.004E−01	1.794E−01	1.660E−01	1.616E−01	1.585E−01	1.567E−01
Mg	12	24.305	4.156E+01	2.983E+00	7.453E−01	3.309E−01	2.553E−01	2.385E−01	2.136E−01	1.976E−01	1.924E−01	1.887E−01	1.866E−01
Al	13	26.982	4.878E+01	3.501E+00	8.749E−01	3.885E−01	2.998E−01	2.801E−01	2.508E−01	2.321E−01	2.259E−01	2.216E−01	2.191E−01
Si	14	28.086	5.657E+01	4.061E+00	1.015E+00	4.506E−01	3.478E−01	3.249E−01	2.909E−01	2.692E−01	2.621E−01	2.571E−01	2.542E−01
P	15	30.974	6.494E+01	4.662E+00	1.165E+00	5.174E−01	3.994E−01	3.731E−01	3.341E−01	3.092E−01	3.010E−01	2.953E−01	2.919E−01
S	16	32.066	7.388E+01	5.304E+00	1.326E+00	5.888E−01	4.544E−01	4.246E−01	3.802E−01	3.518E−01	3.425E−01	3.360E−01	3.322E−01
Cl	17	35.453	8.341E+01	5.988E+00	1.497E+00	6.648E−01	5.131E−01	4.794E−01	4.294E−01	3.973E−01	3.868E−01	3.794E−01	3.751E−01
Ar	18	39.948	9.351E+01	6.713E+00	1.678E+00	7.454E−01	5.754E−01	5.376E−01	4.815E−01	4.456E−01	4.338E−01	4.255E−01	4.207E−01
K	19	39.098	1.042E+02	7.480E+00	1.870E+00	8.305E−01	6.411E−01	5.990E−01	5.365E−01	4.964E−01	4.833E−01	4.741E−01	4.687E−01
Ca	20	40.078	1.154E+02	8.288E+00	2.072E+00	9.203E−01	7.104E−01	6.637E−01	5.944E−01	5.501E−01	5.355E−01	5.253E−01	5.194E−01
Sc	21	44.959	1.273E+02	9.138E+00	2.284E+00	1.015E+00	7.834E−01	7.319E−01	6.555E−01	6.066E−01	5.905E−01	5.793E−01	5.727E−01
Ti	22	47.880	1.397E+02	1.003E+01	2.507E+00	1.114E+00	8.598E−01	8.034E−01	7.195E−01	6.659E−01	6.482E−01	6.359E−01	6.287E−01
V	23	50.942	1.527E+02	1.096E+01	2.740E+00	1.217E+00	9.398E−01	8.781E−01	7.865E−01	7.278E−01	7.085E−01	6.951E−01	6.872E−01
Cr	24	51.996	1.662E+02	1.194E+01	2.983E+00	1.326E+00	1.023E+00	9.562E−01	8.564E−01	7.925E−01	7.715E−01	7.569E−01	7.483E−01
Mn	25	54.938	1.804E+02	1.295E+01	3.237E+00	1.438E+00	1.110E+00	1.038E+00	9.293E−01	8.600E−01	8.372E−01	8.213E−01	8.120E−01
Fe	26	55.847	1.951E+02	1.401E+01	3.501E+00	1.556E+00	1.201E+00	1.122E+00	1.005E+00	9.302E−01	9.055E−01	8.884E−01	8.783E−01

(Continued)

TABLE C.2   Rutherford Scattering Cross Sections (Barns/Steradian) of the Elements for 1 MeV ¹H (Protons) (Continued)

Element	Atomic no. (Z_2)	Mass (M_2)	Scattering Angle (°)										
			175	170	165	160	150	140	135	120	90	60	30
Co	27	58.933	9.472E−01	9.581E−01	9.766E−01	1.003E+00	1.084E+00	1.210E+00	1.295E+00	1.678E+00	3.776E+00	1.511E+01	2.104E+02
Ni	28	58.690	1.019E+00	1.030E+00	1.050E+00	1.079E+00	1.166E+00	1.302E+00	1.393E+00	1.804E+00	4.061E+00	1.625E+01	2.263E+02
Cu	29	63.546	1.093E+00	1.105E+00	1.127E+00	1.157E+00	1.251E+00	1.396E+00	1.494E+00	1.936E+00	4.356E+00	1.743E+01	2.427E+02
Zn	30	65.390	1.169E+00	1.183E+00	1.206E+00	1.239E+00	1.338E+00	1.494E+00	1.599E+00	2.072E+00	4.662E+00	1.865E+01	2.598E+02
Ga	31	69.723	1.249E+00	1.263E+00	1.288E+00	1.323E+00	1.429E+00	1.596E+00	1.708E+00	2.212E+00	4.978E+00	1.991E+01	2.774E+02
Ge	32	72.610	1.331E+00	1.346E+00	1.372E+00	1.409E+00	1.523E+00	1.700E+00	1.820E+00	2.357E+00	5.304E+00	2.122E+01	2.955E+02
As	33	74.922	1.415E+00	1.432E+00	1.459E+00	1.499E+00	1.620E+00	1.808E+00	1.935E+00	2.507E+00	5.641E+00	2.257E+01	3.143E+02
Se	34	78.960	1.502E+00	1.520E+00	1.549E+00	1.591E+00	1.719E+00	1.920E+00	2.054E+00	2.661E+00	5.988E+00	2.395E+01	3.336E+02
Br	35	79.904	1.592E+00	1.610E+00	1.641E+00	1.685E+00	1.822E+00	2.034E+00	2.177E+00	2.820E+00	6.345E+00	2.538E+01	3.536E+02
Kr	36	83.800	1.684E+00	1.704E+00	1.737E+00	1.784E+00	1.928E+00	2.152E+00	2.303E+00	2.983E+00	6.713E+00	2.685E+01	3.740E+02
Rb	37	85.468	1.779E+00	1.800E+00	1.834E+00	1.884E+00	2.036E+00	2.273E+00	2.433E+00	3.151E+00	7.091E+00	2.837E+01	3.951E+02
Sr	38	87.620	1.877E+00	1.898E+00	1.935E+00	1.988E+00	2.148E+00	2.398E+00	2.566E+00	3.324E+00	7.480E+00	2.992E+01	4.168E+02
Y	39	88.906	1.977E+00	2.000E+00	2.038E+00	2.094E+00	2.262E+00	2.526E+00	2.703E+00	3.501E+00	7.879E+00	3.152E+01	4.390E+02
Zr	40	91.224	2.080E+00	2.103E+00	2.144E+00	2.203E+00	2.380E+00	2.657E+00	2.844E+00	3.683E+00	8.288E+00	3.315E+01	4.618E+02
Nb	41	92.906	2.185E+00	2.210E+00	2.253E+00	2.314E+00	2.500E+00	2.792E+00	2.988E+00	3.870E+00	8.708E+00	3.483E+01	4.852E+02
Mo	42	95.940	2.293E+00	2.319E+00	2.364E+00	2.428E+00	2.624E+00	2.929E+00	3.135E+00	4.061E+00	9.138E+00	3.655E+01	5.091E+02
Tc	43	98.000	2.403E+00	2.431E+00	2.478E+00	2.545E+00	2.750E+00	3.071E+00	3.286E+00	4.257E+00	9.578E+00	3.831E+01	5.336E+02
Ru	44	101.070	2.516E+00	2.545E+00	2.594E+00	2.665E+00	2.880E+00	3.215E+00	3.441E+00	4.457E+00	1.003E+01	4.012E+01	5.588E+02
Rh	45	102.906	2.632E+00	2.662E+00	2.714E+00	2.788E+00	3.012E+00	3.363E+00	3.599E+00	4.662E+00	1.049E+01	4.196E+01	5.844E+02
Pd	46	106.420	2.750E+00	2.782E+00	2.836E+00	2.913E+00	3.148E+00	3.514E+00	3.761E+00	4.871E+00	1.096E+01	4.385E+01	6.107E+02
Ag	47	107.868	2.871E+00	2.904E+00	2.960E+00	3.041E+00	3.286E+00	3.669E+00	3.926E+00	5.086E+00	1.144E+01	4.577E+01	6.375E+02
Cd	48	122.411	2.995E+00	3.029E+00	3.088E+00	3.172E+00	3.427E+00	3.826E+00	4.095E+00	5.304E+00	1.194E+01	4.774E+01	6.650E+02
In	49	114.820	3.121E+00	3.157E+00	3.218E+00	3.305E+00	3.572E+00	3.987E+00	4.268E+00	5.528E+00	1.244E+01	4.975E+01	6.930E+02
Sn	50	118.710	3.250E+00	3.287E+00	3.350E+00	3.442E+00	3.719E+00	4.152E+00	4.444E+00	5.756E+00	1.295E+01	5.180E+01	7.215E+02
Sb	51	121.750	3.381E+00	3.420E+00	3.486E+00	3.581E+00	3.869E+00	4.320E+00	4.623E+00	5.988E+00	1.347E+01	5.390E+01	7.507E+02
Te	52	127.600	3.515E+00	3.555E+00	3.624E+00	3.723E+00	4.022E+00	4.491E+00	4.806E+00	6.225E+00	1.401E+01	5.603E+01	7.804E+02

53	I	126.904	3.651E+00	3.693E+00	3.765E+00	3.867E+00	4.179E+00	4.665E+00	4.993E+00	6.467E+00	1.455E+01	5.821E+01	8.107E+02
54	Xe	131.290	3.790E+00	3.834E+00	3.908E+00	4.015E+00	4.338E+00	4.843E+00	5.183E+00	6.713E+00	1.511E+01	6.042E+01	8.416E+02
55	Cs	132.905	3.932E+00	3.977E+00	4.054E+00	4.165E+00	4.500E+00	5.024E+00	5.377E+00	6.964E+00	1.567E+01	6.268E+01	8.731E+02
56	Ba	137.327	4.076E+00	4.123E+00	4.203E+00	4.317E+00	4.665E+00	5.208E+00	5.574E+00	7.220E+00	1.625E+01	6.498E+01	9.051E+02
57	La	138.906	4.223E+00	4.272E+00	4.354E+00	4.473E+00	4.833E+00	5.396E+00	5.775E+00	7.480E+00	1.683E+01	6.732E+01	9.377E+02
58	Ce	140.115	4.373E+00	4.423E+00	4.509E+00	4.631E+00	5.004E+00	5.587E+00	5.980E+00	7.745E+00	1.743E+01	6.971E+01	9.709E+02
59	Pr	140.908	4.525E+00	4.577E+00	4.665E+00	4.792E+00	5.178E+00	5.781E+00	6.187E+00	8.014E+00	1.803E+01	7.213E+01	1.005E+03
60	Nd	144.240	4.680E+00	4.734E+00	4.825E+00	4.956E+00	5.355E+00	5.979E+00	6.399E+00	8.288E+00	1.865E+01	7.460E+01	1.039E+03
61	Pm	145.000	4.837E+00	4.893E+00	4.987E+00	5.123E+00	5.535E+00	6.180E+00	6.614E+00	8.567E+00	1.928E+01	7.710E+01	1.074E+03
62	Sm	150.360	4.997E+00	5.054E+00	5.152E+00	5.292E+00	5.718E+00	6.384E+00	6.833E+00	8.850E+00	1.991E+01	7.965E+01	1.109E+03
63	Eu	151.965	5.159E+00	5.219E+00	5.320E+00	5.464E+00	5.904E+00	6.592E+00	7.055E+00	9.138E+00	2.056E+01	8.224E+01	1.146E+03
64	Gd	157.250	5.325E+00	5.386E+00	5.490E+00	5.639E+00	6.093E+00	6.803E+00	7.281E+00	9.430E+00	2.122E+01	8.488E+01	1.182E+03
65	Tb	158.925	5.492E+00	5.555E+00	5.663E+00	5.817E+00	6.285E+00	7.017E+00	7.510E+00	9.727E+00	2.189E+01	8.755E+01	1.219E+03
66	Dy	162.500	5.663E+00	5.728E+00	5.838E+00	5.997E+00	6.480E+00	7.235E+00	7.743E+00	1.003E+01	2.257E+01	9.026E+01	1.257E+03
67	Ho	164.930	5.835E+00	5.903E+00	6.017E+00	6.180E+00	6.678E+00	7.456E+00	7.979E+00	1.034E+01	2.325E+01	9.302E+01	1.296E+03
68	Er	167.260	6.011E+00	6.080E+00	6.197E+00	6.366E+00	6.879E+00	7.680E+00	8.219E+00	1.065E+01	2.395E+01	9.582E+01	1.335E+03
69	Tm	168.934	6.189E+00	6.260E+00	6.381E+00	6.555E+00	7.083E+00	7.907E+00	8.463E+00	1.096E+01	2.466E+01	9.866E+01	1.374E+03
70	Yb	173.040	6.370E+00	6.443E+00	6.567E+00	6.746E+00	7.290E+00	8.138E+00	8.710E+00	1.128E+01	2.538E+01	1.015E+02	1.414E+03
71	Lu	174.967	6.553E+00	6.628E+00	6.756E+00	6.940E+00	7.499E+00	8.372E+00	8.961E+00	1.161E+01	2.611E+01	1.045E+02	1.455E+03
72	Hf	178.490	6.739E+00	6.817E+00	6.948E+00	7.137E+00	7.712E+00	8.610E+00	9.215E+00	1.194E+01	2.685E+01	1.074E+02	1.496E+03
73	Ta	180.948	6.927E+00	7.007E+00	7.142E+00	7.337E+00	7.928E+00	8.851E+00	9.473E+00	1.227E+01	2.761E+01	1.104E+02	1.538E+03
74	W	183.850	7.119E+00	7.201E+00	7.340E+00	7.539E+00	8.146E+00	9.095E+00	9.734E+00	1.261E+01	2.837E+01	1.135E+02	1.580E+03
75	Re	186.207	7.312E+00	7.396E+00	7.539E+00	7.745E+00	8.368E+00	9.342E+00	9.999E+00	1.295E+01	2.914E+01	1.166E+02	1.623E+03
76	Os	190.200	7.509E+00	7.595E+00	7.742E+00	7.952E+00	8.593E+00	9.593E+00	1.027E+01	1.330E+01	2.992E+01	1.197E+02	1.667E+03
77	Ir	192.220	7.708E+00	7.796E+00	7.947E+00	8.163E+00	8.820E+00	9.847E+00	1.054E+01	1.365E+01	3.071E+01	1.229E+02	1.711E+03
78	Pt	195.080	7.909E+00	8.000E+00	8.154E+00	8.377E+00	9.051E+00	1.010E+01	1.081E+01	1.401E+01	3.152E+01	1.261E+02	1.756E+03
79	Au	196.967	8.113E+00	8.207E+00	8.365E+00	8.593E+00	9.285E+00	1.037E+01	1.109E+01	1.437E+01	3.233E+01	1.293E+02	1.801E+03
80	Hg	200.590	8.320E+00	8.416E+00	8.578E+00	8.812E+00	9.521E+00	1.063E+01	1.138E+01	1.473E+01	3.315E+01	1.326E+02	1.847E+03
81	Tl	204.383	8.529E+00	8.627E+00	8.794E+00	9.033E+00	9.761E+00	1.090E+01	1.166E+01	1.511E+01	3.399E+01	1.360E+02	1.894E+03

(Continued)

TABLE C.2   Rutherford Scattering Cross Sections (Barns/Steradian) of the Elements for 1 MeV ¹H (Protons) (Continued)

Element	Atomic no. (Z_2)	Mass (M_2)	Scattering Angle (°)										
			175	170	165	160	150	140	135	120	90	60	30
Pb	82	207.200	8.741E+00	8.842E+00	9.012E+00	9.258E+00	1.000E+01	1.117E+01	1.195E+01	1.548E+01	3.483E+01	1.393E+02	1.941E+03
Bi	83	208.980	8.956E+00	9.059E+00	9.233E+00	9.485E+00	1.025E+01	1.144E+01	1.225E+01	1.586E+01	3.569E+01	1.428E+02	1.988E+03
Po	84	209.000	9.173E+00	9.278E+00	9.457E+00	9.715E+00	1.050E+01	1.172E+01	1.254E+01	1.625E+01	3.655E+01	1.462E+02	2.036E+03
At	85	210.000	9.392E+00	9.500E+00	9.684E+00	9.948E+00	1.075E+01	1.200E+01	1.284E+01	1.663E+01	3.743E+01	1.497E+02	2.085E+03
Rn	86	222.000	9.615E+00	9.725E+00	9.913E+00	1.018E+01	1.100E+01	1.228E+01	1.315E+01	1.703E+01	3.831E+01	1.533E+02	2.135E+03
Fr	87	223.000	9.840E+00	9.953E+00	1.014E+01	1.042E+01	1.126E+01	1.257E+01	1.345E+01	1.743E+01	3.921E+01	1.568E+02	2.185E+03
Ra	88	226.025	1.007E+01	1.018E+01	1.038E+01	1.066E+01	1.152E+01	1.286E+01	1.377E+01	1.783E+01	4.012E+01	1.605E+02	2.235E+03
Ac	89	227.028	1.030E+01	1.042E+01	1.062E+01	1.091E+01	1.178E+01	1.316E+01	1.408E+01	1.824E+01	4.103E+01	1.641E+02	2.286E+03
Th	90	232.038	1.053E+01	1.065E+01	1.086E+01	1.115E+01	1.205E+01	1.345E+01	1.440E+01	1.865E+01	4.196E+01	1.678E+02	2.338E+03
Pa	91	231.036	1.077E+01	1.089E+01	1.110E+01	1.140E+01	1.232E+01	1.375E+01	1.472E+01	1.907E+01	4.290E+01	1.716E+02	2.390E+03
U	92	238.029	1.100E+01	1.113E+01	1.134E+01	1.165E+01	1.259E+01	1.406E+01	1.505E+01	1.949E+01	4.385E+01	1.754E+02	2.443E+03
Np	93	237.048	1.124E+01	1.137E+01	1.159E+01	1.191E+01	1.287E+01	1.437E+01	1.537E+01	1.991E+01	4.480E+01	1.792E+02	2.496E+03
Pu	94	244.000	1.149E+01	1.162E+01	1.184E+01	1.217E+01	1.315E+01	1.468E+01	1.571E+01	2.034E+01	4.577E+01	1.831E+02	2.550E+03

TABLE C.3   Rutherford Scattering Cross Sections (Barns/Steradian) of the Elements for 1 MeV ⁴He

Element	Atomic no. (Z_2)	Mass (M_2)	30	60	90	120	135	140	150	160	165	170	175
Li	3	6.941	1.036E+01	7.145E−01	1.524E−01	5.317E−02	3.466E−02	3.114E−02	2.609E−02	2.299E−02	2.199E−02	2.130E−02	2.090E−02
Be	4	9.012	1.844E+01	1.293E+00	2.971E−01	1.143E−01	8.201E−02	7.497E−02	6.465E−02	5.815E−02	5.603E−02	5.456E−02	5.370E−02
B	5	10.811	2.883E+01	2.036E+00	4.812E−01	1.945E−01	1.429E−01	1.316E−01	1.148E−01	1.042E−01	1.007E−01	9.833E−02	9.691E−02
C	6	12.011	4.152E+01	2.942E+00	7.033E−01	2.899E−01	2.152E−01	1.987E−01	1.743E−01	1.588E−01	1.537E−01	1.501E−01	1.481E−01
N	7	14.007	5.653E+01	4.020E+00	9.730E−01	4.096E−01	3.074E−01	2.848E−01	2.512E−01	2.298E−01	2.228E−01	2.179E−01	2.150E−01
O	8	15.999	7.384E+01	5.263E+00	1.284E+00	5.478E−01	4.139E−01	3.842E−01	3.402E−01	3.121E−01	3.028E−01	2.964E−01	2.926E−01
F	9	18.998	9.347E+01	6.676E+00	1.641E+00	7.086E−01	5.389E−01	5.012E−01	4.453E−01	4.095E−01	3.978E−01	3.896E−01	3.848E−01
Ne	10	20.180	1.154E+02	8.248E+00	2.031E+00	8.801E−01	6.705E−01	6.239E−01	5.548E−01	5.107E−01	4.961E−01	4.860E−01	4.800E−01
Na	11	22.990	1.397E+02	9.991E+00	2.469E+00	1.076E+00	8.225E−01	7.661E−01	6.824E−01	6.289E−01	6.113E−01	5.990E−01	5.918E−01
Mg	12	24.305	1.662E+02	1.190E+01	2.943E+00	1.286E+00	9.836E−01	9.165E−01	8.168E−01	7.531E−01	7.321E−01	7.175E−01	7.089E−01
Al	13	26.982	1.951E+02	1.397E+01	3.463E+00	1.518E+00	1.163E+00	1.084E+00	9.674E−01	8.926E−01	8.679E−01	8.508E−01	8.407E−01
Si	14	28.086	2.262E+02	1.620E+01	4.020E+00	1.764E+00	1.353E+00	1.261E+00	1.125E+00	1.039E+00	1.010E+00	9.901E−01	9.784E−01
P	15	30.974	2.597E+02	1.861E+01	4.623E+00	2.033E+00	1.561E+00	1.456E+00	1.300E+00	1.201E+00	1.168E+00	1.145E+00	1.131E+00
S	16	32.066	2.955E+02	2.118E+01	5.263E+00	2.316E+00	1.779E+00	1.660E+00	1.482E+00	1.369E+00	1.332E+00	1.306E+00	1.290E+00
Cl	17	35.453	3.336E+02	2.392E+01	5.950E+00	2.623E+00	2.017E+00	1.882E+00	1.682E+00	1.554E+00	1.512E+00	1.482E+00	1.465E+00
Ar	18	39.948	3.740E+02	2.682E+01	6.680E+00	2.950E+00	2.270E+00	2.119E+00	1.895E+00	1.751E+00	1.704E+00	1.671E+00	1.651E+00
K	19	39.098	4.167E+02	2.988E+01	7.441E+00	3.285E+00	2.528E+00	2.359E+00	2.109E+00	1.949E+00	1.897E+00	1.860E+00	1.838E+00
Ca	20	40.078	4.617E+02	3.311E+01	8.247E+00	3.642E+00	2.803E+00	2.616E+00	2.339E+00	2.162E+00	2.103E+00	2.063E+00	2.039E+00
Sc	21	44.959	5.091E+02	3.652E+01	9.102E+00	4.025E+00	3.100E+00	2.894E+00	2.588E+00	2.393E+00	2.328E+00	2.284E+00	2.257E+00
Ti	22	47.880	5.587E+02	4.008E+01	9.994E+00	4.422E+00	3.406E+00	3.181E+00	2.845E+00	2.631E+00	2.560E+00	2.511E+00	2.482E+00
V	23	50.942	6.107E+02	4.381E+01	1.093E+01	4.838E+00	3.728E+00	3.481E+00	3.114E+00	2.880E+00	2.803E+00	2.749E+00	2.717E+00
Cr	24	51.996	6.649E+02	4.771E+01	1.190E+01	5.269E+00	4.060E+00	3.792E+00	3.392E+00	3.137E+00	3.053E+00	2.995E+00	2.960E+00
Mn	25	54.938	7.215E+02	5.177E+01	1.292E+01	5.722E+00	4.410E+00	4.118E+00	3.685E+00	3.408E+00	3.317E+00	3.253E+00	3.216E+00
Fe	26	55.847	7.804E+02	5.600E+01	1.397E+01	6.190E+00	4.771E+00	4.455E+00	3.987E+00	3.687E+00	3.589E+00	3.520E+00	3.479E+00

(Continued)

TABLE C.3 Rutherford Scattering Cross Sections (Barns/Steradian) of the Elements for 1 MeV ⁴He (Continued)

Element	Atomic no. (Z_2)	Mass (M_2)	30	60	90	120	135	140	150	160	165	170	175
Co	27	58.933	8.416E+02	6.039E+01	1.507E+01	6.679E+00	5.149E+00	4.809E+00	4.303E+00	3.980E+00	3.874E+00	3.800E+00	3.756E+00
Ni	28	58.690	9.051E+02	6.495E+01	1.621E+01	7.183E+00	5.537E+00	5.171E+00	4.628E+00	4.280E+00	4.166E+00	4.086E+00	4.039E+00
Cu	29	63.546	9.709E+02	6.967E+01	1.739E+01	7.711E+00	5.945E+00	5.553E+00	4.970E+00	4.597E+00	4.475E+00	4.389E+00	4.339E+00
Zn	30	65.390	1.039E+03	7.456E+01	1.861E+01	8.254E+00	6.365E+00	5.945E+00	5.321E+00	4.922E+00	4.791E+00	4.699E+00	4.645E+00
Ga	31	69.723	1.109E+03	7.962E+01	1.988E+01	8.818E+00	6.800E+00	6.352E+00	5.686E+00	5.260E+00	5.120E+00	5.022E+00	4.965E+00
Ge	32	72.610	1.182E+03	8.484E+01	2.119E+01	9.398E+00	7.249E+00	6.771E+00	6.062E+00	5.608E+00	5.458E+00	5.354E+00	5.293E+00
As	33	74.922	1.257E+03	9.023E+01	2.253E+01	9.997E+00	7.711E+00	7.203E+00	6.448E+00	5.966E+00	5.807E+00	5.696E+00	5.631E+00
Se	34	78.960	1.335E+03	9.579E+01	2.392E+01	1.062E+01	8.189E+00	7.650E+00	6.849E+00	6.336E+00	6.167E+00	6.050E+00	5.981E+00
Br	35	79.904	1.414E+03	1.015E+02	2.535E+01	1.125E+01	8.679E+00	8.107E+00	7.258E+00	6.715E+00	6.536E+00	6.412E+00	6.338E+00
Kr	36	83.800	1.496E+03	1.074E+02	2.682E+01	1.191E+01	9.185E+00	8.580E+00	7.682E+00	7.107E+00	6.918E+00	6.786E+00	6.709E+00
Rb	37	85.468	1.580E+03	1.134E+02	2.834E+01	1.258E+01	9.703E+00	9.064E+00	8.116E+00	7.509E+00	7.309E+00	7.170E+00	7.088E+00
Sr	38	87.620	1.667E+03	1.197E+02	2.989E+01	1.327E+01	1.024E+01	9.563E+00	8.562E+00	7.922E+00	7.711E+00	7.564E+00	7.478E+00
Y	39	88.906	1.756E+03	1.260E+02	3.149E+01	1.398E+01	1.078E+01	1.007E+01	9.020E+00	8.345E+00	8.123E+00	7.969E+00	7.878E+00
Zr	40	91.224	1.847E+03	1.326E+02	3.312E+01	1.470E+01	1.134E+01	1.060E+01	9.490E+00	8.780E+00	8.547E+00	8.384E+00	8.288E+00
Nb	41	92.906	1.941E+03	1.393E+02	3.480E+01	1.545E+01	1.192E+01	1.114E+01	9.971E+00	9.226E+00	8.980E+00	8.810E+00	8.709E+00
Mo	42	95.940	2.036E+03	1.462E+02	3.652E+01	1.621E+01	1.251E+01	1.169E+01	1.047E+01	9.684E+00	9.426E+00	9.247E+00	9.141E+00
Tc	43	98.000	2.135E+03	1.532E+02	3.828E+01	1.700E+01	1.312E+01	1.225E+01	1.097E+01	1.015E+01	9.882E+00	9.694E+00	9.583E+00
Ru	44	101.070	2.235E+03	1.604E+02	4.009E+01	1.780E+01	1.373E+01	1.283E+01	1.149E+01	1.063E+01	1.035E+01	1.015E+01	1.004E+01
Rh	45	102.906	2.338E+03	1.678E+02	4.193E+01	1.862E+01	1.437E+01	1.342E+01	1.202E+01	1.112E+01	1.083E+01	1.062E+01	1.050E+01
Pd	46	106.420	2.443E+03	1.754E+02	4.382E+01	1.946E+01	1.501E+01	1.403E+01	1.256E+01	1.162E+01	1.131E+01	1.110E+01	1.097E+01
Ag	47	107.868	2.550E+03	1.831E+02	4.574E+01	2.031E+01	1.568E+01	1.464E+01	1.311E+01	1.213E+01	1.181E+01	1.159E+01	1.146E+01
Cd	48	122.411	2.660E+03	1.909E+02	4.772E+01	2.119E+01	1.636E+01	1.528E+01	1.369E+01	1.266E+01	1.233E+01	1.209E+01	1.196E+01
In	49	114.820	2.772E+03	1.990E+02	4.972E+01	2.208E+01	1.704E+01	1.592E+01	1.426E+01	1.319E+01	1.284E+01	1.260E+01	1.246E+01
Sn	50	118.710	2.886E+03	2.072E+02	5.177E+01	2.299E+01	1.775E+01	1.658E+01	1.485E+01	1.374E+01	1.337E+01	1.312E+01	1.297E+01
Sb	51	121.750	3.003E+03	2.156E+02	5.387E+01	2.393E+01	1.847E+01	1.725E+01	1.545E+01	1.430E+01	1.392E+01	1.365E+01	1.350E+01
Te	52	127.600	3.122E+03	2.241E+02	5.600E+01	2.488E+01	1.920E+01	1.794E+01	1.606E+01	1.486E+01	1.447E+01	1.420E+01	1.403E+01

Z	Element	Mass											
53	I	126.904	1.458E+01	1.475E+01	1.503E+01	1.544E+01	1.669E+01	1.863E+01	1.994E+01	2.584E+01	5.818E+01	2.328E+02	3.243E+03
54	Xe	131.290	1.514E+01	1.531E+01	1.561E+01	1.603E+01	1.732E+01	1.935E+01	2.071E+01	2.683E+01	6.040E+01	2.417E+02	3.366E+03
55	Cs	132.905	1.570E+01	1.588E+01	1.619E+01	1.663E+01	1.797E+01	2.007E+01	2.148E+01	2.783E+01	6.265E+01	2.507E+02	3.492E+03
56	Ba	137.327	1.628E+01	1.647E+01	1.679E+01	1.724E+01	1.863E+01	2.081E+01	2.227E+01	2.885E+01	6.496E+01	2.599E+02	3.620E+03
57	La	138.906	1.687E+01	1.706E+01	1.739E+01	1.787E+01	1.931E+01	2.156E+01	2.307E+01	2.989E+01	6.730E+01	2.693E+02	3.751E+03
58	Ce	140.115	1.746E+01	1.767E+01	1.801E+01	1.850E+01	1.999E+01	2.232E+01	2.389E+01	3.095E+01	6.968E+01	2.788E+02	3.884E+03
59	Pr	140.908	1.807E+01	1.828E+01	1.863E+01	1.914E+01	2.069E+01	2.310E+01	2.472E+01	3.203E+01	7.210E+01	2.885E+02	4.019E+03
60	Nd	144.240	1.869E+01	1.891E+01	1.927E+01	1.980E+01	2.139E+01	2.389E+01	2.557E+01	3.313E+01	7.457E+01	2.984E+02	4.156E+03
61	Pm	145.000	1.932E+01	1.954E+01	1.992E+01	2.046E+01	2.211E+01	2.469E+01	2.643E+01	3.424E+01	7.708E+01	3.084E+02	4.296E+03
62	Sm	150.360	1.996E+01	2.019E+01	2.058E+01	2.114E+01	2.285E+01	2.551E+01	2.730E+01	3.537E+01	7.963E+01	3.186E+02	4.438E+03
63	Eu	151.965	2.061E+01	2.085E+01	2.125E+01	2.183E+01	2.359E+01	2.634E+01	2.819E+01	3.652E+01	8.222E+01	3.289E+02	4.582E+03
64	Gd	157.250	2.127E+01	2.152E+01	2.193E+01	2.253E+01	2.435E+01	2.719E+01	2.910E+01	3.770E+01	8.485E+01	3.395E+02	4.729E+03
65	Tb	158.925	2.194E+01	2.220E+01	2.262E+01	2.324E+01	2.512E+01	2.804E+01	3.001E+01	3.888E+01	8.752E+01	3.502E+02	4.878E+03
66	Dy	162.500	2.262E+01	2.289E+01	2.333E+01	2.396E+01	2.590E+01	2.891E+01	3.095E+01	4.009E+01	9.024E+01	3.610E+02	5.029E+03
67	Ho	164.930	2.332E+01	2.358E+01	2.404E+01	2.470E+01	2.669E+01	2.980E+01	3.189E+01	4.131E+01	9.299E+01	3.720E+02	5.182E+03
68	Er	167.260	2.402E+01	2.429E+01	2.476E+01	2.544E+01	2.749E+01	3.069E+01	3.285E+01	4.256E+01	9.579E+01	3.832E+02	5.338E+03
69	Tm	168.934	2.473E+01	2.502E+01	2.550E+01	2.619E+01	2.830E+01	3.160E+01	3.383E+01	4.382E+01	9.863E+01	3.946E+02	5.496E+03
70	Yb	173.040	2.545E+01	2.575E+01	2.624E+01	2.696E+01	2.913E+01	3.253E+01	3.481E+01	4.510E+01	1.015E+02	4.061E+02	5.657E+03
71	Lu	174.967	2.619E+01	2.649E+01	2.700E+01	2.774E+01	2.997E+01	3.346E+01	3.582E+01	4.640E+01	1.044E+02	4.178E+02	5.820E+03
72	Hf	178.490	2.693E+01	2.724E+01	2.777E+01	2.852E+01	3.082E+01	3.441E+01	3.683E+01	4.772E+01	1.074E+02	4.297E+02	5.985E+03
73	Ta	180.948	2.768E+01	2.800E+01	2.854E+01	2.932E+01	3.169E+01	3.538E+01	3.786E+01	4.905E+01	1.104E+02	4.417E+02	6.152E+03
74	W	183.850	2.845E+01	2.878E+01	2.933E+01	3.013E+01	3.256E+01	3.635E+01	3.891E+01	5.040E+01	1.134E+02	4.539E+02	6.322E+03
75	Re	186.207	2.922E+01	2.956E+01	3.013E+01	3.095E+01	3.345E+01	3.734E+01	3.997E+01	5.178E+01	1.165E+02	4.662E+02	6.494E+03
76	Os	190.200	3.001E+01	3.036E+01	3.094E+01	3.179E+01	3.435E+01	3.835E+01	4.104E+01	5.317E+01	1.197E+02	4.787E+02	6.668E+03
77	Ir	192.220	3.081E+01	3.116E+01	3.176E+01	3.263E+01	3.526E+01	3.936E+01	4.213E+01	5.458E+01	1.228E+02	4.914E+02	6.845E+03
78	Pt	195.080	3.161E+01	3.198E+01	3.259E+01	3.348E+01	3.618E+01	4.039E+01	4.323E+01	5.600E+01	1.260E+02	5.043E+02	7.024E+03
79	Au	196.967	3.243E+01	3.280E+01	3.343E+01	3.435E+01	3.711E+01	4.144E+01	4.435E+01	5.745E+01	1.293E+02	5.173E+02	7.205E+03
80	Hg	200.590	3.325E+01	3.364E+01	3.429E+01	3.522E+01	3.806E+01	4.249E+01	4.548E+01	5.892E+01	1.326E+02	5.304E+02	7.389E+03
81	Tl	204.383	3.409E+01	3.448E+01	3.515E+01	3.611E+01	3.902E+01	4.356E+01	4.663E+01	6.040E+01	1.359E+02	5.438E+02	7.574E+03

(Continued)

TABLE C.3   Rutherford Scattering Cross Sections (Barns/Steradian) of the Elements for 1 MeV ^{4}He (Continued)

Element	Atomic no. (Z_2)	Mass (M_2)	175	170	165	160	150	140	135	120	90	60	30
							Scattering Angle (°)						
Pb	82	207.200	3.494E+01	3.534E+01	3.602E+01	3.701E+01	3.999E+01	4.465E+01	4.779E+01	6.190E+01	1.393E+02	5.573E+02	7.763E+03
Bi	83	208.980	3.580E+01	3.621E+01	3.691E+01	3.792E+01	4.097E+01	4.574E+01	4.896E+01	6.342E+01	1.427E+02	5.710E+02	7.953E+03
Po	84	209.000	3.667E+01	3.709E+01	3.780E+01	3.883E+01	4.196E+01	4.685E+01	5.015E+01	6.496E+01	1.462E+02	5.848E+02	8.146E+03
At	85	210.000	3.754E+01	3.798E+01	3.871E+01	3.976E+01	4.297E+01	4.797E+01	5.135E+01	6.651E+01	1.497E+02	5.988E+02	8.341E+03
Rn	86	222.000	3.844E+01	3.888E+01	3.963E+01	4.071E+01	4.399E+01	4.911E+01	5.256E+01	6.809E+01	1.532E+02	6.130E+02	8.538E+03
Fr	87	223.000	3.933E+01	3.979E+01	4.056E+01	4.166E+01	4.502E+01	5.026E+01	5.379E+01	6.968E+01	1.568E+02	6.273E+02	8.738E+03
Ra	88	226.025	4.024E+01	4.071E+01	4.149E+01	4.263E+01	4.606E+01	5.142E+01	5.504E+01	7.129E+01	1.604E+02	6.418E+02	8.940E+03
Ac	89	227.028	4.117E+01	4.164E+01	4.244E+01	4.360E+01	4.711E+01	5.260E+01	5.630E+01	7.292E+01	1.641E+02	6.565E+02	9.144E+03
Th	90	232.038	4.210E+01	4.258E+01	4.340E+01	4.459E+01	4.818E+01	5.379E+01	5.757E+01	7.457E+01	1.678E+02	6.714E+02	9.351E+03
Pa	91	231.036	4.304E+01	4.353E+01	4.437E+01	4.558E+01	4.925E+01	5.499E+01	5.886E+01	7.624E+01	1.716E+02	6.864E+02	9.560E+03
U	92	238.029	4.399E+01	4.450E+01	4.536E+01	4.659E+01	5.034E+01	5.621E+01	6.016E+01	7.793E+01	1.754E+02	7.015E+02	9.771E+03
Np	93	237.048	4.495E+01	4.547E+01	4.635E+01	4.761E+01	5.144E+01	5.744E+01	6.147E+01	7.963E+01	1.792E+02	7.169E+02	9.985E+03
Pu	94	244.000	4.592E+01	4.645E+01	4.735E+01	4.864E+01	5.256E+01	5.868E+01	6.280E+01	8.135E+01	1.831E+02	7.324E+02	1.020E+04

Appendix D: Proton-Induced Gamma Emissions from Light Elements ($3 \leq Z \leq 26$)

Table D.1 lists nuclear reactions, resonance energies, energies of emitted γ-rays, resonance cross sections, and resonance widths. The number of γ-ray energies is limited to those that are of the greatest aid in identifying the resonance. Therefore, they are generally the three highest energies, but where an exceptionally intense low-energy γ-ray occurs, it has been used to substitute one of the high-energy lines. Where the identification of the γ-ray transition in a compound nucleus is definite, the γ-ray energy listed was obtained by taking the difference between the excitation energies of the two states involved in the transition, instead of using the directly measured value. The cross section given is the total cross section in millibarns (mb) or the resonance strength (electron volts) at the resonance peak. Where more than one primary γ-ray is emitted, the tabulated value of the cross section is the sum of all such individual primary γ-ray cross sections. For those resonances that are too narrow for such cross-section measurements, the integrated cross section, $\int \sigma\, dE$, has been tabulated where this measurement has been made. In these instances the abbreviation "eV-b" for "electronvolt barn" has been inserted in the cross-section column. The width column gives the measured full width at half-maximum of the resonance in the laboratory system of coordinates. In most cases, this is the observed overall experimental width. However, for certain narrow well-known resonances, the tabulated width is the actual intrinsic resonance width; that is, it is a processed number that results after the factors such as beam width and Doppler broadening have been removed from the actual experimental value. The data are taken from Appendix 12 of *Handbook of Modern Ion Beam Materials Analysis,* 2nd ed., eds. Y. Wang and M. Nastasi, MRS Publisher, Warrendale, PA, 2009.

TABLE D.1 (p,γ) and (p,αγ) Resonances by Element

Element	Reaction	Ep (keV)	σ (mb) or Strength (eV)	Resonance Width (keV)	Eγ (MeV)	Relative Intensity
Li	^{7}Li(p,γ)^{8}Be	441	5.9 mb	12.2	17.64	63
					14.75	37
		1030		168	18.15	40
					15.25	60
		2060		310	16.15	100
Be	^{9}Be(p,γ)^{10}B	319	0.14 eV	120	6.15	21
					5.15	55
					4.75	11
					1.02	58
					0.72	84
		992	10.4 eV	82	7.5	~100
		1083	0.78 eV	2.65	6.85	85
					4.45	4.5
					3.01	9.5
					2.4	15
					0.72	96
B	^{11}B(p,γ)^{12}C	163	0.157 mb	5.2	16.11	3.5
					11.68	96.5
					4.43	96.5
	^{10}B(p,γ)^{11}C	675	0.050 mb	322	12.15	~100
		1146	0.0055 mb	450	9.7	
		1180	0.0075 mb	570	9.4	
	^{11}B(p,γ)^{12}C	1390	0.053 mb	1270	17.23	66
					12.80	34
					4.43	34

Element	Reaction	E (keV)	σ		Eγ (MeV)	(%)
C	$^{12}C(p,\gamma)^{13}N$	457	0.127 mb	35	2.36	100
	$^{13}C(p,\gamma)^{14}N$	551	1.44 mb	25	8.06	80
		1152	0.56 mb	4.1	8.62	23
					4.67	24
					2.42	40
					2.31	62
		1320	0.062 mb	440	8.71	~90
		1462	0.074 mb	17	5.83	18
					5.10	53
					3.08	84
	$^{12}C(p,\gamma)^{13}N$	1540	0.037 mb	9	8.98	~100
		1689	0.035 mb	67	3.51	95
					2.36	5
					1.14	5
	$^{13}C(p,\gamma)^{14}N$	1748	340 mb	0.075	9.17	86
					6.45	6
					2.73	6
					2.74	9
N	$^{14}N(p,\gamma)^{15}O$	278	0.014 eV	1.06	6.79	23
					6.18	58
					1.38	58
	$^{15}N(p,\gamma)^{16}O$	335	0.007 mb	110	12.44	~100
	$^{15}N(p,\alpha\gamma)^{12}C$	335	0.03 mb	110	4.43	100
		429	1560 mb	0.12	4.43	100
	$^{15}N(p,\gamma)^{16}O$	429	0.001 mb	0.103	6.40	60
					6.13	60
		710		40	6.72	

(Continued)

TABLE D.1   (p,γ) and (p,αγ) Resonances by Element (Continued)

Element	Reaction	Ep (keV)	σ (mb) or Strength (eV)	Resonance Width (keV)	Eγ (MeV)	Relative Intensity
	$^{15}N(p,\alpha\gamma)^{12}C$	897	800 mb	1.7	4.43	100
	$^{15}N(p,\alpha\gamma)^{12}C$	1028	15 mb	140	4.43	100
	$^{15}N(p,\alpha\gamma)^{12}C$	1028	1 mb	140	13.09	~100
	$^{14}N(p,\gamma)^{15}O$	1058	0.37 mb	3.9	8.28	53
					5.24	42
					3.04	42
	$^{15}N(p,\alpha\gamma)^{12}C$	1210	425 mb	22.5	4.43	100
	$^{14}N(p,\gamma)^{15}O$	1550	0.09 eV	34	6.18	36
					5.18	64
	$^{15}N(p,\alpha\gamma)^{12}C$	1640	340 mb	68	4.43	100
	$^{14}N(p,\gamma)^{15}O$	1742	0.06 eV	8	8.92	50
					5.18	39
	$^{15}N(p,\alpha\gamma)^{12}C$	1806	0.52 mb	4.2	8.98	94
		1979	35 mb	23	4.43	100
O	$^{18}O(p,\gamma)^{19}F$	630	0.10 eV	2.0	8.39	42
		841	1.4 eV	48	8.68	30
		1167	0.29 eV	0.05	6.32	47
					2.58	47
					0.197	>50
					0.110	>35
		1398	0.08 eV	3.6	9.32	30
		1684	0.025 eV	8	8.24	32
		1768	1.2 eV	3.8	9.67	22
		1928	2.8 eV	0.3	9.62	41
F	$^{19}F(p,\alpha\gamma)^{16}O$				γ1	γ2

Reaction	Energy	Cross section		6.13	6.72
$^{19}\text{F}(p,\gamma)^{22}\text{Ne}$	224	0.2 mb	0.94	100	
	340.5	102 mb	2.22	96.5	0.5
	484	32 mb	0.86	79	1
	594	7 mb	24	~100	
	668	57 mb	6.4	81	0.3
	832	19 mb	6.5		
	872	661 mb	4.3	68	24
	935	180 mb	7.7	76	3
	1088	13 mb	0.14		
	1136	15 mb	2.5		
	1280	29 mb	19	74	8
	1347	89 mb	4.7	55	14
	1371	300 mb	11.8	87	8
	1692		35		
	1350			1.63	
$^{22}\text{Ne}(p,\gamma)^{23}\text{Na}$	638	5.6 eV	0.065	9.406	73
				0.440	12
	851	14 eV	0.006	9.606	27
				9.166	42
				5.693	21
				0.440	>47
$^{20}\text{Ne}(p,\gamma)^{21}\text{Na}$	1169	1.6 eV	0.016	3.544	92
				1.828	6
				0.332	8

Ne

(Continued)

TABLE D.1 **(p,γ) and (p,αγ) Resonances by Element (*Continued*)**

Element	Reaction	Ep (keV)	σ (mb) or Strength (eV)	Resonance Width (keV)	Eγ (MeV)	Relative Intensity
Na	^{23}Na(p,γ)^{24}Mg	308.75	0.84 eV	<0.002	10.618	28
					7.748	46
					4.239	35
					1.369	>60
		512	0.73 eV	<0.05	10.813	71
					7.943	9
					1.369	>80
					10.9	
		592	1.9 eV	0.03	10.970	11
		677	5.1 eV	<0.07	8.100	45
					7.104	21
					4.239	35
					1.369	>52
	^{23}Na(p,αγ)^{20}Ne	1011	0.37 eV	<0.5	1.634	100
		1011	55 eV	<0.1	1.634	100
	^{23}Na(p,γ)^{24}Mg	1164	160 eV	1.2		
		1164	1.8 eV	1.4		
Mg	^{24}Mg(p,γ)^{25}Al	419	0.042 eV	<0.044	2.674	25
					2.222	31
					0.452	43
	^{26}Mg(p,γ)^{27}Al	454	0.0715 eV	<0.08	7.864	48
					7.695	16
		823	0.64 eV	1.3	0.843	>62
		840	1.9 eV	0.240	2.610	77
					0.452	83

Element	Reaction					
Al	$^{27}\text{Al}(p,\gamma)^{28}\text{Si}$	1549	5.8 eV	0.018	9.761	20
					8.748	30
					7.552	27
					1.013	>40
		406	0.00863 eV	<0.04	7.360	72
					2.835	>72
					1.779	>80
		632	2.6 eV	0.0067	10.416	74
					7.581	15
					1.779	>94
		992	24 eV	0.070	10.763	76
					4.744	10
					1.779	>93
Si	$^{29}\text{Si}(p,\gamma)^{30}\text{P}$	417	1.04 eV	~0.1	5.9	93
	$^{30}\text{Si}(p,\gamma)^{31}\text{P}$	620	4.4 eV	0.068	7.897	
P	$^{31}\text{P}(p,\gamma)^{32}\text{S}$	541	0.480 eV	<<1	7.159	80
					2.230	~100
		811	1.0 eV	<0.42	7.420	57
					4.956	38
					2.230	>84
		1251	4.3 eV	1.50	7.846	32
					3.852	42
					2.230	>90
S	$^{34}\text{S}(p,\gamma)^{35}\text{Cl}$	929	1 eV	0.014	7.274	72
					6.055	20
					1.219	24

(Continued)

TABLE D.1 (p,γ) and (p,αγ) Resonances by Element (Continued)

Element	Reaction	Ep (keV)	σ (mb) or Strength (eV)	Resonance Width (keV)	Eγ (MeV)	Relative Intensity
	$^{32}S(p,p'\gamma)^{32}S$	1211	9.7 eV	0.0105	4.385	97
					3.163	87
					2.23	
Cl	$^{37}Cl(p,\gamma)^{38}Ar$	3094	61 eV	0.34		
		766	4.3 eV	<1	8.820	44
					2.168	>60
Ar	$^{40}Ar(p,\gamma)^{41}K$	1087	2.1 eV	0.008	8.86	
Ca	$^{40}Ca(p,\gamma)^{41}Sc$	1842	0.280 mb		2.883	
Ti	$^{48}Ti(p,\gamma)^{49}V$	1007				7.582
		1013	<1.2 eV			7.650
Cr	$^{52}Cr(p,\gamma)^{53}Mn$	1005	0.89 eV	<0.1	2.5–4.7	
Ni	$^{58}Ni(p,\gamma)^{59}Cu$	1424	1.7 eV-b	<0.045	4.82	27
					4.33	50

Appendix E: Characteristic X-Ray Energies and Relative Intensities

The following data are taken from Appendix 16 of *Handbook of Modern Ion Beam Materials Analysis*, 2nd ed., eds. Y. Wang and M. Nastasi, MRS Publisher, Warrendale, PA, 2009.

TABLE E.1   Binding Energies (eV) for the K, L, and M Shells

Z	K	L1	L2	L3	M1	M2	M3	M4	M5
1	13.598								
2	24.587								
3	54.75	5.3							
4	111.9	8							
5	188	12.6	4.7						
6	284.1	10	9						
7	400.5	15	8.9	9.7					
8	532	23.7	6.8	7.4					
9	685.4	34	8.4	8.7					
10	870.1	48.47	21.66	21.56					
11	1072.1	63.3	31.1	31	0.7				
12	1305	89.4	51.5	51.3	2.1				
13	1559.6	117.7	73.15	72.72	0.7	0.5			
14	1838.9	148.7	99.5	98.9	7.6	3			
15	2145.5	189.3	136.2	135.3	16.2	9.6	10.1		
16	2472	229.2	165.4	164.2	15.8	7.8	8.2		
17	2822.4	270.2	201.6	200	17.5	6.7	6.7		
18	3202.9	326	250.55	248.5	29.24	15.93	15.76		
19	3607.4	377.1	296.3	293.6	33.9	18.1	17.8	0	0
20	4038.1	437.8	350	346.4	43.7	25.8	25.5	0	0
21	4492.8	500.4	406.7	402.2	53.8	33.8	31.5	6.6	0
22	4966.4	563.7	461.5	455.5	60.3	35.6	32.2	3.7	0
23	5465.1	628.2	520.5	512.9	66.5	40	35	2.2	0

24	5989.2	694.6	583.7	574.5	74.1	45.9	39.9	2.9	2.2
25	6539	769	651.4	640.3	83.9	53.1	46.4	3.5	2.7
26	7113	846.1	721.1	708.1	92.9	58.1	52	3.9	3.1
27	7708.9	925.6	793.6	778.6	100.7	63.2	57.7	2.7	3.3
28	8332.8	1008.1	871.9	854.7	111.8	71.2	69.7	3.9	3.3
29	8978.9	1096.1	951	931.1	119.8	75.3	72.8	1.8	1.5
30	9658.6	1193.6	1042.8	1019.7	135.9	88.6	85.6	7.9	8
31	10367.1	1297.7	1142.3	1115.4	158.1	106.8	102.9	20.7	15.7
32	11103.1	1414.3	1247.8	1216.7	180	127.9	120.8	29.2	28.5
33	11866.7	1526.5	1358.6	1323.1	203.5	146.4	140.5	41.7	40.9
34	12657.8	1653.9	1476.2	1435.8	231.5	168.2	161.9	57.4	56.4
35	13473.7	1782	1596	1549.9	256.5	189.3	181.5	70.1	69
36	14325.6	1921	1727.2	1674.9	292.1	222.1	214.4	95.04	93.82
37	15199.7	2065.1	1863.9	1804.4	322.1	247.4	238.5	111.8	110.3
38	16104.6	2216.3	2006.8	1939.6	357.5	279.8	269.1	135	133.1
39	17038.4	2372.5	2155.5	2080	393.6	312.4	300.3	159.6	157.4
40	17997.6	2531.6	2306.7	2222.3	430.3	344.2	330.5	182.4	180
41	18985.6	2697.7	2464.7	2370.5	468.4	378.4	363	207.4	204.6
42	19999.5	2865.5	2625.1	2520.2	504.6	409.7	392.3	230.3	227
43	21044	3042.5	2793.2	2676.9	544	444.9	425	256.4	252.9
44	22117.2	3224	2966.9	2837.9	585	482.8	460.6	283.6	279.4
45	23219.9	3411.9	3146.1	3003.8	627.1	521	496.2	311.7	307
46	24350.3	3604.3	3330.3	3173.3	669.9	559.1	531.5	340	334.7
47	25514	3805.8	3523.7	3351.1	717.5	602.4	571.4	372.8	366.7
48	26711.2	4018	3727	3537.5	770.2	650.7	616.5	410.5	403.7

(Continued)

TABLE E.1   Binding Energies (eV) for the K, L, and M Shells (Continued)

Z	K	L1	L2	L3	M1	M2	M3	M4	M5
49	27939.9	4237.5	3938	3730.1	825.6	702.2	664.3	450.8	443.1
50	29200.1	4464.7	4156.1	3928.8	883.8	756.4	714.4	493.3	484.8
51	30491.2	4698.3	4380.4	4132.3	943.7	811.9	765.6	536.9	527.5
52	31813.8	4939.2	4612	4341.4	1006	869.7	818.7	582.5	572.1
53	33169.4	5188.1	4852.1	4557.1	1072.1	930.5	874.6	631.3	619.4
54	34561.4	5452.8	5103.7	4782.2	1148.4	999	937	690.6	674.7
55	35984.6	5714.3	5359.4	5011.9	1217.1	1065	997.6	739.5	725.5
56	37440.6	5988.8	5623.6	5247	1292.8	1136.7	1062.2	796.1	780.7
57	38924.6	6266.3	5890.6	5482.5	1361.3	1204.4	1123.4	848.5	831.7
58	40443	6548.8	6164.2	5723.4	1434.6	1272.8	1185.4	901.3	883.3
59	41990.6	6834.8	6440.4	5964.3	1511	1337.4	1242.2	951.1	931
60	43568.9	7126	6721.5	6207.9	1575.3	1402.8	1297.4	999.9	977.7
61	45184	7427.9	7012.8	6459.3	1648.6	1471.4	1356.9	1051.5	1026.9
62	46834.2	7736.8	7311.8	6716.2	1722.8	1540.7	1419.8	1106	1080.2
63	48519	8052	7617.1	6976.9	1800	1613.9	1480.6	1160.6	1130.9
64	50239.1	8375.6	7930.3	7242.8	1880.8	1688.3	1544	1217.2	1185.2
65	51995.7	8708	8251.6	7514	1967.5	1767.7	1611.3	1275	1241.2
66	53788.5	9045.8	8580.8	7790.1	2046.8	1841.8	1675.6	1332.5	1294.9
67	55617.7	9394.2	8917.8	8071.1	2128.3	1922.8	1741.2	1391.5	1351.4
68	57485.5	9751.3	9264.3	8357.9	2216.7	2005.8	1811.8	1453.3	1409.3
69	59389.6	10115.7	9616.9	8648	2306.8	2089.8	1884.5	1514.6	1467.7
70	61332.3	10486.4	9978.2	8943.6	2398.1	2173	1949.8	1576.3	1527.8
71	63313.8	10870.4	10348.6	9244.1	2491.2	2263.5	2023.6	1639.4	1588.5

72	65350.8	11270.7	10739.4	9560.7	2600.9	2365.4	2107.6	1716.4	1661.7
73	67416.4	11681.5	11136.1	9881.1	2708	2468.7	2194	1793.2	1735.1
74	69525	12099.8	11544	10206.8	2819.6	2574.9	2281	1871.6	1809.2
75	71676.4	12526.7	11958.7	10535.3	2931.7	2681.6	2367.3	1948.9	1882.9
76	73870.8	12968	12385	10870.2	3048.5	2792.2	2457.2	2030.8	1960.1
77	76111	13418.5	12824.1	11215.2	3173.7	2908.7	2550.7	2116.1	2040.4
78	78394.8	13880.1	13272.6	11563.7	3297.2	3026.7	2645.7	2201.7	2121.4
79	80724.9	14352.8	13733.6	11918.7	3424.9	3147.8	2743	2291.1	2205.7
80	83102.3	14839.3	14208.7	12283.9	3561.6	3278.5	2847.1	2384.9	2294.9
81	85530.4	15346.7	14697.9	12657.5	3704.1	3415.7	2956.6	2485.1	2389.3
82	88004.5	15860.8	15200	13035.2	3850.7	3554.2	3066.4	2585.6	2484
83	90525.9	16387.5	15711.1	13418.6	3999.1	3696.3	3176.9	2687.6	2579.6
84	93099.9	16927.9	16238	13810.6	4153.5	3844.3	3293.4	2793.6	2679.2
85	95724	17481.5	16777.3	14208	4311.7	3995.8	3410.5	2901.8	2780.7
86	98397.2	18048.7	17329.7	14611.4	4474.3	4151.5	3530.5	3012.3	2884.2
87	101129.9	18634.1	17900.5	15025.6	4645.7	4316	3657.3	3129.7	2994.9
88	103916.2	19236.7	18484.3	15444.4	4822	4485	3786.6	3248.4	3104.9
89	106756.3	19845.9	19083	15871.2	5000.6	4656.8	3916.7	3370.1	3219.7
90	109649.1	20472.1	19693.2	16300.3	5182.3	4830.4	4046.1	3490.8	3332
91	112596.1	21111.4	20313.7	16729.1	5366.9	5002.7	4173.8	3606.4	3439.4
92	115600.6	21757.4	20947.6	17166.3	5548	5182.2	4303.4	3727.6	3551.7

Note: The energy of a particular x-ray diagram line is the difference between the binding energies of the levels involved. For example, the KL3 ($K\alpha_1$) line of iron has an energy of $7113 - 708 = 6405$ eV, and the L_3M_5 ($L\alpha_1$) line of gold has an energy of $11918.7 - 2205.7 = 9713$ eV.

TABLE E.2   K X-Ray Energies and Relative Intensities for Elements 6 ≤ Z ≤ 60

Line	C-6		N-7		O-8		F-9		Ne-10	
KL_2 ($K\alpha_2$)	0.275	0.3333	0.392	0.3335	0.525	0.3239	0.677	0.3282	0.848	0.3302
KL_3 ($K\alpha_1$)	0.284	0.6667	0.391	0.6665	0.525	0.6470	0.677	0.6551	0.849	0.6580

Line	Na-11		Mg-12		Al-13		Si-14		P-15	
KL_2 ($K\alpha_2$)	1.041	0.3308	1.253	0.3313	1.486	0.3268	1.739	0.3230	2.009	0.3188
KL_3 ($K\alpha_1$)	1.041	0.6604	1.254	0.6620	1.487	0.6536	1.740	0.6418	2.010	0.6286

Line	S-16		Cl-17		Ar-18		K-19		Ca-20	
KL_2 ($K\alpha_2$)	2.307	0.3122	2.621	0.3069	2.952	0.3011	3.311	0.2989	3.688	0.2962
KL_3 ($K\alpha_1$)	2.308	0.6179	2.622	0.6073	2.954	0.5971	3.314	0.5909	3.692	0.5854
KM_2 ($K\beta_3$)	2.464	0.0192	2.816	0.0248	3.187	0.0309	3.589	0.0343	4.012	0.0369
KM_3 ($K\beta_1$)	2.464	0.0382	2.816	0.0490	3.187	0.0607	3.590	0.0686	4.013	0.0739

Line	Sc-21		Ti-22		V-23		Cr-24		Mn-25	
KL_2 ($K\alpha_2$)	4.086	0.2977	4.505	0.2972	4.945	0.2971	5.405	0.2974	5.888	0.2970
KL_3 ($K\alpha_1$)	4.091	0.5875	4.511	0.5854	4.952	0.5844	5.415	0.5840	5.899	0.5828
KM_2 ($K\beta_3$)	4.459	0.0359	4.931	0.0369	5.425	0.0372	5.943	0.0373	6.486	0.0383
KM_3 ($K\beta_1$)	4.461	0.0713	4.934	0.0728	5.430	0.0736	5.949	0.0740	6.493	0.0755

Line	Fe-26		Co-27		Ni-28		Cu-29		Zn-30	
KL_2 ($K\alpha_2$)	6.392	0.2972	6.915	0.2973	7.461	0.2972	8.028	0.2975	8.616	0.2973
KL_3 ($K\alpha_1$)	6.405	0.5820	6.930	0.5812	7.478	0.5806	8.048	0.5797	8.639	0.5785
KM_2 ($K\beta_3$)	7.055	0.0389	7.646	0.0394	8.262	0.0398	8.904	0.0402	9.570	0.0409
KM_3 ($K\beta_1$)	7.061	0.0765	7.651	0.0772	8.263	0.0781	8.906	0.0787	9.573	0.0799

Line	Ga-31		Ge-32		As-33		Se-34		Br-35	
KL_2 $(K\alpha_2)$	9.225	0.2971	9.855	0.2957	10.508	0.2946	11.182	0.2933	11.878	0.2926
KL_3 $(K\alpha_1)$	9.252	0.5775	9.886	0.5743	10.544	0.5717	11.222	0.5683	11.924	0.5647
KM_2 $(K\beta_3)$	10.260	0.0414	10.975	0.0421	11.720	0.0429	12.490	0.0437	13.284	0.0443
KM_3 $(K\beta_1)$	10.264	0.0810	10.982	0.0825	11.726	0.0839	12.496	0.0855	13.292	0.0865

Line	Kr-36		Rb-37		Sr-38		Y-39		Zr-40	
KL_2 $(K\alpha_2)$	12.598	0.2916	13.336	0.2907	14.098	0.2899	14.883	0.2893	15.691	0.2884
KL_3 $(K\alpha_1)$	12.651	0.5628	13.395	0.5598	14.165	0.5569	14.958	0.5544	15.775	0.5520
KM_2 $(K\beta_3)$	14.103	0.0442	14.952	0.0448	15.825	0.0453	16.726	0.0459	17.653	0.0465
KM_3 $(K\beta_1)$	14.111	0.0865	14.961	0.0876	15.835	0.0886	16.738	0.0898	17.667	0.0909
KN_2 $(K\beta_2^{II})$	14.311	0.0044	15.185	0.0052	16.084	0.0058	17.013	0.0063	17.968	0.0067
KN_3 $(K\beta_2^{I})$	14.312	0.0085	15.186	0.0100	16.085	0.0114	17.016	0.0122	17.972	0.0130

Line	Nb-41		Mo-42		Tc-43		Ru-44		Rh-45	
KL_2 $(K\alpha_2)$	16.521	0.2880	17.374	0.2874	18.251	0.2868	19.150	0.2863	20.074	0.2861
KL_3 $(K\alpha_1)$	16.615	0.5499	17.479	0.5480	18.367	0.5463	19.279	0.5444	20.216	0.5428
KM_2 $(K\beta_3)$	18.607	0.0471	19.590	0.0476	20.599	0.0481	21.634	0.0485	22.699	0.0490
KM_3 $(K\beta_1)$	18.623	0.0919	19.607	0.0929	20.619	0.0937	21.657	0.0947	22.724	0.0953
KN_2 $(K\beta_2^{II})$	18.950	0.0069	19.961	0.0072	21.001	0.0075	22.070	0.0078	23.168	0.0080
KN_3 $(K\beta_2^{I})$	18.956	0.0135	19.967	0.0140	21.007	0.0146	22.076	0.0151	23.174	0.0156

Line	Pd-46		Ag-47		Cd-48		In-49		Sn-50	
KL_2 $(K\alpha_2)$	21.020	0.2860	21.990	0.2858	22.984	0.2857	24.002	0.2855	25.044	0.2856
KL_3 $(K\alpha_1)$	21.177	0.5414	22.163	0.5398	23.174	0.5380	24.210	0.5362	25.271	0.5341
KM_2 $(K\beta_3)$	23.791	0.0494	24.912	0.0498	26.060	0.0503	27.238	0.0506	28.444	0.0511
KM_3 $(K\beta1)$	23.819	0.0962	24.943	0.0971	26.095	0.0978	27.276	0.0985	28.486	0.0991

(Continued)

TABLE E.2   K X-Ray Energies and Relative Intensities for Elements $6 \leq Z \leq 60$ (Continued)

Line	Pd-46		Ag-47		Cd-48		In-49		Sn-50	
KN_2 ($K\beta_2''$)	24.296	0.0082	25.451	0.0085	26.640	0.0089	27.858	0.0092	29.106	0.0096
KN_3 ($K\beta_2'$)	24.300	0.0159	25.458	0.0165	26.646	0.0172	27.865	0.0179	29.114	0.0186

Line	Sb-51		Te-52		I-53		Xe-54		Cs-55	
KL_2 ($K\alpha_2$)	26.111	0.2853	27.202	0.2853	28.317	0.2848	29.458	0.2843	30.625	0.2834
KL_3 ($K\alpha_1$)	26.359	0.5321	27.472	0.5299	28.612	0.5284	29.779	0.5266	30.973	0.5254
KM_2 ($K\beta_3$)	29.679	0.0513	30.944	0.0516	32.239	0.0518	33.562	0.0520	34.920	0.0522
KM_3 ($K\beta_1$)	29.726	0.0996	30.995	0.1001	32.295	0.1004	33.624	0.1008	34.987	0.1011
KN_2 ($K\beta_2''$)	30.387	0.0099	31.697	0.0101	33.039	0.0104	34.397	0.0106	35.812	0.0108
KN_3 ($K\beta_2'$)	30.396	0.0193	31.717	0.0198	33.050	0.0202	34.405	0.0207	35.823	0.0212

Line	Ba-56		La-57		Ce-58		Pr-59		Nd-60	
KL_2 ($K\alpha_2$)	31.817	0.2835	33.034	0.2835	34.279	0.2815	35.550	0.2844	36.847	0.2844
KL_3 ($K\alpha_1$)	32.194	0.5231	33.442	0.5213	34.720	0.5167	36.026	0.5187	37.361	0.5177
KM_2 ($K\beta_3$)	36.304	0.0524	37.720	0.0526	39.170	0.0525	40.653	0.0531	42.166	0.0533
KM_3 ($K\beta_1$)	36.378	0.1016	37.801	0.1019	39.258	0.1016	40.748	0.1027	42.271	0.1031
KN_2 ($K\beta_2''$)	37.249	0.0111	38.719	0.0114	40.220	0.0114	41.754	0.0116	43.326	0.0117
KN_3 ($K\beta_2'$)	37.261	0.0218	38.733	0.0223	40.236	0.0222	41.773	0.0227	43.344	0.0229

Note: Only high-intensity K x-ray lines are included. The two columns of data for each element contain the energy of the transition line in kiloelectron volts and the relative intensity of each line, respectively. The intensities do not sum to 1.0 since some of the weak lines are not listed.

TABLE E.3 L X-Ray Energies and Relative Intensities for Elements 26 ≤ Z ≤ 92

Line	Fe-26		Co-27		Ni-28		Cu-29		Zn-30	
$L_3M_5(L\alpha_1)$	0.704	0.8137	0.775	0.8268	0.849	0.8364	0.928	0.8474	1.009	0.8499
$L_2M_4(L\beta_1)$	0.717	0.9103	0.790	0.9248	0.866	0.9358	0.948	0.9481	1.032	0.9507
$L_2N_4(L\gamma_1)$	—	—	—	—	—	—	—	—	—	—
$L_3N_5(L\beta_2)$	—	—	—	—	—	—	—	—	—	—
$L_1M_3(L\beta_3)$	0.794	0.6375	0.868	0.6375	0.938	0.6369	1.023	0.6358	1.108	0.6352

Line	Ga-31		Ge-32		As-33		Se-34		Br-35	
$L_3M_5(L\alpha_1)$	1.096	0.8524	1.186	0.8545	1.282	0.8559	1.379	0.8570	1.480	0.8580
$L_2M_4(L\beta_1)$	1.122	0.9535	1.216	0.9557	1.317	0.9575	1.419	0.9589	1.526	0.9601
$L_2N_4(L\gamma_1)$	—	—	—	—	—	—	—	—	—	—
$L_3N_5(L\beta_2)$	—	—	—	—	—	—	—	—	—	—
$L_1M_3(L\beta_3)$	1.195	0.6345	1.294	0.6187	1.386	0.6068	1.492	0.5950	1.601	0.5817

Line	Kr-36		Rb-37		Sr-38		Y-39		Zr-40	
$L_3M_5(L\alpha_1)$	1.587	0.8587	1.694	0.8593	1.806	0.8596	1.922	0.8601	2.042	0.8468
$L_2M_4(L\beta_1)$	1.638	0.9610	1.752	0.9617	1.872	0.9622	1.996	0.9629	2.124	0.9492
$L_2N_4(L\gamma_1)$	—	—	—	—	—	—	—	—	2.302	0.0150
$L_3N_5(L\beta_2)$	—	—	—	—	—	—	—	—	2.219	0.0134
$L_1M_3(L\beta_3)$	1.707	0.5685	1.827	0.5588	1.947	0.5507	2.072	0.5458	2.201	0.5412

Line	Nb-41		Mo-42		Tc-43		Ru-44		Rh-45	
$L_3M_5(L\alpha_1)$	2.166	0.8356	2.293	0.8272	2.424	0.8185	2.558	0.8097	2.696	0.8001
$L_2M_4(L\beta_1)$	2.257	0.9356	2.395	0.9266	2.538	0.9170	2.683	0.9073	2.834	0.8969
$L_2N_4(L\gamma_1)$	2.462	0.0290	2.623	0.0390	2.792	0.0494	2.964	0.0599	3.144	0.0707

(Continued)

TABLE E.3   L X-Ray Energies and Relative Intensities for Elements 26 ≤ Z ≤ 92 (Continued)

Line	Nb-41		Mo-42		Tc-43		Ru-44		Rh-45	
$L_3N_5(L\beta_2)$	2.367	0.0258	2.518	0.0346	2.674	0.0437	2.836	0.0529	3.001	0.0627
$L_1M_3(L\beta_3)$	2.335	0.5382	2.473	0.5345	2.618	0.5304	2.763	0.5268	2.916	0.5230

Line	Pd-46		Ag-47		Cd-48		In-49		Sn-50	
$L_3M_5(L\alpha_1)$	2.838	0.7892	2.984	0.7815	3.133	0.7746	3.287	0.7680	3.444	0.7624
$L_2M_4(L\beta_1)$	2.990	0.8841	3.151	0.8758	3.316	0.8681	3.487	0.8611	3.662	0.8546
$L_2N_4(L\gamma_1)$	3.328	0.0842	3.519	0.0930	3.716	0.1006	3.920	0.1080	4.131	0.1147
$L_3N_5(L\beta_2)$	3.172	0.0740	3.348	0.0818	3.528	0.0886	3.713	0.0951	3.904	0.1009
$L_1M_3(L\beta_3)$	3.073	0.5206	3.234	0.5164	3.402	0.5122	3.573	0.5077	3.750	0.5013

Line	Sb-51		Te-52		I-53		Xe-54		Cs-55	
$L_3M_5(L\alpha_1)$	3.605	0.7573	3.769	0.7525	3.937	0.7482	4.111	0.7442	4.286	0.7403
$L_2M_4(L\beta_1)$	3.843	0.8491	4.029	0.8439	4.220	0.8391	4.422	0.8348	4.620	0.8306
$L_2N_4(L\gamma_1)$	4.347	0.1205	4.570	0.1259	4.800	0.1309	5.036	0.1353	5.280	0.1397
$L_3N_5(L\beta_2)$	4.100	0.1060	4.301	0.1106	4.507	0.1149	4.720	0.1187	4.936	0.1224
$L_1M_3(L\beta_3)$	3.933	0.4956	4.121	0.4899	4.314	0.4835	4.516	0.4774	4.717	0.4715

Line	Ba-56		La-57		Ce-58		Pr-59		Nd-60	
$L_3M_5(L\alpha_1)$	4.467	0.7362	4.651	0.7326	4.840	0.7335	5.034	0.7324	5.230	0.7322
$L_2M_4(L\beta_1)$	4.828	0.8262	5.043	0.8228	5.262	0.8241	5.489	0.8229	5.722	0.8222
$L_2N_4(L\gamma_1)$	5.531	0.1441	5.789	0.1477	6.052	0.1465	6.322	0.1477	6.602	0.1486
$L_3N_5(L\beta_2)$	5.156	0.1259	5.384	0.1293	5.613	0.1279	5.850	0.1287	6.090	0.1290
$L_1M_3(L\beta_3)$	4.927	0.4663	5.143	0.4613	5.363	0.4610	5.593	0.4583	5.829	0.4556

Line	Pm-61		Sm-62		Eu-63		Gd-64		Tb-65	
$L_3M_5(L\alpha_1)$	5.431	0.7314	5.636	0.7304	5.846	0.7302	6.059	0.7282	6.275	0.7287
$L_2M_4(L\beta_1)$	5.956	0.8215	6.206	0.8209	6.456	0.8206	6.714	0.8186	6.979	0.8195
$L_2N_4(L\gamma_1)$	6.891	0.1493	7.180	0.1501	7.478	0.1505	7.788	0.1524	8.104	0.1517
$L_3N_5(L\beta_2)$	6.336	0.1295	6.587	0.1299	6.842	0.1301	7.102	0.1315	7.368	0.1306
$L_1M_3(L\beta_3)$	6.071	0.4532	6.317	0.4504	6.571	0.4476	6.832	0.4435	7.097	0.4417

Line	Dy-66		Ho-67		Er-68		Tm-69		Yb-70	
$L_3M_5(L\alpha_1)$	6.495	0.7280	6.720	0.7277	6.948	0.7270	7.181	0.7261	7.414	0.7256
$L_2M_4(L\beta_1)$	7.249	0.8190	7.528	0.8183	7.810	0.8180	8.103	0.8172	8.401	0.8168
$L_2N_4(L\gamma_1)$	8.418	0.1522	8.748	0.1529	9.089	0.1533	9.424	0.1540	9.779	0.1545
$L_3N_5(L\beta_2)$	7.638	0.1309	7.912	0.1306	8.188	0.1311	8.472	0.1313	8.758	0.1315
$L_1M_3(L\beta_3)$	7.370	0.4390	7.653	0.4360	7.940	0.4327	8.231	0.4296	8.537	0.4264

Line	Lu-71		Hf-72		Ta-73		W-74		Re-75	
$L_3M_5(L\alpha_1)$	7.654	0.7240	7.898	0.7225	8.145	0.7185	8.396	0.7155	8.651	0.7126
$L_2M_4(L\beta_1)$	8.708	0.8155	9.021	0.8139	9.341	0.8094	9.670	0.8062	10.008	0.8032
$L_2N_4(L\gamma_1)$	10.142	0.1558	10.514	0.1574	10.892	0.1584	11.283	0.1599	11.684	0.1610
$L_3N_5(L\beta_2)$	9.048	0.1325	9.346	0.1334	9.649	0.1341	9.959	0.1351	10.273	0.1359
$L_1M_3(L\beta_3)$	8.847	0.4218	9.163	0.4174	9.488	0.4130	9.819	0.4082	10.159	0.4032

Line	Os-76		Ir-77		Pt-78		Au-79		Hg-80	
$L_3M_5(L\alpha_1)$	8.910	0.7095	9.173	0.7058	9.441	0.7032	9.711	0.6999	9.987	0.6972
$L_2M_4(L\beta_1)$	10.354	0.7998	10.706	0.7953	11.069	0.7927	11.439	0.7893	11.823	0.7867
$L_2N_4(L\gamma_1)$	12.094	0.1623	12.509	0.1632	12.939	0.1640	13.379	0.1649	13.828	0.1656
$L_3N_5(L\beta_2)$	10.596	0.1368	10.918	0.1373	11.249	0.1379	11.582	0.1385	11.923	0.1391
$L_1M_3(L\beta_3)$	10.511	0.3980	10.868	0.3936	11.234	0.3885	11.610	0.3831	11.992	0.3780

(Continued)

TABLE E.3 L X-Ray Energies and Relative Intensities for Elements 26 ≤ Z ≤ 92 (Continued)

Line	Tl-81		Pb-82		Bi-83		Po-84		At-85	
$L_3M_5(L\alpha_1)$	10.266	0.6945	10.549	0.6918	10.836	0.6891	11.128	0.6867	11.424	0.6840
$L_2M_4(L\beta_1)$	12.210	0.7836	12.611	0.7809	13.021	0.7781	13.441	0.7756	13.873	0.7728
$L_2N_4(L\gamma_1)$	14.288	0.1668	14.762	0.1678	15.244	0.1688	15.740	0.1698	16.248	0.1710
$L_3N_5(L\beta_2)$	12.268	0.1397	12.620	0.1402	12.977	0.1409	13.338	0.1413	13.705	0.1422
$L_1M_3(L\beta_3)$	12.390	0.3725	12.794	0.3665	13.211	0.3602	13.635	0.3544	14.071	0.3479

Line	Rn-86		Fr-87		Ra-88		Ac-89		Th-90	
$L_3M_5(L\alpha_1)$	11.724	0.6818	12.029	0.6791	12.338	0.6765	12.650	0.6742	12.966	0.6719
$L_2M_4(L\beta_1)$	14.316	0.7702	14.770	0.7676	15.233	0.7651	15.712	0.7623	16.200	0.7599
$L_2N_4(L\gamma_1)$	16.768	0.1721	17.301	0.1732	17.845	0.1742	18.405	0.1755	18.977	0.1765
$L_3N_5(L\beta_2)$	14.077	0.1426	14.459	0.1433	14.839	0.1441	15.227	0.1446	15.620	0.1451
$L_1M_3(L\beta_3)$	14.518	0.3414	14.977	0.3352	15.450	0.3288	15.929	0.3223	16.426	0.3158

Line	Pa-91		U-92	
$L_3M_5(L\alpha_1)$	13.291	0.6700	13.613	0.6677
$L_2M_4(L\beta_1)$	16.700	0.7579	17.218	0.7558
$L_2N_4(L\gamma_1)$	19.559	0.1778	20.163	0.1789
$L_3N_5(L\beta_2)$	16.022	0.1459	16.425	0.1467
$L_1M_3(L\beta_3)$	16.938	0.3098	17.454	0.3031

Note: Only high-intensity L x-ray lines are included. The two columns of data for each element contain the energy of the transition line in kiloelectron volts and the relative intensity of each line, respectively. The intensities do not sum to 1.0 since some of the weak lines are not listed.

Appendix F: Useful Schematic Illustrations in Ion Channeling Analysis

The following are several useful schematic illustrations in ion channeling experiments and data interpretation. (From Chapter 12 and Appendix 17 of *Handbook of Modern Ion Beam Materials Analysis,* 2nd ed., eds. Y. Wang and M. Nastasi, MRS Publisher, Warrendale, PA, 2009.)

REFERENCES

Carstanjen, H.-D. 1980. Interstitial positions and vibrational amplitudes of hydrogen in metals investigated by fast ion channeling. *Physica Status Solidi A* 59:11.

Howe, L. M., Swanson, M. L., and Davies, J. A. 1983. *Methods of experimental physics,* vol. 21, p. 275. New York: Academic Press. (General review.)

Swanson, M. L. 1982. The study of lattice defects by channeling. *Reports on Progress in Physics* 45:47. (Review, including channeling studies of lattice defects.)

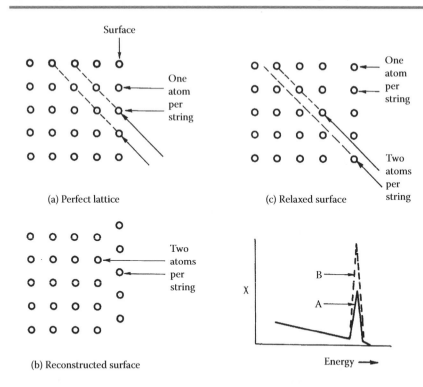

FIGURE F.1   Schematic illustration of measurements of channeling surface peaks to study displacements of surface atoms: (a) perfect lattice, (b) reconstructed surface, and (c) relaxed surface. (From Swanson, M. L. 1982. *Reports on Progress in Physics* 45:47.)

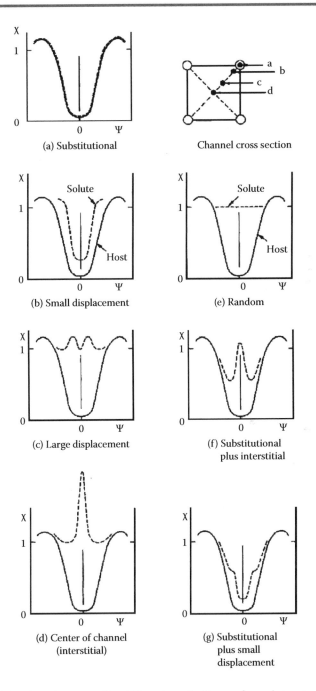

FIGURE F.2   Angular scans for different projections of a solute atom into an axial channel. (From Howe, L. M., Swanson, M. L., and Davies, J. A. 1983. *Methods of Experimental Physics*, vol. 21, p. 275. New York: Academic Press.)

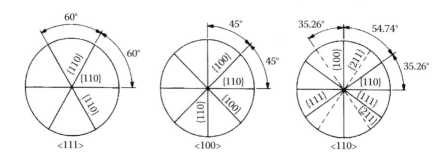

FIGURE F.3   Low-index planes around the ⟨111⟩, ⟨100⟩, and ⟨110⟩ axes in a cubic structure. To assist in orienting crystals, the major axes and planes follow, listed in the order of increasing spacing along the rows and planes; thus, from left to right, the channeling dips progress from strong to weak (e.g., ⟨110⟩ channels give the lowest minimum yields for the fcc and diamond, whereas {110} planes give the lowest dips for the bcc and diamond structures).

Structure	Axes <uvw>	Planes {hkl}
fcc	110, 100, 111	111, 100, 110
bcc	111, 100, 110	110, 100, 211
diamond	110, 111, 100	110, 111, 100

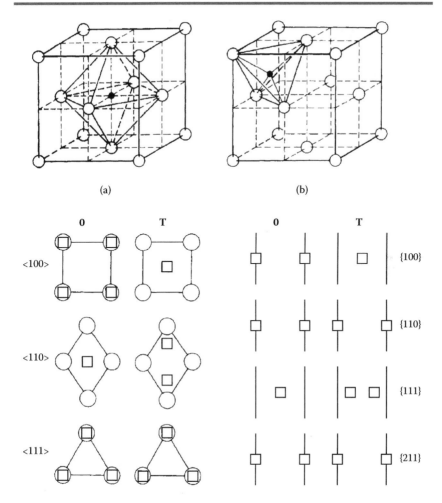

FIGURE F.4   Top: (a) octahedral and (b) tetrahedral interstitial sites in an fcc crystal. Bottom: Projections of octahedral and tetrahedral interstitial sites onto the major axial and planar channels of an fcc structure. (From Carstanjen, H.-D. 1980. *Physica Status Solidi A* 59:11.)

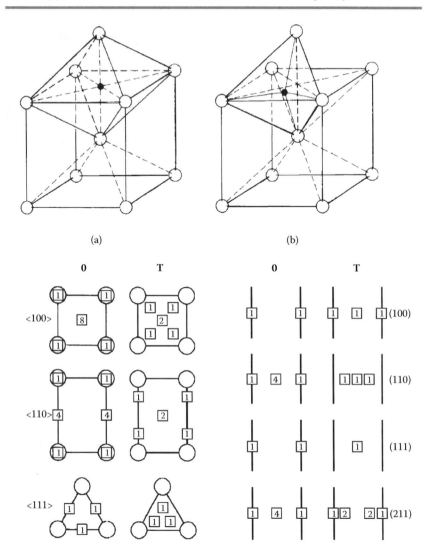

FIGURE F.5   Top: (a) octahedral and (b) tetrahedral interstitial sites in a bcc crystal. Bottom: projections of octahedral and tetrahedral interstitial sites onto the major axial and planar channels of a bcc structure. (From Carstanjen, H.-D. 1980. *Physica Status Solidi A* 59:11.)

Appendix G: Data Analysis Software for Ion Beam Analysis

Corteo	http://www.lps.umontreal.ca/~schiette/index.php?n=Recherche.Corteo
	Monte Carlo (MC) simulation program for ion beam analysis spectra
DataFurnace	http://www.surrey.ac.uk/ati/ibc/research/ion_beam_analysis/ndf.htm
	Fitting code (not a simulator) that extracts elemental depth profiles from Rutherford backscattering and related ion beam analysis spectra
FLUX	http://members.home.nl/p.j.m.smulders/FLUX/HTML/
	Program for simulating the trajectories of high-energy ions in single crystals in channeling or near-channeling directions
GeoPIXE	http://www.nmp.csiro.au/GeoPIXE.html
	Software package that performs real-time particle-induced x-ray emission (PIXE)/synchrotron x-ray fluorescence (SXRF) spectral deconvolution using dynamic analysis
GUPIX/GUPIXWIN	http://pixe.physics.uoguelph.ca/gupix/main/
	Software package for fitting PIXE spectra from thin, thick, intermediate, and layered specimens
IBANDL	http://www-nds.iaea.org/ibandl.htm
	Library of experimental data on nuclear cross sections and excitation functions relevant to ion beam analysis (IBA); resulted from the merger of SigmaBase and NRABASE
IBIS	http://www.kfki.hu/~ionhp/
	Collection of resources for the ion beam community (called the Ion Beam Information System) that includes links to several programs and other useful sites, as well as an archive of documents
IDF	http://idf.schemas.itn.pt
	Definition of the IBA data format IDF, including documentation and examples
LibCPIXE	http://sourceforge.net/projects/cpixe
	Library for the simulation of PIXE spectra
MSTAR	https://www-nds.iaea.org/stopping/
	Collection of data on stopping powers of light and heavy ions
NDF	http://www.itn.pt/facilities/lfi/ndf/uk_lfi_ndf.htm
	General-purpose code for the analysis of Rutherford backscattering spectrometry (RBS), elastic recoil detection analysis (ERDA), PIXE, nonresonant nuclear reaction analysis (NRA), and neutron depth profiling (NDP) data for any ion, any target, any geometry, and any number of spectra
QXAS	http://qxas.software.Informer.com/
	DOS-based software package (called Quantitative X-ray Analysis System) for quantitative x-ray fluorescence analysis
RUMP	http://www.genplot.com
	Graphical data analysis and plotting package that provides comprehensive analysis and simulation of RBS and elastic recoil detection (ERD) spectra

SigmaCalc   http://www.surreyibc.ac.uk/sigmacalc/
                   Service providing evaluated (recommended) differential cross sections for ion
                       beam analysis
SIMNRA   http://www.rzg.mpg.de/~mam/
                   Windows-based program for the simulation of backscattering spectra for ion
                       beam analysis with megaelectron volt ions, mainly intended for non-
                       Rutherford backscattering, NRA, and ERDA
SRIM   http://www.srim.org
                   Group of programs that calculate the stopping and range of ions in matter
                       using a quantum mechanical treatment of ion–atom collisions; Web site also
                       includes ion beam analysis software called IBA.

Source: From Chapter 14 in the *Handbook of Modern Ion Beam Materials Analysis,* 2nd ed., eds. Y.
            Wang and M. Nastasi. Warrendale: MRS Publishing, 2009.

Appendix H: Physical Constants, Combinations, and Conversions

PHYSICAL CONSTANTS

Avogadro constant	$N_A = 6.022 \times 10^{23}$	Atoms (molecules)/mol
Boltzmann constant	$k = 8.617 \times 10^{-5}$	eV/K
Gas constant	$R = 8.314$	J/(mol K)
Fundamental charge	$e = 1.602 \times 10^{-19}$	C
Gravitational constant	$G = 6.673 \times 10^{-11}$	N m^2/kg^2
Magnetic constant	$\mu_0 = 4\pi \times 10^{-7}$	N/A^2
Rest mass:		
Of electron	$M_e = 9.109 \times 10^{-31}$	kg
	$= 0.511$	MeV/c^2
Of proton	$M_p = 1.673 \times 10^{-27}$	kg
	$= 938.3$	MeV/c^2
Of neutron	$M_n = 1.675 \times 10^{-27}$	kg
	$= 939.6$	MeV/c^2
Of deuteron	$M_d = 3.344 \times 10^{-27}$	kg
	$= 1.876 \times 10^3$	MeV/c^2
Of alpha	$M_\alpha = 6.645 \times 10^{-27}$	kg
	$= 3.727 \times 10^3$	MeV/c^2
Unified mass unit	$M_\mu = 1.661 \times 10^{-27}$	kg
	$= 931.5$	MeV/c^2
Speed of light in a vacuum	$c = 2.998 \times 10^8$	m/s
Planck constant	$h = 6.626 \times 10^{-34}$	J s
	$= 4.136 \times 10^{-21}$	MeV s
$h/2\pi$	$\hbar = 6.582 \times 10^{-16}$	eV s

COMBINATIONS

Bohr magneton	$m_B = e\hbar/2m_e = 9.274 \times 10^{-24}$	J/T
Bohr radius	$a_0 = \hbar^2/2m_e^2 = 5.292 \times 10^{-2}$	nm
Bohr velocity	$v_0 = e^2/\hbar = 2.188 \times 10^6$	m/s
Classical electron radius	$r_e = e^2/mc^2 = 2.818 \times 10^{-6}$	nm
Compton wavelength	$\lambda_c = h/2e = 2.426 \times 10^{-12}$	m
Coulomb constant	$k = 1/(4\pi\varepsilon_0) = 8.988 \times 10^9$	N m^2/C^2
Electronic charge	$e^2 = 1.440$	eV nm
	$= 1.440 \times 10^{-13}$	MeV cm
Fine structure constant	$\alpha = e^2/\hbar c = 7.297 \times 10^{-3}$	
Hydrogen binding energy	$e^2/2a_0 = 13.606$	eV

CONVERSIONS

1 Å	=	10^{-8} cm	1 T	=	10^4 G
	=	10^{-1} nm	1 Ci	=	3.700×10^{10} Bq
1 μm	=	10^{-4} cm		=	3.700×10^{10} decays/s
	=	1×10^4 Å	1 R	=	2.58×10^{-4} C/kg (air)
1 barn	=	1×10^{-24} cm^2	1 rad	=	1 cGy
1 atm	=	101.3 kPa		=	1×10^{-2} Gy
	=	1.013 bar	1 rem	=	1 cSv
	=	760 mmHg		=	1×10^{-2} Sv
	=	14.7 psi	1 mrem	=	10 μSv
1 Torr	=	1 mmHg			
	=	1.333 mbar	Speed		
1 year	=	365.24 days	of 1 MeV alpha	=	6.94×10^6 m/s
	=	3.156×10^7 s	of 1 MeV electron	=	2.82×10^8 m/s
1 C	=	6.242×10^{18} electrons	of 1 MeV proton	=	1.38×10^7 m/s
1 eV	=	1.602×10^{-19} J	of 1 MeV deuteron	=	9.78×10^6 m/s
1 eV/particle	=	23.06 kcal/mol			
	=	96.53 kJ/mol			

Index